ASSET INTEGRITY MANAGEMENT FOR OFFSHORE AND ONSHORE STRUCTURES

ASSET INTEGRITY MANAGEMENT FOR OFFSHORE AND ONSHORE STRUCTURES

Mohamed A. El-Reedy, PhD
Structural Consultant Engineer

ELSEVIER

Gulf Professional Publishing
An imprint of Elsevier

Gulf Professional Publishing is an imprint of Elsevier
50 Hampshire Street, 5th Floor, Cambridge, MA 02139, United States
The Boulevard, Langford Lane, Kidlington, Oxford, OX5 1GB, United Kingdom

Library of Congress Cataloging-in-Publication Data
A catalog record for this book is available from the Library of Congress

British Library Cataloguing-in-Publication Data
A catalogue record for this book is available from the British Library

ISBN: 978-0-12-824540-8

For information on all Gulf Professional publications
visit our website at https://www.elsevier.com/books-and-journals

Publisher: Charlotte Cockle
Senior Acquisitions Editor: Katie Hammon
Editorial Project Manager: Aleksandra Packowska
Production Project Manager: Prem Kumar Kaliamoorthi
Cover Designer: Christian Bilbow

Typeset by STRAIVE, India

This book is dedicated to the spirits of my mother and my father, and to my wife and my children Maey, Hisham, and Mayar.

Contents

About the author

Mohamed A. El-Reedy's background is in structural engineering. His main area of research is the reliability of concrete and steel structures. He has provided consulting to different engineering companies and oil and gas industries in Egypt and to international companies such as the International Egyptian Oil Company (IEOC) and British Petroleum (BP). Moreover, he provides different concrete and steel structure design packages for residential buildings, warehouses, and telecommunication towers and electrical projects with WorleyParsons Egypt. He has participated in liquefied natural gas (LNG) and natural gas liquid (NGL) projects with international engineering firms. Currently, Dr. El-Reedy is responsible for reliability, inspection, and maintenance strategy for onshore concrete structures and offshore steel structure platforms. He has performed these tasks for hundreds of structures in the Gulf of Suez in the Red Sea.

Dr. El-Reedy has consulted with and trained executives at many organizations, including the Arabian American Oil Company (ARAMCO), BP, Apachi, Abu Dhabi Marine Operating Company (ADMA), the Abu Dhabi National Oil Company, King Saudi's Interior ministry, Qatar Telecom, the Egyptian General Petroleum Corporation, Saudi Arabia Basic Industries Corporation (SAPIC), the Kuwait Petroleum Corporation, and Qatar Petrochemical Company (QAPCO). He has taught technical courses on repair and maintenance for reinforced concrete structures and advanced materials in the concrete industry worldwide, especially in the Middle East.

Dr. El-Reedy has written numerous publications and has presented many papers at local and international conferences sponsored by the American Society of Civil Engineers, the American Society of Mechanical Engineers, the American Concrete Institute, the American Society for Testing and Materials, and the American Petroleum Institute. He has published many research papers in international technical journals and has authored over 10 books on total quality management, quality management and quality assurance, economic management of engineering projects, repair and protection of reinforced concrete structures, offshore structures design, project management, and others. He received his bachelor's degree from Cairo University in 1990, his master's degree in 1995, and his PhD from Cairo University in 2000.

Preface

Most oil and gas companies first began to develop asset integrity management approaches at the end of 1980 and the beginning of 1990. At that time, around 40% of the facilities worldwide were aging. Many oil and gas projects began in the 1950s and the numbers increased in the 1960s, so many facilities were over 30 years old, and doubts were arising about the direction of maintenance approaches from the point of view of equipment economics and reliability.

Mechanical engineers had been pioneers in maintenance philosophies and approaches, due mainly to their dealings with rotating equipment; however, due to the differing nature of structures and other static equipment, they were a step behind in their development in this area. The run-to-failure or corrective maintenance approaches were proving to be expensive in general and could affect production and increase the risk of investments; therefore there was a new direction to develop a system to maintain the liquid inside the equipment with zero leakage. From that philosophy came the approach of asset integrity management.

This term refers to managing the integrity between different entities with different scopes, to achieve general maintenance of an asset; thus this task cannot be done by one entity. However, you will find many companies and other entities on the market who will tell you that they are doing asset integrity management. Take care: this is largely a marketing issue. Even the best companies can achieve management of only two or three elements but not all, as the integrity of the asset should be managed between the owner, the operator (if different), the surveillance and inspection company for onshore and offshore, integrity experts to diagnose the problems, the engineering firm for evaluation and assessment using complicated analyses like pushover or finite element, the engineering firm or integrity experts to define the structure of the static facilities and its fitness for service, the owner or operator integrity team so the owner can define a maintenance plan, and software developers to collect and archive all these data.

So, in the end, the owner is responsible for managing the integrity between many entities to maintain the asset.

This book is a blend of practical coverage and theoretical background of integrity management elements for pipelines, tanks, and offshore and concrete structures. The corrosion phenomena is discussed in detail, as it is the main driver of any deterioration. The structure probability of failure and its reliability as part of integrity management is also discussed. Inspection methods and up-to-date techniques are presented in detail, and assessment and evaluation of offshore and onshore structures, tanks, and pipelines are illustrated. Ways to protect structures from corrosion are also examined in this book. The integrity management system and procedures, in addition to qualitative and quantitative risk assessment with real-world examples, are presented in the last chapter.

In this book I have attempted to cover all aspects of structure integrity management systems from a practical point of view with a strong theoretical background, which is my approach for all my books.

Mohamed A. El-Reedy, PhD
www.elreedyman.com
Cairo, Egypt

CHAPTER 1

Corrosion effects on offshore and onshore structures

1.1 Introduction

Corrosion is the main factor affecting deterioration of structures offshore and onshore, and specifically in plants near the coast, which includes most oil and gas plants worldwide. The main driver in structural integrity management is corrosion; repair projects are even referred to as brown field projects, as an indication of corrosion effects. Thus it is mandatory to know the basics and the features of corrosion, which are the main topics of this chapter.

Many textbooks examine corrosion phenomena through in-depth presentations of chemical reactions and electrical theories. However, the objective of this chapter is to describe corrosion in a simple form that matches the requirements of engineering aspects, in order to discuss the effects of corrosion on onshore and offshore structures.

In general, the main factors in any type of corrosion are the environmental conditions surrounding the structure. The variability and characteristics of these factors cannot be defined easily. Practically, the main characteristics of the environment are that it is variable and its condition can change over time.

The local environment at the metal surface is the microenvironment that affects the metal. The real corrosion damage is mainly affected by the local environment reactivity. Therefore local effects like flow, pH cells, deposits, and galvanic effects should be considered in experimental investigations with the nominal environmental conditions to define the lifetime prediction.

In general, corrosion has begun when rust is present on any steel surface. The main two drivers of corrosion are chemical reactions and electrical current. Once corrosion of the steel occurs, in a hole that contains water, electrons are obtained according to Eq. (1.1), which represents an anodic reaction. The nature of offshore structures is that they are exposed continuously to saline water throughout their lifetime. Therefore water is the main factor that causes corrosion in offshore structures. It is thus important to look at the reactions of iron (Fe) with water (H_2O). The reaction between steel

Asset Integrity Management for Offshore and Onshore Structures
https://doi.org/10.1016/B978-0-12-824540-8.00001-6

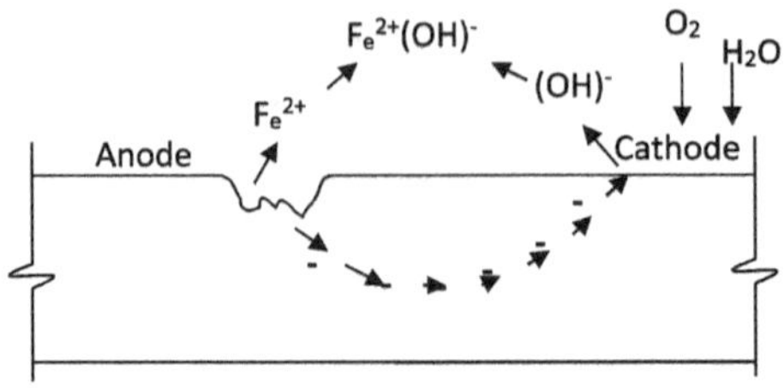

Fig. 1.1 Corrosion process on the steel surface.

ions and water in the presence of oxygen accumulates a solution at anodic areas in an amount that is chemically equivalent to the reaction at cathodic areas as well; these reactions are presented in Fig. 1.1. In general, the corrosion process on the steel surface conforms to the anodic chemical reaction at the anodic area as per the chemical reaction in Eq. (1.1).

$$Fe \rightarrow Fe^{2+} + 2e^- \tag{1.1}$$

If the electrons accumulate on one part of the steel rather than the other part, another chemical reaction will occur as a combination of the electrodes with oxygen and water. This is called a cathodic reaction and its equation is given as:

Cathodic Reaction:

$$2e^- + H_2O + \frac{1}{2}O_2 \rightarrow 2OH^- \tag{1.2}$$

In Eq. (1.2), it is clear that the presence of (OH^-) occurs due to a cathodic reaction. The hydroxide ions affect the alkalinity and slightly reduce the effect of chlorides. Based on that, it is obvious that the presence of water and oxygen is the main cause for corrosion.

As shown in the preceding equations and Fig. 1.1, the first step in the corrosion process is the anodic and cathodic reaction, as the hydroxide ions (OH^-) will react with ferrous iron (Fe^{2+}) as result of chemical Eq. (1.1). This reaction will produce ferrous hydroxide, which will react with oxygen and water again and produce ferric hydroxide. The equations of the chemical reactions are as follows:

$$Fe^{2+} + 2OH^- \rightarrow Fe(OH)_2 \tag{1.3}$$

$$4Fe(OH)_2 + O_2 + 2H_2O \rightarrow 4Fe(OH)_3 \tag{1.4}$$

$$2Fe(OH)_3 \rightarrow Fe_2O_3 \cdot H_2O + 2H_2O \tag{1.5}$$

From these chemical reactions, the ferrous hydroxides ($Fe(OH)_2$) will react with oxygen and water to produce ferric hydroxides ($Fe(OH)_3$) and this component is rust; its chemical name is hydrated ferric oxide and the chemical term is $Fe_2O_3 \cdot H_2O$.

Saturated $Fe(OH)_3$ is almost neutral in pH. Magnetic hydrous ferrous ferrite, $Fe_3O_4 \_ nH_2O$, can form a black intermediate layer between hydrous Fe_2O_3 and FeO. This is the reason behind the fact that rust films normally consist of three layers of iron oxides in different states of oxidation.

1.2 Corrosion effects on offshore structure

1.2.1 Corrosion in seawater

Many industries are affected by seawater, such as shipping, FPSO, offshore oil and gas facilities, power plants, and coastal industrial plants. Seawater is used in many applications, such as cooling purposes, firefighting systems, oil-field water injection, and desalination plants. For many years a great deal of research and development has been carried out, studying corrosion problems in these systems; however, after all these studies have been done on the behavior of materials in seawater, failures are still occurring.

Dissolved materials concentrations in the sea vary greatly with location and time, because rivers dilute seawater, as do rain and melting ice, and seawater can be concentrated by evaporation. The most important properties of seawater are as follows:

- Remarkably constant ratios of the concentrations of the major constituents worldwide
- Sodium chloride (salt) exists with high salt concentration
- High electrical conductivity
- Relatively high and constant pH
- Buffering capacity
- Solubility of gases, of which oxygen and carbon dioxide in particular are important in the context of corrosion
- Organic compounds exist
- Biological life is represented as microfouling, such as bacteria and slime, and macrofouling, such as seaweed, mussels, barnacles, and many kinds of fish. All these biological effects depend on depth, temperature, intensity of light, and the availability of nutrients. The salinity of seawater is its only main specification.

In 1902, salinity was defined as the total amount of solid material, weighed by grams, contained in 1 kg of seawater. This was expressed by Eq. (1.6), where S is the salinity and Cl is chlorinity, which are expressed as parts per thousand (‰).

$$S(\%) = 0.03 + 1.805Cl(\%) \tag{1.6}$$

Based on this equation, at zero chlorinity the salinity is 0.03%, which required more investigation. In 1969, the United Nations Scientific, Education and Cultural Organization (UNESCO) defined a more accurate relationship between chlorinity and salinity. The proposed equation is as follows:

$$S\,(‰) = 1.80655Cl\,(‰) \tag{1.7}$$

It can be noted that there are no major differences between these two equations and their results are the same at salinity equal to 35%.

In 1978, another salinity definition was raised in terms of conductivity ratio, which is given by this equation:

$$S = 0.0080 - 0.1692K^{0.5} + 25.3853K + 14.0941K^{1.5} - 7.0261K^2 + 2.7081K^{2.5} \tag{1.8}$$

where S is the salinity of a sample of seawater, K is the seawater sample electrical conductivity at 15°C and 1 standard atmosphere pressure, to that of a potassium chloride (KCl) solution in which the mass fraction of KCl is 0.0324356, at the same temperature and pressure.

Note that K is equal to a practical salinity of 35.

It is worth mentioning that this definition (‰) is not used; however, an old value of 35‰ corresponds to a new value of 35. In general, the electrical conductivity falls within the range 32‰ to 35‰.

It is found that when corrosion occurs, the anodic reaction rate is exactly equal to the cathodic reaction rate. In an environment having good conductivity, as in seawater or seabed mud, the corroding metal displays a single potential that lies between E_c and E_a. In Fig. 1.2A, this condition is met where the anodic and cathodic curves cross. The potential at the crossover point is referred to as the corrosion potential, E_{cor}. The current, I_{cor}, is referred to as the corrosion current, and it is an electrical current that represents the corrosion rate. In practice, a corroding metal does not take up potential E_a or E_c, but spontaneously moves to E_{cor}.

While the shape of the individual $E - \log I$ curves may vary, as it depends on environmental conditions, the manner in which the diagrams, so-called polarization diagrams, are interpreted in terms of E_{cor} and I_{cor} remains the same.

Fig. 1.2 Evan's diagram.

Fig. 1.2A represents the Evans diagram, showing the polarization curves for separate cathodic and anodic reactions. When the cathodic and anodic current densities are equal, the two curves shall intersect at the point that defines the corrosion rate in terms of a mean corrosion current density (i_{cor}). The electrode potential of the couple at this point is termed the corrosion potential (E_{cor}).

In some time at the metal surface, there is a difference between the electrode potentials developed at the anodic and cathodic sites. It may be the amount is significant in ohmic drop under conditions where corrosion macrocells are formed. This happens if the anodic and cathodic areas are separated by a medium of high electrolytic resistance. Therefore the Evans diagram will be modified as shown in Fig. 1.2D. Thus the mean corrosion rate I_{cor} is now reduced and the corrosion potential varies with the location between the limits anode E_{cor} and cathode E_{cor}, the positions of local anodes being indicated by the region of low corrosion potential. If the anodic reaction is steeply polarized, such as owing to the presence of a passive film, then i_{cor} is small and E_{cor} assumes a value close to the reversible potential of the cathodic reaction, as shown in Fig. 1.2C. On the other hand, if the cathodic reaction is steeply polarized, such as owing to limited oxygen availability, the situation is as shown in Fig. 1.2B, with i_{cor} again small but the corrosion potential close to the reversible potential of the anodic reaction.

1.2.2 Steel corrosion in seawater

The corrosion of steel in seawater and also in seabed mud can be adequately represented by Eq. (1.1), although the process normally proceeds to the precipitation of ferric hydroxide. In the case of clean steel in seawater, the anodic process occurs with greater facility than the cathodic. In consequence, the corrosion reaction can go no faster than the rate of cathodic oxygen reduction. The latter usually proves to be controlled by the rate of arrival of the oxygen at the metal surface, which in turn is controlled by the linear water flow rate and the dissolved oxygen concentration in the bulk seawater.

Fig. 1.3 shows the corrosion in the bracing of an offshore structure.

This may be represented on a polarization diagram as shown in Fig. 1.4. At first, the cathodic kinetics get faster as the potential becomes more negative from Ec. This has the effect of depleting the oxygen immediately adjacent to the metal surface, thus causing the reaction to be more difficult. Ultimately, a point is reached where the surface concentration of oxygen has dropped to zero and oxygen can then only be reduced as and when it reaches the surface. Further lowering of the potential cannot increase the

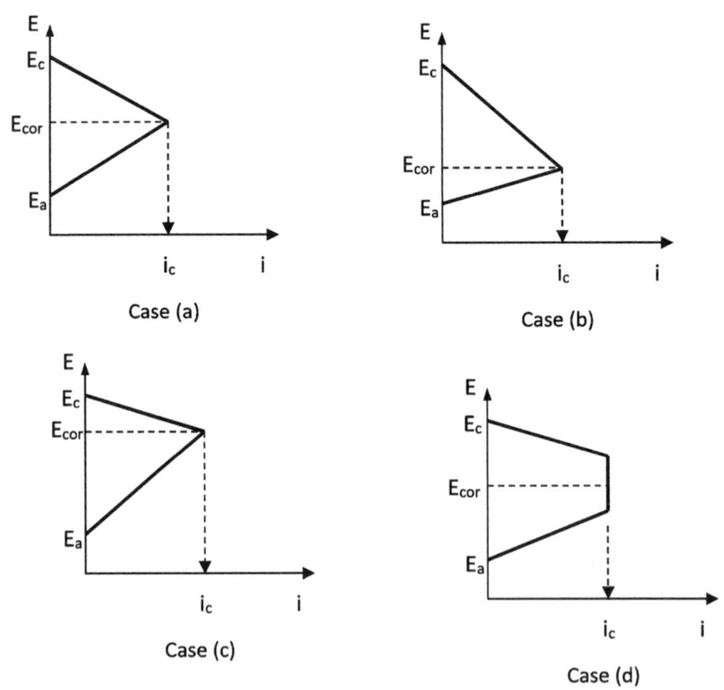

Fig. 1.3 Corrosion on a jacket structure.

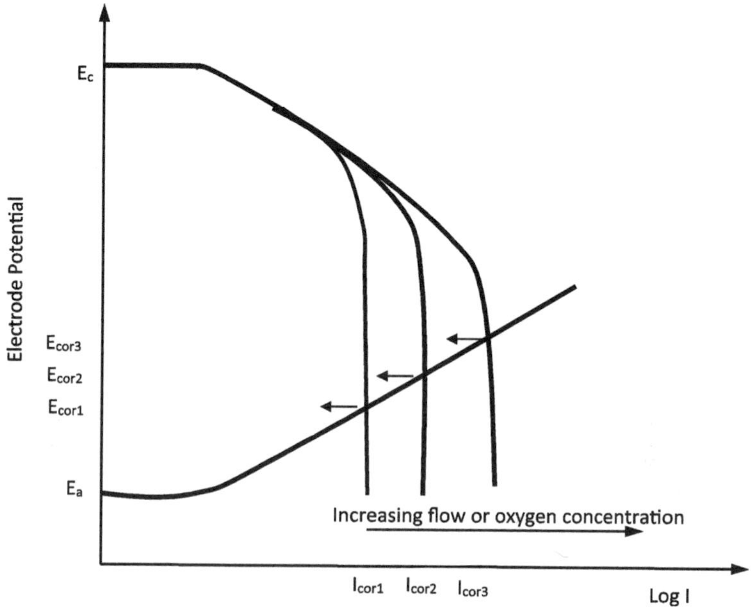

Fig. 1.4 Polarization diagram with increasing oxygen concentration.

cathodic reaction rate, because the kinetics are now governed by potential-independent diffusion processes. A plateau, or limiting, current is observed.

Fig. 1.4 shows that the corrosion rate is then equal to this limiting current. It is worth mentioning that the limiting current can be increased by increasing the oxygen flow, either by raising the bulk oxygen concentration or increasing the flow rate; both cause the corrosion rate to increase.

To a first approximation, it can be stated that the corrosion rate of clean steel in aerated seawater under turbulent flow conditions is directly proportional to the bulk oxygen concentration and the linear velocity.

In 1994 Ashworth used a technique to estimate the maximum corrosion rates of clean steel in North Sea water at 7°C, as illustrated in Table 1.1.

In practice, corrosion products and marine fouling build up on steel as it corrodes in seawater. These generally produce lower corrosion rates.

In 1994 Rowlands suggested that the corrosion rate of fully immersed steel is fairly rapid in the first few months of exposure, but it falls progressively with time. A value of 0.13 mm/annual-year is considered a good estimation in any part of the world. However, pits may grow at 3 to 10 times more than that rate.

Table 1.1 Estimated maximum corrosion rates of clean steel in North Sea water 7°C.

	O_2 concern				
Linear flow rate (m/s)	**6ppm**	**7ppm**	**8ppm**	**9ppm**	**10ppm**
0	0.080	0.094	0.107	0.120	0.134
0.3	0.091	0.107	0.123	0.138	0.154
0.4	0.096	0.111	0.128	0.144	0.160
0.6	0.104	0.121	0.138	0.156	0.174
1	0.120	0.140	0.160	0.179	0.199
2	0.160	0.187	0.213	0.240	0.266
4	0.240	0.280	0.320	0.360	0.400

In the case of offshore structures in shallow waters, they are simultaneously exposed to a number of discrete corrosive environments such as the marine atmosphere, the splash zone, the tidal zone, the fully submerged zone, and the mud zone. There are different results for those discussed in the corrosion rate data presented in 1949 in the work of Humble. The peak corrosion rates are found immediately below the mean low tide zone and in the splash zone. The typical mean corrosion rate in the splash zone, given quiet sea conditions, is estimated as 0.25 to 0.75 mm/a.

Fig. 1.4 shows the polarization diagram representing control of the corrosion rate by controlling the rate of oxygen arrival at the surface and the effect of increasing oxygen availability.

The peak corrosion rate is often attributed to galvanic action between steel in contact with the oxygen-rich surface waters, which is the cathodic area, and the steel at somewhat greater depth exposed to waters of lesser oxygen content, which is the anodic area. Really, it is difficult to conceive that the change in oxygen concentration with depth is sufficiently great enough to cause the effect. The corrosion effect in the topside of offshore structures is presented in Fig. 1.5.

The corrosion in an FPSO vessel of one of its tanks is shown in Fig. 1.6.

1.3 Pipelines

As per a study done by the National Association of Corrosion Engineers (NACE), the cost due to corrosion in a pipeline is in the range of $5.4 billion and $8.6 billion in the United States alone. Therefore corrosion in pipelines is one of the major challenges for any pipeline integrity engineer, as it is the main reason for pipeline failure in the oil and gas industry.

Fig. 1.5 Corrosion on the platform topside.

Fig. 1.6 Corrosion in FPSO tank.

The corrosion in the pipeline reduces the thickness at the same time that the pipeline is under pressure, so if the pressure is the same with time, the strength will be reduced dramatically due to the reduction in wall thickness. This reduction can cause a leak, which affects the reliability of the pipeline, as any leak will directly affect the production.

Note that low-carbon steel has been associated with susceptibility to oxidation in the presence of electrolytes, water, and carbon dioxide.

Pipeline corrosion will occur due to internal or external factors; internal corrosion is due to the fluid inside it, which is a function of the material composition of the fluid and also its velocity is considered one of the main factors. The external corrosion is due to its contact with the soil; Baker (2008) mentioned that controlling the external corrosion is mainly by coating and utilizing a cathodic protection system.

The concept of cathodic protection is producing an electric current in the pipeline to compensate for the electron movement from the anode to the cathode as per the corrosion fundamentals, as discussed in the previous section. The cathodic protection system function is to create a cathodic field over the pipeline, which implies that the anodes in the exposed surface are nonreactive. The pipe will act like a cathode, as there is a lack of movement of electrons. In addition, cathodic protection leads to the development of deposits that protect the steel since they are alkaline in nature.

Baker (2008) suggested two main types of cathodic protection, which are sacrificial anode protection by connecting the pipeline with an external metal that has a relatively higher activity than steel. This metal will be located away from the pipeline but within the electrolyte, which is the soil. As a result, there will be an electric current that transfers electrons from this metal to the pipeline and it is depilated with time. Therefore the sacrificial metal undergoes corrosion, thereby protecting the oil and gas pipeline from corrosion.

The second method is the impressed-current anode type, by producing a direct current between the pipeline and anode from an external electrical source.

Choosing between these two types depends on many factors, such as the cost and its operability and maintenance.

Another scenario involving pipelines is that the corrosion will happen inside, due to a chemical reaction between the carbon steel and the hydrocarbon inside it (Fig. 1.7).

- In the oil and gas industry, the corrosion usually occurs due to the result of chemical reactions between CO_2 or H_2S and steel, with water usually

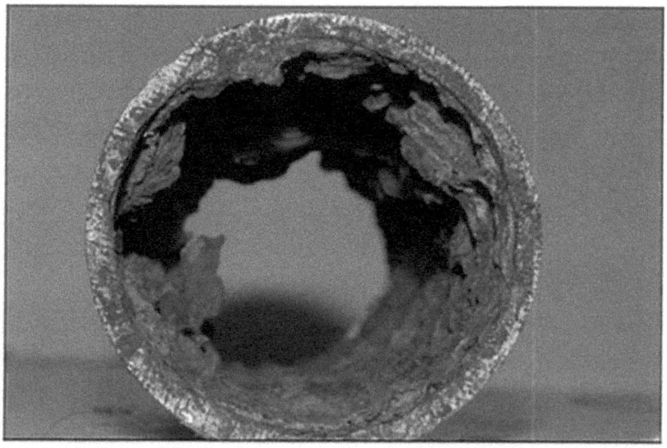

Fig. 1.7 Internal pipeline corrosion.

acting as a catalyst for the reaction. In general, the most common types of corrosion that cause a pitting corrosion in the oil and gas industry are sweet, sour, oxygen, crevice, and microbiologically induced corrosion.

1.4 Onshore structure

In the past, it was stated that concrete is a more durable material than steel structures; around 1980–1990 the statement changed to "concrete is a durable material but needs some precautions." It was shown by researchers Cady and Weyers (1984); Kilareski (1980); Mori and Ellingwood (1994a,b); Takewaka and Matsumoto (1988); and Thoft-Christensen (1995) that in general any reinforced concrete structure will be exposed during its lifetime to environmental stress such as corrosion or expansive aggregate reactions or others, which attacks the concrete or steel reinforcement. Based on this, it has been shown by statistics worldwide that the corrosion of reinforced steel bars is the most common problem facing structural engineers in the last 30 years.

To prevent corrosion of the steel reinforcement bars in a concrete structure, the main step is to understand the corrosion phenomenon, so that we can suggest the best method of protection and repair methods as well.

During the winter the ice accumulates over bridges in cold-weather countries like the United States, Canada, and others, and salt is used to melt the ice as a normal practice. The salt is mainly sodium chloride, which is a main source of steel corrosion. Therefore chloride attack is the main challenge in these countries and the cost is high every year for maintenance and repairs, especially for the bridge decks and piers.

Some countries like the Middle East, especially in the Gulf area as these countries are the main producers of oil and gas worldwide, have very hot weather and high humidity. Their ground water table also has a high content of salt. Thus the main challenge is the corrosion in the steel bars. Therefore in the Middle East a great deal of money is spent due to corrosion of steel bars.

The corrosion process occurs slowly and propagates with time, and the deterioration rate varies from one structure to another, depending on many factors. Corrosion of steel reinforcing bars directly affects the structure's safety. The corrosion depends mainly on the concrete quality and the environmental conditions surrounding the structure. Therefore the main factors affecting the structure safety over time are the corrosion rate, the location of the member in the structure, the type of structure member, and its redundancy.

The main challenge for an integrity management system is to predict the time period in which to do inspection or minor repairs, considering that minimizing the time interval between inspections will cost but will maintain structural safety. On the other hand, increasing the time interval between inspections can decrease the cost of inspections during the structure lifetime, but it will affect structure safety and can lead to a major repair, which is expensive. So, the main task of the integrity engineer is to define the optimum inspection time intervals.

Many accidents worldwide have occurred due to a lack of maintenance strategy. For example, a large part of a bridge in New York fell, causing a motorcycle rider to lose control of his vehicle and die. It is obvious that bad maintenance strategy, lack of understanding of the corrosion process, and not considering risk during maintenance planning, which is the main core of the integrity management system, caused a death.

In 1960 to 1970, there was worldwide acceptance of using seawater in concrete mixtures as it increases the concrete strength, which really is true. In this time period calcium chloride additives were also used as an accelerator for concrete setting; these additives were used in 1970. For example, in the Middle East in the past pure water was very expensive, and they were just starting in their building growth, so they used seawater extensively; after that, they found severe deterioration in reinforced concrete structures, causing many problems and requiring expensive rehabilitation and rebuilding.

The effects of corrosion not only create a serious problem for structural engineers but also affect national economies. Therefore engineers must have an in-depth understanding of the nature of the corrosion process and the reasons for its occurrence. Understanding the problem is the first step in solving the problem.

Many textbooks examine the corrosion process in concrete. There are two main factors: chemical reactions and electrical currents. The mix between them to match with the structural engineering requirements was illustrated by El-Reedy (2007), to describe corrosion in an easy-to-understand way so that structure engineers can take action in a timely manner. On the other hand, it is required to select the best protection method based on the environmental conditions, type of corrosion, and nature of the project.

As a rule of thumb, if the piece of steel is in dry air or is submerged in water, the piece of steel will not corrode. However, if it is submerged in water and removed and then exposed to air, and this is repeated over a cyclic period, this piece of steel will corrode.

In the case of steel bars embedded in concrete, the concrete is a porous material containing water in the voids during the pouring and after that, during the curing process or due to rainy weather or any weather with high relative humidity. Thus the concrete will contain humidity, which is a main cause of corrosion. This is contrary to the old statement from textbooks, articles, and lectures that "a concrete structure is durable and has no need for inspection like steel structures"; nowadays, this statement has been changed to "concrete is a durable material but …."

In reality, the reinforced concrete structure is durable. The reason why the steel bars in concrete do not corrode is that concrete is alkaline in nature and alkalinity counteracts acidity. As a result of its alkalinity, concrete can protect the steel bar from corrosion. The concrete is alkaline because inside its microvoids there is a high concentration of the oxides calcium, sodium, and magnesium. After mixing water with Portland cement, a dehydration process occurs and produces very alkaline oxides and hydroxides. The measure of alkalinity and acidity is pH, which is hydrogen ion concentration. A strong acid has pH equal to 1 (or less), a strong alkali has pH equal to 14 (or more), and a neutral solution has pH equal to 7. Concrete has a pH of 12–13; steel starts to rust at a pH of 8–9.

These oxides produce hydroxides from the cement reaction with water that have a high alkalinity, with pH of 12–13. This alkalinity will produce a passive layer around the steel reinforcement surface; this layer is dense and prevents corrosion, and if the alkalinity is maintained around the steel bars corrosion will not occur during its lifetime. In this case, the old statement is valid, as the reinforced concrete structure is durable and does not need inspections.

But, unfortunately, this concrete alkalinity around the steel bars will not remain for a long time. Two main factors always have an effect in breaking the alkalinity, which are the carbonation process and the chlorides attached to the concrete from the inside or outside. The corrosion process in this case is the same as that discussed in Section 1.1 and the same chemical reactions are valid. After the passive layer is broken down, the pH changes until it reaches 8–9, which means it has reached acidity; rust will appear instantly on the steel bar's surface. The chemical reaction processes are almost similar in carbonation, sulfate, and chloride attacks. When corrosion of the reinforced steel bars in concrete occurs, they melt in the water-containing voids.

The chemical reaction in brief is the transformation of steel from ferrous hydroxide ($Fe(OH)_2$), that will react with oxygen (O_2) and water (H_2O) to produce ferric hydroxide ($Fe(OH)_3$). The ferric hydroxide is the hydrated ferric oxide, which is rust, and its chemical term is $Fe_2O_3{\cdot}H_2O$.

It has been found that ferric hydroxide is a major factor in concrete deterioration and spalling of concrete cover, because its volume can increase from two times to ten times or more that of its original volume. Because the steel bars are confined by concrete, if the steel bar volume increases, it will produce high stresses and the weak points will start to crack. These cracks propagate over time until the concrete cover breaks down. Then during a visual inspection rust can be seen, with its brown color, on the steel bar.

There are three types of corrosion of steel bars, which are black, bacteria, and pit formation. Their characteristics are described in the following sections.

1.4.1 Black corrosion

A concrete foundation or any member immersed in water sometimes develops black corrosion; this type of corrosion usually occurs if oxygen is not available or if there is a long distance between the location of the anode and cathode. This type of corrosion is called black corrosion because the steel bars have a black or green color on their surface. The main problem for this corrosion is the lack of the presence of any warning of cracks or spalling of the concrete cover and it usually exists in offshore structures such as harbors, since part of the concrete is submerged in the water.

You can recognize this corrosion type by its smell, similar to rotten eggs, due to the production of hydrogen sulfide, and the pitting is smooth with black color in the case of steel bars exposed to soil saturated with water, such as in foundations. It is important that any honeycomb or crack should be repaired as soon as possible before rust begins.

1.4.2 Pit formation

The uniform corrosion in steel reinforcing bars starts with the formation of a small pit and then these pits increase with time and combine to form uniform corrosion. This is clearly seen in steel reinforcement bars exposed to carbonation or chloride effects. There are many chemical reactions describing the formation of pits in various textbooks, and sometimes it is very complicated. However, from a practical point of view, pit corrosion is very simple, especially in cases of chloride attacks.

As chloride attaches to some parts of the steel reinforcement, the voids of the cement mortar are located around the steel reinforcement bars in the concrete, or sulfide is present inside the steel bars. Therefore the passive layer is more vulnerable to chloride attack, because the electrochemical potential difference attracts chloride ions. Therefore corrosion starts and acids are formed as hydrogen from the existing MnS inside the steel bars and HCl from the chloride ions, if they are present. The chemical reactions are as follows:

$$Fe^{2+} + H_2O\ FeOH^+ + H^+ \tag{1.9}$$

$$MnS + 2H^+\ H_2S + Mn^{2+} \tag{1.10}$$

Rust will form over the pit, with concentration of the acid (H^+) as per Eq. 1.9 and without oxygen, so that the iron stays in solution; the acidity will prevent the formation of a protective oxide layer, thus accelerating corrosion. The pitting formation is shown in Fig. 1.8.

The pit corrosion phenomenon is well known to be a galvanic action between a cathode, which is a relatively large area of passivity on the steel, and an anode, which is a small pit. Inside these pits there is a high chloride concentration with a low pH value.

For pitting to exist and continue, there must be enough oxygen, as it is the main reason for the anode polarization. The corrosion potential of steel reinforcement to have pitting varies between the passive state and that of the anodic pitting areas, which is normally in the range of -200 to $-500\,mV$, which will be examined later in Chapter 5.

Fig. 1.8 Pitting corrosion.

1.4.3 Reasons for steel-in-concrete corrosion

The steel corrosion process has been discussed in previous sections, along with methods of occurrence and main causes. Fig. 1.9 presents the corrosion of a reinforced concrete slab. Fig. 1.10 shows corrosion in the steel bars for a ring beam under a steel storage tank.

In general, there are three main reasons for corrosion of steel in concrete, which are chloride attack, sulfide attack, and carbon dioxide penetration, which is called the carbonation process. These processes cause corrosion

Fig. 1.9 Corrosion of steel bars in concrete slab.

Fig. 1.10 Corrosion on steel for ring beam under storage tank.

of steel but do not influence the corrosion of steel in concrete directly. However, there are additional reasons for corrosion to occur, such as the presence of chloride or other chemicals inside the concrete and the number of voids in the concrete that affect the steel bars. Moreover, some other acids, like sulfate, in soil will also attack the concrete and react with the concrete components and also damage the concrete alkalinity around the steel bars, and then corrosion can start on the steel.

The main causes of corrosion of steel in concrete are chloride attack and carbonation transformation.

1.4.3.1 Carbonation

The carbonation process is a major enemy that attacks the concrete surface until reaching the steel bars, by converting the alkalinity of the concrete due to the existing hydroxides to acid by chemical reaction. Like many other gases, carbon dioxide dissolves in water to form an acid. But in this case the carbonic acid does not attack the cement paste, but rather neutralizes the alkalis in the pore water, mainly forming calcium carbonate according to the following chemical reactions:

$$CO_2 + H_2O \rightarrow H_2CO_3 \tag{1.11}$$

$$H_2CO_3 + Ca(OH)_2 \rightarrow CaCO_3 + 2H_2O \tag{1.12}$$

The calcium hydroxyl existing in the concrete is the main source of alkalinity and it maintains a pH level of 12–13; after carbonates attack the concrete surface and propagate inside the concrete, calcium carbonate forms as per Eq. (1.12). From this, the value of pH will be reduced to reach the level that causes corrosion of the steel reinforcement.

Therefore it is obvious that the carbonation process occurs rapidly if the concrete cover is not very thick. On the other hand, the concrete cover may be thick enough, conforming with the codes and standards requirements, but there may be voids in the concrete due to bad compaction, so the carbonation transformation process can occur due to the existence of pore opening voids in the concrete, which are the main cause of the rapid penetration of CO_2 inside the concrete. The carbonation process can occur if the alkalinity in the concrete voids is relatively small. This occurs when the cement content is less than the minimum values in the codes and standards/normal practice, and the water-to-cement (w/c) ratio is high; it can also be due to a bad curing process during construction.

The carbonation propagates inside concrete according to the diffusion theory. The diffusion rate is in inverse proportion to the concrete cover thickness, which is the distance between the steel reinforcement bars and the concrete surface:

$$\frac{ds}{dt} = \frac{D_r}{s} \tag{1.13}$$

where s is the distance from the steel bars to the concrete surface, t is the time, and D_r is the diffusion rate, which depends on the concrete quality. If the concrete is affected by carbonation or chlorides, the value of alkalinity will decrease from pH 11–13 until it reached a pH equal to or less than 8. At this low level of alkalinity, the passive layer protection will not protect steel, so this is when the corrosion process starts.

There are many factors affecting the capability of concrete to resist the carbonation propagation inside the concrete. The carbonation propagation time depends on the thickness of the concrete cover and on the quality of the concrete in general, based on the water-cement ratio and also the curing after the concrete pouring.

High-quality concrete, as defined by the percentage of cement, water-cement ratio w/c, compaction, and curing process, will have resistance high enough to prevent CO_2 propagating inside the concrete.

The concrete properties can be enhanced to decrease the voids by using new materials such as silica fume, which has a very fine particle size, around 0.1–0.2 μm. These fine particles will reduce the voids in concrete. In the 1980s, the Japanese created a noncompacted concrete that contained a high percentage of silica fume with a special concrete mix design. Using these materials, a high-strength concrete can be produced that can reach concrete compressive strengths up to 150 MPa.

The updated codes and standards, like EC2 and BS, provide a table that correlates between the environmental conditions around the structure and defines the minimum concrete cover, minimum cement content, and maximum water-cement ratio, and also the minimum concrete grade.

For an aging structure, the carbonation process is the main reason for corrosion of the reinforcing bars, especially for those that were poorly constructed, or structures containing a small amount of cement content in the concrete mix, less than the standard, or those located in polluted areas, such as urban regions. The carbonation process rarely happens in modern concrete bridges and new buildings, due to good quality control concepts and adherence. With good quality control, the percentage of water to cement is low, cement content is sufficient, and the compaction and concrete curing are very good. Also, in these structures, if the concrete cover maintains a match to the codes and specifications, it will be sufficient to prevent the spread of carbonation to the steel reinforcement throughout the structure's lifetime.

For reinforced concrete structures that are exposed directly to seawater or are near to it and those affected by sea water spray, the chlorides in this case will penetrate through the concrete cover until they reach the steel reinforcement bars; the chlorides cause a steel corrosion that is faster than the tin case of the carbonation process. The reason for the quick speed of the cycle of dry and wet on the surface of concrete is the process of carbonation, as the propagation of carbon dioxide gas is permitted in the dry cycle and dissolved in the wet cycle. Therefore this is considered a problem in some country regions like coastal cities, which are regularly exposed to cycles of wet and dry.

If, after investigation, it is found that corrosion has occurred because of inadequate concrete cover thickness, then the concrete around the steel reinforcement is carbonated and is causing the corrosion. In general, if the concrete cover is less than that required by the code or standard, the carbonation transformation process will speed up. This situation may lead to the process of corrosion starting at around 5 years after the construction year. If the concrete is of good quality, the carbonation will have difficulty in propagating and the inadequate thickness of the concrete cover will minimize the occurrence of carbonation as well.

Carbonation propagation rate

The propagation of carbon dioxide inside concrete follows Fick's law: the diffusion rate is in inverse proportion to the distance from the concrete surface, as in Eq. (1.14). Since the carbonation process is causing a change in the characteristics of pore voids in concrete, it reduces the base and also the value of pH. These equations present an approximate time of carbonation propagation.

Table 1.2 shows the different equations that define the relationship between the carbonation propagation rate and the concrete quality and environmental condition factors, such as relative humidity, w/c ratio, strength, and CaO content, which depends on the cement type and its alkali content surrounding the structure. If there is a crack, it will cause a change in concrete composition and moisture levels with depth and will also lead to a deviation from the perfect diffusion equation, so the rate will be higher than that in these equations.

Thus the suitable equation that should be used depends on the environmental conditions and all these factors, so each country and region should have its own study to define the optimum equation to use. For example, consider the basic equation:

$$C_d = At^{0.5} \tag{1.14}$$

where C_d is the carbonation depth in millimeters, A is a coefficient, and t is the time (years).

As an indication, in applying this equation, the depth of carbonation is about 16 mm in 16 years in the case of weak concrete and 4 mm in 20 years for good-quality concrete. Therefore it can be expected that the diffusion rate ranges between 0.25 and 1 mm/year.

Parrott (1987) performed research focusing on the time of the start of corrosion on steel reinforced bars in concrete. In bridges, the main driver of corrosion is chloride in cold countries; however, due to traffic with carbon dioxide it was found that the carbonation did not affect the bridges, even after 20 years, as from a practical point of view it is usually the case that bridges are constructed with high quality control and there are many precautions in compaction, curing process, and concrete mix design.

As discussed earlier, the design engineers during the design phase can control the carbonation process from the design phase by considering the proper selection of the thickness of the concrete cover and concrete mix by lowering the w/c ratio based on the surrounding environmental conditions.

Table 1.2 A selection of carbonation depth equations.

Equation	Parameters
$d = A\ t^n$	d, carbonation depth t, time in years A, diffusion coefficient n, exponent (approximately ½)
$d = ABC\ t^{0.5}$	$A = 1.0$ for external exposure $B = 0.07$ to 1.0 depending on surface finish $C = R\ (wc - 0.25)/(0.3(1.15 + 3wc))^{1/2}$ R = coefficient of neutralization, a function of mix design and additives
$d = A(\mathrm{B}wc - c)\ t^{0.5}$	A is a function of curing B and C are a function of fly ash used
$d = 0.43(wc - 0.4)(12(t - 1))^{0.5} + 0.1$	28 days cured
$d = 0.53\ (wc - 0.3)\ (12\mathrm{t})^{0.5} + 0.2$	Uncured
$d = (2.6(wc - 0.3)^2 + 0.16)t^{0.5}$	Sheltered
$d = ((wc - 0.3)^2 + 0.07)\ t^{0.5}$	Unsheltered
$d = 10.3\ \mathrm{e}^{-0.123f28}$ at 3 years	Unsheltered $f\mathrm{X}$ = strength at day X
$d = 3.4\ \mathrm{e}^{-0.34f28}$ at 3 years	Sheltered
$d = 680\ (f28 + 25)^{-1.5} - 0.6$ at 2 years	
$d = \mathrm{A} + \mathrm{B}/f28^{0.5} + \mathrm{c}/(\mathrm{CaO} - 46)^{0.5}$	CaO is alkali content expressed as CaO
$d = (0.508/f35^{0.5} - 0.047)(365\,\mathrm{t})^{0.5}$	
$d = 0.846(10wc/(10f7)^{0.5} - 0.193 - 0.076wc)(12\ \mathrm{t})^{0.5} - 0.95$	
$d = A((T - t_i)t^{0.75}\ \mathrm{C}_1/\mathrm{C}_2)^{0.5}$	t_i = induction time T = temperature in °K $C_1 = CO_2$ concentration $C_2 = CO_2$ bound by concrete

From Parrott, L.J., 1987. Steel corrosion in concrete: how does it occur? Mater. Protect. 6, 19–23.

It is thus recommended that the integrity teams from different disciplines should review the engineering document and specification from an integrity point of view. In the meantime, quality-control procedures should be applied to the construction of reinforcement concrete structures to maintain their durability.

Schiessl (1988) presents in Fig. 1.11 the relationship between the carbonation depth and corresponding time per year in different environmental

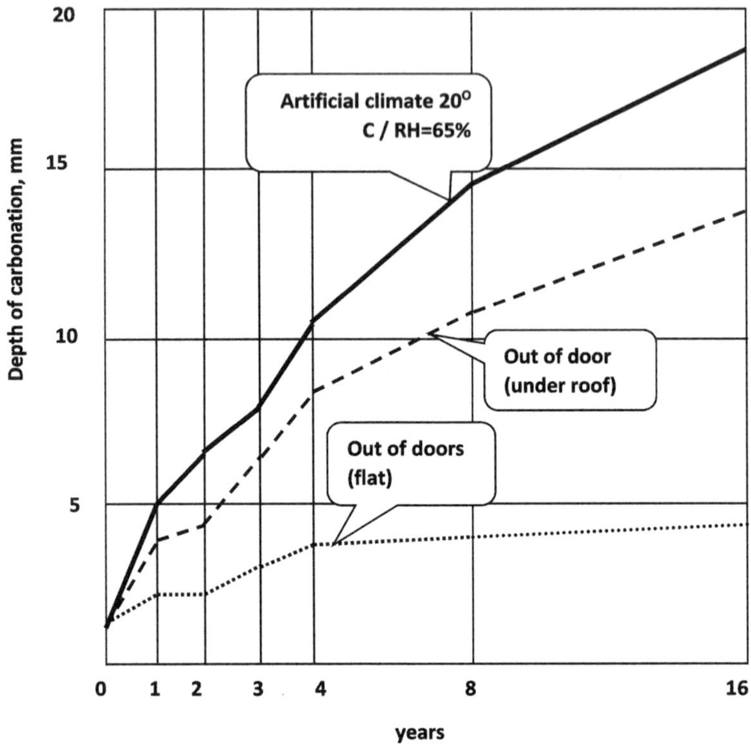

Fig. 1.11 Relation between carbonation depth with time at different environmental condition.

conditions. It was found that a relative humidity around 65% represents the worst conditions for fast propagation of carbonation inside the concrete.

1.4.3.2 Chloride attack

The chloride causes a dramatic problem to steel bars and 70% of the corrosion cases are due to chloride. The chloride can affect the steel bars in two ways: the chloride may be inside the concrete, or from outside with a marine structure or any building near the sea or ocean. Chloride being inside concrete can happen due to the following reasons (this can occur due to bad quality control, as the codes and standards provide many limits for chloride in all concrete components):

- the concrete was mixed using seawater
- using a calcium chloride additive to accelerate concrete setting time

- aggregate contains chlorides due to not being washed well
- using additives with higher chloride content than that defined in the specification
- the water in the concrete mix had a higher chloride content, greater than specified in codes and standards.

The chlorides can attack concrete externally when a concrete structure is exposed directly to seawater spray or is continuously exposed to seawater or salt water, such as when storing brackish water or with the use of salt to melt ice, which is common in countries with cold weather. In the oil and gas field, in some cases concrete tanks are used to store a brackish for water flood process system, colling system or firefighting, and the chloride will directly affect the tank surface. The chloride inside can occur as with concrete mix water, having a greater number of chloride ions, or any of its components having a higher chloride concentration.

It is very important to define the reason for chloride from inside or from outside, and this can be done by doing a test, as discussed in Chapter 5 as the repair methodology in both cases and precautions are different.

Chloride propagation

Chloride propagation inside concrete is similar to carbonation propagation, but its calculation is more complicated because salt water is absorbed quickly by dry concrete. In some cases, the movement of saline water is due to capillary rise phenomena and the results of chemical reactions between the chlorides and the concrete components absorbed in the pore voids of the concrete. The main factor is the chloride concentration, which varies. Due to these differences, the use of carbonation propagation definitions for chloride will not provide accurate data. In addition, defining chloride concentration on a concrete surface is difficult because its value varies with time from 0% to 100%, depending on whether it is a dry or wet surface, evaporation, and many other factors.

For integrity management, we need to make the decision not to go deeply into research of this topic. There are tests that represent the penetration of chloride inside concrete, as will be discussed in the next chapters. Using this test, there is a figure to represent the chloride concentration from the concrete surface until around 700 mm inside the concrete. Based on the tests, it is found that that the concentration of chlorides decreases rapidly after a short distance from the spread inside the concrete. From the graph of chloride concentration with depth, it can be obtained that the concentration has the shape of a simple parabolic curve, as described in the research in 1990 by Poulson with the Nordic Concrete Research Group.

It is important to know that the propagation of chlorides inside concrete depends on:

- the diffusion factor
- factors influencing the start of the chlorides
- the capillary rise phenomenon
- the absorption coefficient, which affects the outset at the first 5 mm from the surface.

1.4.4 Corrosion rates

The main question in structural integrity is the optimum time to start inspection. This mainly depends on the corrosion rate. As a general rule, when corrosion starts there is a reduction in the steel bar area that dramatically affects the structure safety, so if the corrosion rate is high then the probability of failure will be high and the reliability will be low. Therefore the maintenance program is based on that concept.

Corrosion rates are thus very important factors to know or predict. The corrosion rate value is affected by different factors; if these factors can be controlled, we can maintain a slower corrosion rate with the lowest value. In this case, the corrosion will occur but will not cause a serious problem to the structure over time.

The main factor that affects the corrosion rate is the presence of oxygen, especially in the cathodic area. So if the oxygen is nil, the corrosion rate will be slower. Different methods can be used to prevent the propagation of oxygen inside the concrete. This begins with a very simple method that costs nothing: apply concrete quality control on the site by having good concrete compaction to obtain a dense concrete cover with minimum voids, to prevent or reduce oxygen propagation inside the concrete.

The main items in quality control onsite involve taking care of the concrete compaction in order to obtain a dense concrete cover, so that oxygen propagation will be very slow or preventive. From a practical point of view, this is not so simple to apply and can be different from one construction site to another as it depends on the skills and knowledge of quality on the contractor and owner side, and whether quality control is truly one of the project targets or not. When preventing the propagation of oxygen inside concrete, the oxygen in the steel bars in the cathode area will be less, so the corrosion rate will be at a minimum value. It is also found that when the concrete strength is high, the concrete has a high resistance to electrical current, which will prevent or slow down electron movement from the anode to the cathode area and this will decrease the corrosion rate as well.

The passive protection layer will be stable if the aqueous solution contains a high quantity of hydroxide ions (OH^-). This layer is affected by chloride ions (CL^-) or carbonates reducing the hydroxide in the water solution in the concrete pore void, causing the formation of voids and pitting in steel bars.

In general, the steel corrosion rate in concrete is expected to change as a function of time. In attempts to model this time-dependent corrosion behavior, it is convenient to distinguish and clearly understand the following corrosion process stages:

1. After pouring the concrete, the steel bars are protected by concrete as in alkaline, as shown in Fig. 1.12.
2. The carbonation or chloride propagation takes a time, tp, which is the propagation time until it reaches the steel bars, as shown in Fig. 1.13, and the pH will reach 8; this time period can be calculated from Eq. (1.14).
3. The corrosion begins and takes a time, t_c, which is the corrosion time, and there is a reduction of the steel reinforced area, which will increase the structure probability of failure.

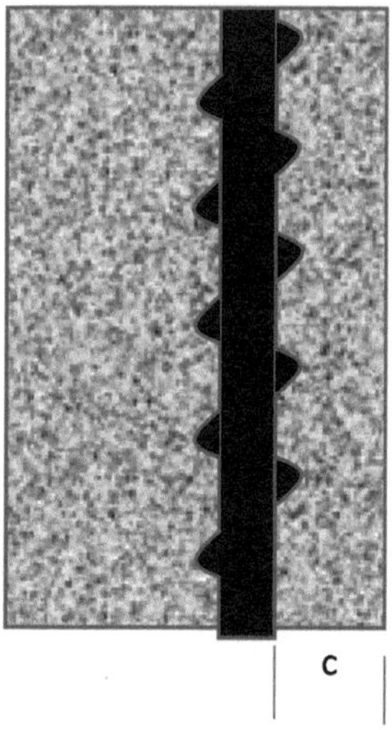

Fig. 1.12 Day one after construction.

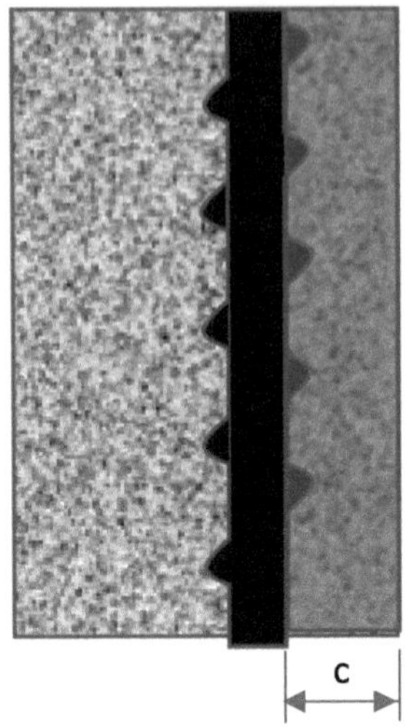

Fig. 1.13 Propagation of carbonation or chloride.

4. During the time period until spalling of the concrete cover, t_s, the structure is under a high value of probability of failure, as there is a reduction in the steel bar area and also there is a reduction of the concrete section, so action should be taken. This time period can be calculated from Eq. (1.15).

These stages present the main tools for defining the maintenance strategy or the integrity management system program. If you want to do nothing until reaching stage 5, a corrective action or "run to failure" strategy is implied. This can actually work and is valid in some cases; this is a main strategy in some facilities and currently it is used in most of the concrete structures in residential buildings. However, with industrial standards, an integrity management system is typically in place, so the strategy is to be proactive at stages 1 and 2.

Carbonation, chloride, and bacteria propagation and the transformation process from alkalinity to acidity are the main causes of undermining the passive protection of steel reinforcement and thus allowing the starting of corrosion.

When the corrosion process starts, the deterioration of steel depends mainly on the corrosion rate. This corrosion rate, as already discussed, depends on many factors; in the case of carbonation, the rate of corrosion is greatly influenced by moisture-relative humidity, for example. The carbonation propagation is much lower during a decrease in the relative humidity inside the concrete voids, at less than 75%. The rate of corrosion increases significantly when the relative humidity increases to 95%. Based on Tutti's studies in 1982, it was found that by reducing the temperature, the corrosion rate would decrease: a reduction from 5 to 10 times in the corrosion rate with a temperature reduction of 10 °C, and vice versa with temperature increases.

It can be seen that the corrosion rate in water was 0.0678 mm/year and 0.0073 mm/year for the uncoated and coated bars, respectively. With the NaCL and Na_2So_4 solutions, the corrosion rate was 0.098 mm/year and 0.0130 coating for uncoated and coated steel bars, respectively.

El-Abiary et al. (1992) stated that the corrosion rate in most research is between 0.015 and 0.09 mm/year.

1.4.5 Corrosion's effect on spalling of concrete cover

As the volume of a steel bar increases due to corrosion, so the pressure increases, and cracks propagate through the cover. Experimental results have shown that, although at the early stages of crack propagation several cracks appear, by the end of the test a single crack finally breaks the cover (Nguyen et al., 2006; Ohtsu and Yosimura, 1997) on the weakest side of the concrete element. When this crack appears, the internal stresses relax, which stops the propagation of other internal cracks (Nguyen et al., 2006).

Most problems that occur because of corrosion of steel in concrete are due not only to the reduction in the steel section but also to failure of the concrete cover. Various efforts have been made to estimate the amount of corrosion that will cause spalling of the concrete cover.

As per Broomfield in 1997, the cracking will start with less than 0.1 mm of steel bar section loss, but in some cases far less than 0.1 mm has been needed. This mainly depends on the way that the oxide is distributed, the concrete existing strength, the applied stresses on it, and the bar distribution and shape, as it can cause a stress concentration that leads to a rapid occurrence of crack propagation; this can happen with steel bars in the corner, as there is not much confinement so the cracks appear quickly. For any column, the corrosion will start on the corner at the bottom of the column, as it is the first part to experience cracking due to corrosion. The inspection team should be aware of that during inspection. From the corrosion rate measurements, about 10 μm section loss or 30 μm rust growth is sufficient to cause cracking.

As per El-Abiary et al. (1992), the time period between the starting of corrosion and spalling of concrete can be calculated by this equation:

$$t_s = \frac{0.08.c}{d.C_r} \tag{1.15}$$

where t_s is time from corrosion start until spalling of the concrete cover in years, c is the concrete cover in millimeters, d is the steel bar diameter, and C_r is the average of corrosion rate.

In 1988 Morinaga, S. provided a study presenting the conditions when the concrete cover cracks as a result of the corrosion of reinforcing steel bars. Morinaga performed two experiments. One of them was done to define the tensile stress when the concrete cover cracks. In this test, a hollow concrete cylinder was used as a specimen, with the internal diameter of the hollow cylinder simulating the reinforcing steel bar diameter and the wall thickness of the hollow cylinder as the thickness of the concrete cover. Then oil pressure was applied to the inner surface of the hollow cylinder. Then the influence of bar diameter, cover thickness, and tensile strength of the concrete on the maximum oil pressure at the specimen failure was examined and investigated.

The second test was performed to define the tensile strain when the concrete cover cracks. In this test a piece of reinforcing steel bar was embedded at the center of the concrete cylinder. The specimen was immersed in a salt solution and a direct electric current applied to the reinforcing steel to force it to corrode electrolytically. After the concrete cylinder was cracked, the reinforcing steel was weighed and the amount of corrosion products was calculated.

After that, Morinaga merged the results of these two experimental tests and provided the equation to estimate the amount of corrosion when the concrete cover cracks. This equation is as follows:

$$Q_{cr} = 0.602\left(1 + \frac{2c}{d}\right)^{0.85} d \tag{1.16}$$

where Q_{cr} is the amount of corrosion when concrete cracks ($\times 10^{-4}\,\mathrm{g/cm^2}$); C is the concrete cover thickness in millimeters; and d is reinforcing steel diameter in millimeters.

1.4.6 Bond strength between concrete and corroded steel bars

There has been much research and development in this area, but in general the studies agree that the bond strength is increased in low levels of corrosion but is decreased rapidly in the case of a high level of corrosion, as per a study done by Valenet (2012).

An experimental research study done by Coccia et al. (2016) revealed that low corrosion percentages, such as lower than 0.5%–0.6% in mass loss, will cause an increase of the bond strength up to about 50%–60% with respect to the sound bars; higher corrosion entities cause a sharp bond reduction; as a matter of fact, for a mass loss of about 1.5% the maximum bond stress is reduced about 40%.

1.4.7 Influence steel bar corrosion on the shear strength

The effect of corroded stirrups on the shear strength was studied by Xue et al. (2012) and they concluded that if the percentage maximum local mass loss of the stirrups was below 35%, stirrup corrosion had little influence on the load-carrying mechanism and the modified truss theory could be applied. The shear capacity can be evaluated simply by allowing for the reduction of Vs due to the sectional area loss of the stirrups. The influence of stirrup corrosion on Vc can be ignored. If the percentage maximum local mass loss of stirrup area is higher than 35%, the shear capacity decreases gradually as the corrosion level of the stirrups increases. However, there is little difference in critical crack behavior between the corroded specimens and the sound specimen.

With regard to corroded longitudinal steel bars, Xue and Seki (2010) studied the influence of longitudinal bar corrosion on shear behavior of RC beams and they found that the bond behavior between longitudinal bars and concrete plays a very important role in the load-carrying mechanism of RC beams. As longitudinal bars corrode, the deterioration of bond strength may reduce the stiffness of the RC beams, and furthermore result in a transition of the load-carrying mechanism.

The shear behavior of RC beams with corroded longitudinal reinforcing steel bars is influenced not only by the corrosion level of the longitudinal bars, but also by the shear-span-to-effective-depth ratio. When the shear-span-to-effective-depth ratio is above 3.0, the prediction of the shear capacity of RC beams with corroded longitudinal bars using the current shear equation will result in an unsafe overvaluation.

References

Baker, M., 2008, November. Pipeline Corrosion: Final Report. U.S. Department of Transportation: Pipeline and Hazardous Materials Safety Administration.

Cady, P.D., Weyers, R.E., 1984. Deterioration rates of concrete bridge decks. J. Transp. Eng. 110 (1), 34–444.

Coccia, S., Imperatore, S., Rinaldi, Z., 2016. Influence of corrosion on the bond strength of steel rebars in concrete. Mater. Struct. 49 (1), 537–551.

El-Abiary, S., et al., 1992. Deteriorated Concrete Structures. University Publishing House, Cairo, Egypt.

El-Reedy, M.A., 2007. Steel-Reinforced Concrete Structures: Assessment and Repair of Corrosion, second ed. CRC Press.

Kilareski, W.P., 1980. Corrosion-induced deterioration of reinforced concrete—an overview. Mater. Perform. Natl. Assoc. Corros. Eng. 19 (3), 48–50.

Mori, Y., Ellingwood, B.R., 1994a. Maintaining reliability of concrete structures. I: role of inspection/repair. J. Struct. Eng. ASCE 120 (3), 824–845.

Mori, Y., Ellingwood, B.R., 1994b. Maintaining reliability of concrete structures. I: role of inspection/repair. J. Struct. Eng. ASCE 120 (3), 824–845.

Nguyen, Q.T., Millard, A., Care, S., L'Hostis, V., Berthaud, Y., 2006. Fracture of concrete caused by the reinforcement corrosion products. J. Phys. IV 136, 109–120.

Ohtsu, M., Yosimura, S., 1997. Analysis of crack propagation and crack initiation due to corrosion of reinforcement. Constr. Build. Mater. 11 (7–8), 437–442.

Parrott, L.J., 1987. Steel corrosion in concrete: how does it occur? Mater. Protect. 6, 19–23.

Schiessl, P., 1988. Corrosion of steel in concrete. Report of the Technical Committee 60-CSC.

Takewaka, K., Matsumoto, S., 1988. Quality and cover thickness of concrete based on the estimation of chloride penetration in marine environments. In: Malhotra, V.M. (Ed.), Concrete in Marine Environment, SP-109. American Concrete Institute (ACI), Detroit, MI, pp. 381–400.

Thoft-Christensen, P., 1995. Advanced bridge management systems. Struct. Eng. Rev. 7 (3), 149–266.

Valenet, M., 2012. Bond Strength Between corroded steel rebar and concrete. IACSIT Int. J. Eng. Technol. 4 (5), 653–656.

Xue, X., Hiroshi, S., Song, Y., 2012. Influence of Stirrup Corrosion on Shear Strength of RC Beams Applied Mechanics and Materials. vols. 204–208, pp. 3287–3293.

Xue, X., Seki, H., 2010. Influence of longitudinal bar corrosion on shear behavior of RC beams. Jpn. Concr. Inst. J. Adv. Concr. Technol. 8 (2), 156.

CHAPTER 2

Integrity management versus reliability

2.1 Introduction

The main target in asset integrity management is to maintain the structure reliability over its lifetime so that its full function is preserved and so that there are no negative effects on the industry production process.

The asset integrity management for a maintenance plan is based on a facility's qualitative and quantitative risk assessment, which is directly affected by the facility's or structure's reliability or probability of failure. Therefore this chapter is crucial for understanding the core of the asset integrity management concept, to be able to deal with it professionally.

There are many mature structures worldwide, which is reflected in numerous research studies and textbooks focusing on the reliability of reinforced concrete structures, offshore structure platforms, and pipelines. Also, there is a trend to use various tools to predict the structure service life. This chapter covers updated research and development in the area of reliability and its relationship with integrity management, as it is considered the technical arm of the integrity management system.

The reliability analysis depends on probability theory. In the 17th century, Blaise Pascal and Pierre de Fermat began to formulate probability theory. Renyi (1973) stated that they created mathematical problems to present games of chance. Engineers were very interested and focused on their theories of probability, because they focus on the likelihood of occurrence of a certain event and try to define uncertain measures of random events.

The concept of probability is well known, but there is some difficulty in defining uncertainty and accurately evaluating it for large structural systems, as with platform structures. The developments in computer capabilities have allowed the creation of a numerical solution for large-scale, complex systems that have uncertainties in their behavior.

The main tools for integrity management systems are using a proactive maintenance approach depending on a risk-based inspection. Risk for any structure is the probability of failure multiplied by the consequences in

Asset Integrity Management for Offshore and Onshore Structures
https://doi.org/10.1016/B978-0-12-824540-8.00002-8

the case of its failure. Thus there is a direct relation between asset integrity management and the calculation of its probability of failure and its reliability as the other face of the coin.

The reliability analysis for onshore structures, pipelines, offshore structures, and FPSOs, with the main concept of structure reliability and ways to obtain the structure probability of failure, is discussed in this chapter.

One of the interesting research fields involving structural reliability is the study of sources of randomness in structure, including the variability of resistance and loads. It is important to know that there are two types of variables that affect the reliability of structures:

- basic variables such as load and resistance variables
- nonbasic variables such as the uncertainty in the theoretical equation of design and structural analysis and others.

All the methods that are used to obtain the system probability of failure mainly depend on knowing the probability distributions of loads and strength variables. For example, with a live load having a high variability, many statistical research studies were carried out to study the live loads probability characteristics.

On the other hand, many research works also study the variables that affect the strength of the structure, for example the concrete strength, yield strength, and the reinforcing steel bars cross-sectional area.

To obtain the structure radiality, the first step is to define the failure modes and represent them by their limit state equation, and then the second step is to define the best probability distribution and main statistical parameters like the mean, standard deviation, and the coefficient of variation for each parameter. Then by using any of the reliability analysis methods that will be discussed in the next chapter, the structure probability of failure can be obtained.

The main parameters for any mode of failure are the load and the strength. As shown in Fig. 2.1, the probability of failure will occur when a high applied load value and lower strength value occur at the same time.

2.2 Principle of reliability

The reliability analysis principle can be applied in any structure system related to offshore and onshore structures. During the design stage, the designer will input the values of the loads into the design codes or standards using only one value; however, the value should be represented by a probability distribution, as it may be high or low. When calculating a member

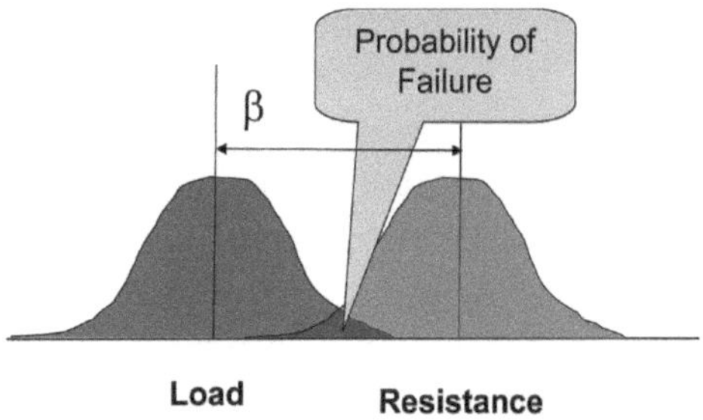

Fig. 2.1 Probability of failure diagram.

resistance capacity from an equation in any code or standard, one value is provided as well, so the calculation is called a deterministic calculation. However, if every parameter is taken as a variable represented by a probability distribution, the member resistance capacity will be represented as a probabilistic distribution and the calculation method is called a probabilistic study. This type of study is applicable in research and in reliability analysis calculations, but this will be superseded in daily design activities by factors for loads and designs, to cover the variability.

Fig. 2.1 shows the master scene of the probability of failure definition and its relationship to reliability. One can find from this figure that the probability of any structure failure occurs in the zone of intersection between the two curves of load and resistance. The definition of structure failure is a partial collapse, total collapse as per Fig. 2.2, or member failure. As shown

Fig. 2.2 Collapse of new building in Dubai.

in Fig. 2.1, the failure happens when the load has its highest value and at the same time the strength of the member is too weak due to deterioration, bad quality control, incorrect design, or a combination of different reasons. This curve is the main principle to define the relationship between the reliability and probability of failure in any structure system. The probability of failure is the main target for structural integrity, as the purpose of integrity is to maintain the structure safe during its lifetime, or to have a reliable structure. With pipelines, tanks, or any vessel the main goal of its structural integrity is to keep the hydrocarbon safely inside it, from this objective there will be an enough safety margin to have any leak or damage to the facility.

Practically speaking, the load usually increases as a result of changes in the mode of operation or many loads affect the structure at the same time; for example, in residential buildings sometimes all the furniture may be stored in one room just while painting or remodeling the other rooms, or the load may increase for a party or celebration.

The strength can be weakened due to bad quality control during construction or installation in the case of offshore structure new buildings. The strength of the structure will weaken with age as a result of deterioration due to corrosion in the steel bars and spalling concrete cover or cracks in a concrete structure member in old buildings, or in the case of corrosion in cutting members and other anomalies with offshore structures.

2.2.1 Structure reliability calculation

There are different methods to obtain the structure reliability or, on the other hand, to calculate the structure probability of failure. This method of calculating the structure probability of failure falls under the concept of a risk quantitative method. As discussed in following chapters, it can be obtained in pipelines or tanks or even onshore structures, but it is most costly to apply it to offshore structures. A great deal of research and development is ongoing in this area, and we will also use it as presented in the last chapter, as it is part of the structure integrity management system.

In 2012, El-Reedy presented in detail different methods that can be used to calculate structure reliability:

(1) First-order reliability method (FORM)
(2) Second-order reliability method (SORM)
(3) Monte-Carlo simulation method
(4) Integration method
(5) Approximate methods

The first three methods are most commonly used for onshore and offshore structural reliability. According to Ang and Tang (1984), Melchers (1983), Frangopol (1985), and Schueller and Stix (1987), the analytical and integration methods are not easy to use and the simulation method is the most appropriate method. The approximate method was initiated by Cornell (1967) by providing the upper and lower bound and there was some discussion and further development by Moses (1982) and by Garson (1980) and Ramachandran and Baker (1985), but this method is not currently applicable.

Many research studies have been carried out on reliability of concrete, steel, pipelines, and offshore structures. Recently, due to the aging of offshore and onshore facilities and structures, there has been an impetus to study structural reliability and remaining life.

One example of a reliability study was performed by Ahmed (1990) concerning the reliability of plane ductile structures against plastic collapse, which considered all possible collapse modes of ductile plane structures. This research was based on a parametric study with quantitative comparison of the relation between the factor of safety and the reliability index. This research was also extended to cover the optimum design of portal frames using a probabilistic design approach; the study examined the effectiveness of different alternatives for strengthening an existing portal frame capable of withstanding a higher load.

Frangopol and Curley (1987) studied the reliability of existing trusses and bridges and they presented a method to define the structural redundancy of deteriorating structures. In 1987 Takaoka, Shiraki, and Matsuho studied the reliability analysis of expressway bridges using random process theory and a simulation technique.

For offshore structure reliability, in 1982 Thoft-Christensen and Baker performed many detailed reliability studies of offshore structures under the actions of extreme environmental conditions, such as wind, waves, and current, and they presented various ways of modeling jacket structures.

El-Reedy and Ahmed (1998) studied the reliability of tubular joints on offshore structures based on an inspection strategy. A reliability analysis of reinforced concrete members after strengthening by epoxy-bonded carbon fiber-reinforced polymer (CFRP) was published by Plevris et al. (1995). Surahman and Rojiani (1983) studied reinforced concrete frames from the point of view of reliability analysis and creating an optimum design approach. Sirvidya and Ranganathan (1995) studied a reliability-based optimal design of a reinforced concrete frame structure. Dagher et al. (1993) presented system reliability concepts in the design of transmission pipelines.

In 1969 Castello and Chu studied singly and doubly reinforced concrete sections and examined the probabilistic relationships between flexural resistance and material strengths. El-Reedy et al. (2000) discussed the reliability of reinforced concrete columns under corrosion attack.

Many studies have considered the corrosion of steel bars in reinforced concrete structures and its effect on the structural reliability: for example, a study was carried out by Frangopol et al. (1997) on the reliability of reinforced concrete girders under corrosion attack. Mori and Ellingwood (1994) discussed comprehensively the reliability of concrete structures and also examined inspection and repair optimization, which is really the core of an integrity management system. A technical report from Det Norske Veritas USA Inc. (DNV) (1996) discussed offshore structure reliability.

Important research was carried out in 1990 by Mirza and MacGregor covering materials and geometric variations and their effects on the reliability of reinforced concrete slender columns bent in single curvature. In this study the columns had hinge connections in both ends, with equal load eccentricities acting at both ends. It was found that the material strengths and geometric properties of the column varied randomly and the resulting variation in the ultimate strength of the column was determined. In this study different combinations of the size of the cross-sectional, the concrete strength, the ratios for longitudinal reinforcement, the slenderness ratio, and the end eccentricity ratio were examined. All these variables were considered to determine their influence on the properties of the slender column strength probability distribution. From this study, it was found that the ratio of longitudinal steel reinforcement, the slenderness ratio, and the end eccentricity ratio had a significant influence on the probability distribution properties of the slender column strength. In addition, the study revealed that the concrete strength variation was a major contributing factor to the slender column strength variability in the case of low eccentricity ratios, whereas the variability in the steel strength had a major influence on the slender column strength variability when the end eccentricity ratios were high. This study is important because it describes the variation in strength of the column in different cases, considering that the column is the low redundancy member in the structure, since probability of failure of the whole structure relies on its probability of failure.

In 2000 a study was carried out by El-Reedy et al. to consider column reliability under different corrosion rates related to different environmental conditions and considering all the preceding variations, in addition to loading effects. Currently, due to more advanced computing power, the most practical reliability calculation is the Monte Carlo simulation technique. In this study, the reliability of a reinforced concrete column was calculated under corrosion attack with different corrosion rates, reinforcement ratios, and eccentricity, to discuss their influence on the reliability of the reinforced concrete column.

The following sections examine the statistical data and probability distributions used to calculate the reliability of a structure. After obtaining the probabilistic calculation of load, the probabilistic calculation of strength is discussed in Chapter 5, along with practical methods that can be used to calculate the structure probability of failure, or that of any element, by focusing on the reinforced concrete column.

With mature structures, the first step is to do a complete evaluation of the structure to determine all the probable events that cause cracks or deterioration of any of the structure members. Therefore the inspection technique and evaluation methodology should be understood very well, as discussed in Chapter 3. Also, this chapter illustrates the nondestructive tests that can be used to evaluate the concrete strength, in addition to the corrosion tests that can provide a full image of the structure from a corrosion point of view.

All the engineers and owners are very interested in knowing the remaining life of any structure, due to business objectives or for preparing a maintenance plan.

2.2.1.1 Reliability analysis using first- and second-order methods

In 1969 Cornell presented a second-moment method as an alternate formula to calculate the reliability index, β, also called the safety index, which was easier to calculate than the probability of failure. The safety index or reliability index was defined as the mean safety margin divided by its standard deviation, the safety margin simply meaning mean strength minus the mean load that affects the element.

If the distributions of the basic design variables are normal, the reliability index can be expressed in terms of the first and second moments, which are the design variables mean and variance, and the probability of failure $P[F]$ can be obtained from the following formula:

$$P[F] = \phi(-\beta) \tag{2.1}$$

where ϕ is the cumulative of the function of the standard normal distribution. Hasofer and Lind (1974), Ditlevesn (1979), and Shinozuka (1983) provided a geometrical definition for the reliability index, which is the shortest distance from the mean of the basic limit state equation variables to the boundary of the failure surface, considering that all variables are measured in standard deviation units.

With variables that follow other probability distributions than the normal distribution, the reliability index can be obtained by using the Rackwitz and Fiessler (1978) method, as this approach is used to obtain a normal approximation of nonnormal distributions at the so-called design points. Thoft-Christensen and Murotsu (1986) stated that in cases in which the limit state function is complicated, the reliability index can be evaluated iteratively. This method is traditionally used in offshore structure reliability since using the simulation is not easy. However, some research uses simulation methods, but only for load and resistance in general and not considering all factors.

2.2.2 Probability in limit states design

In 1976 Allen provided the development of probability-based design criteria that can be used in the general limit state design philosophy. The definition of the limit state is a condition in which a structure or structural element in some way becomes unsafe or not fit for purpose. In most structure systems, the limit states can be categorized into two groups:

(1) ultimate limit states, which are directly related to the structural collapse or failure of part of the structure that may cause a loss of life or major financial loss, and the probability of occurrence is low.

(2) serviceability limit states, which are directly related to the structure being unfit for purpose, such as cracks or deformations on the structure members.

Limit state design is a process consisting of these three steps:

(1) definition of all the modes of failure of the structure or modes that can affect its serviceability;

(2) definition of the acceptable levels of safety against each limit state;

(3) consideration by the designer of each significant limit state.

The limit state equation is a function of the structure resistance R of the structural element and of the load effect L acting on it; R and L are random variables and this limit state equation in most research has two forms, as follows:

$$Z = R/L \text{ or } Z = R - L \tag{2.2}$$

They are essentially the same but the first form is better in the case of using a lognormal distribution, where

R is the resistance random variable,

L is the load random variable,

Z is also a random variable.

whereas R and L are calculated from the formulated probabilistic formula for resistance and load, respectively.

By knowing the probability density function of the resistance R and load effect L, the probability density function of Z can be obtained.

The probability of failure of a structural element (the shaded area represented in Fig. 2.1) is equal to

$$P_f = P[R/L < 1.0] \tag{2.3}$$

and since ln1.0 = 0.0, then Eq. (2.3) becomes

$$P_f = P[\ln(R/L) < 0],$$

$$P_f = \int_0^1 f_z(z)dz = F_z(1.0) \tag{2.4}$$

where $f_z(z)$ is the probability density function of the random variable Z.

If the probability density function of variable z is a normal distribution, the probability of failure will be calculated by the following equation:

$$P_f = P[Z < 1.0] = \phi\left[\frac{1.0 - \mu_z}{\sigma_z}\right] \tag{2.5}$$

where ϕ is the cumulative standard normal distribution function.

The structure safety or its reliability can be measured in terms of a safety index, also called the reliability index, β, as in Fig. 2.1, which was defined in 1969 by Cornell using the second-moment format as the number of standard deviations of the safety factor z by which its mean exceeds 1.0.

Thus the safety index β is

$$\beta = \left[\frac{1 - \mu_z^N}{\sigma_z^N}\right] \tag{2.6}$$

According to Ang and Tang (1984), Z is a normally distributed random variable if R and L are lognormally distributed variables. Based on their study, the live load is represented by a lognormal distribution. But structural member strength has different parameters represented by different probability distributions. Considering that the strength of the structural element is a variable, R, which was represented by a distribution that is not a lognormal distribution. Based on that, the random variable, Z, representing the safety of the structural element, will not be a normal distribution.

2.2.2.1 Equivalent normal distribution

According to Ang et al. (1984), if the random variable Z probability distribution is known and it does not follow a normal distribution, the probability of failure will be obtained by using a numerical integration technique; the equivalent normal distribution for a nonnormal variable will be obtained by making the cumulative probability and also the probability density ordinate of the equivalent normal distribution equal to those of the corresponding nonnormal distribution at the appropriate point on the failure surface, as shown in Fig. 2.3.

By equating the cumulative probabilities at the point of failure, as mentioned previously, with $Z=1.0$,

$$\phi\left(\frac{1.0-\mu_Z^N}{\sigma_Z^N}\right)=F_Z(1.0) \tag{2.7}$$

By substituting Eq. (2.4) into Eq. (2.7), then

$$\phi\left(\frac{1.0-\mu_Z^N}{\sigma_Z^N}\right)=P_f \tag{2.8}$$

where μ_Z^N and σ_Z^N are the mean value and standard deviation, respectively, of the equivalent normal distribution for Z.

The preceding equality then yields

$$\mu_Z^N=1.0-\sigma_Z^N\phi^{-1}\left[P_f\right] \tag{2.9}$$

whereas, equating the corresponding probability density ordinates at $Z=1.0$,

$$\frac{1}{\sigma_Z^N}\varphi\left(\frac{1.0-\mu_Z^N}{\sigma_Z^N}\right)=f_z(1.0) \tag{2.10}$$

Then by solving Eqs. (2.8) and (2.10), the standard deviation of the equivalent normal distribution Z can be calculated using this equation:

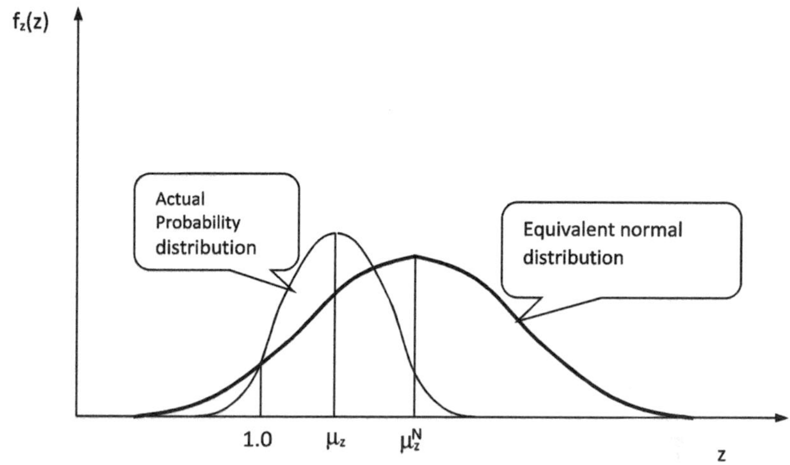

Fig. 2.3 Random variables actual and equivalent normal distribution.

$$\sigma_Z^N = \frac{\varphi\{\phi^{-1}[P_f]\}}{f_z(1.0)} \tag{2.11}$$

and from Eqs. (2.9) and (2.11) one can get μ_Z^N, σ_Z^N.

Then the reliability index is:

$$\beta = \frac{1-\mu_Z^N}{\sigma_Z^N} \tag{2.12}$$

On the other hand, Eq. (2.12) can be used to evaluate the probability of failure P_f by converting the nonnormal distribution variant into a normal distribution variant.

2.2.2.2 Procedure for reliability analysis

Probability-based limit state design customarily employs loads or load effects that are multiplied by load factors and resistance, which are multiplied by resistance factors in a set of checking equations which have the general form:

Factored resistance ≥ Effects of factored load

Ang and Cornell (1974) presented a method to calculate probability-based design provided by classical reliability theory. The loads and resistance are assumed to be statistical variables and the necessary statistical parameters are assumed to be variable. According to Galamobos et al. (1982), this provides a strong theoretical basis for their evaluation of structural reliability, but the many conceptual and operational difficulties in its use led to the creation of first-order, second-moment (FOSM) methods.

Galamobos et al. (1982) introduced a mathematical formula defining the relation between resistance and load effect variables for the specific limit state of interest:

$$g(X_1, X_2, \ldots X_m) = 0 \tag{2.13}$$

in which x_i represents the resistance and load variable failures, and by this equation the failure happens when $g < 0$ for the specific ultimate or serviceability limit state. Then the reliability associated with any given design is checked by identifying a point (X_1, X_2,...X_m) satisfying Eq. (2.13).

$$x_i = \mu_{x_i} - \alpha_i \beta \sigma_{x_i} \tag{2.14}$$

$$\alpha_i = \frac{\sigma_{x_i} \dfrac{\partial g}{\partial x_i}}{\sqrt{\sum \left(\sigma_{x_i} \dfrac{\partial g}{\partial x_i} \right)^2}} \tag{2.15}$$

where μ_{xi} and σ_{xi} are the average and standard deviation, respectively, and β is the reliability index. The partial derivatives in Eq. (2.15) are evaluated at the point defined by Eq. (2.1). Ang and Tang (1984) stated that this point is called the checking or design point. Their study revealed that by doing iteration for the direction cosines, α_i, the minimum value of β is provided, where β is the shortest distance from the origin to the failure surface.

Ang and Tang (1984) introduced the linear performance function, as follows:

$$g(X) = a_0 + \sum_i a_i X_i \tag{2.16}$$

where a_0 and a_i are constants. The shortest distance from the origin to the failure plane surface will be from the following equation, as shown in Fig. 2.4:

$$\beta = \frac{a_0 + \sum_i a_i \mu_{X_i}}{\sqrt{\sum_i (a_i \sigma_{X_i})^2}} \tag{2.17}$$

If the statistical parameters such as mean and standard deviation are available for different variables and their probability distributions are known,

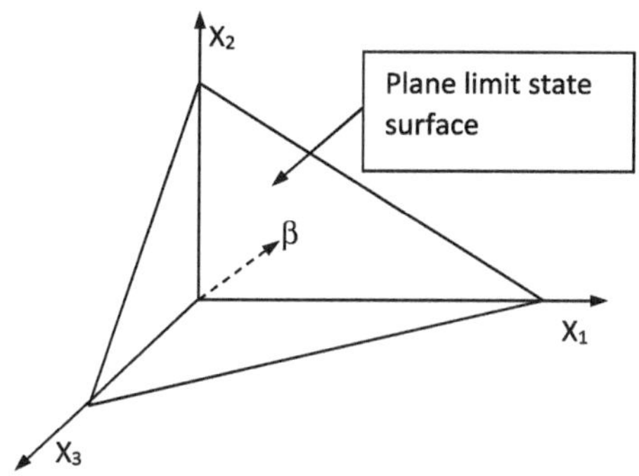

Fig. 2.4 Linear limit state surface in x_1, x_2, x_3 surfaces.

this information can be included in the analysis in an approximate way that does not require multidimensional integration, which is required in traditional reliability analysis.

Corotis and Doshi (1980) and Simiu and Filliben (1975) stated that many structural problems involve random variables that do not follow the normal distribution. In this situation, nonnormal variables are transformed to equivalent normal variables for the solution of Eqs. (2.14)–(2.16).

As per Ang and Tang (1984), this transformation will occur at a point close to that where failure is most likely, meaning at a point on the failure surface; herein μ_{x_i} and σ_{x_i} of the equivalent normal variables are chosen, considering that at the checking point x_i, the cumulative probability distribution and probability density function of the actual and approximately normal variables are equal.

Having determined μ_{x_i} and σ_{x_i} of the equivalent normal distributions, the solution proceeds as in Eqs. (2.15) and (2.16).

2.2.2.3 Reliability index calculation for different codes

The previous approach can be applied to calculate the reliability index as per the code limit state equation, based on the British standard BS 8110. Applying this procedure of BS 8110 for calculating reliability in the case of a reinforced concrete column, the equation of the limit state will be as follows:

$$\varphi R_n \geq 1.4\, D_n + 1.6\, L_n \tag{2.18}$$

where R_n, D_n, and L_n are respective nominal values of resistance and loads.

The reduction factor for strength (φ) will be obtained as its value depends on the percentage of longitudinal steel bars and the ratio between yield strength and concrete compressive strength.

Frequently, structural engineers use "nominal" values of loads and resistances, which are generally different from the mean value. The ratio between the mean value and its nominal is called the bias factor. Galamobos et al. (1982) showed that the ratios between the mean loads and their nominal values are as follows:

$$\frac{\mu_D}{D_n} = 1.05$$

$$\frac{\mu_L}{L_n} = 1.15$$

whereas in the case of axial compression of a tied column,

$$\frac{\mu_R}{R_n} = 0.95$$

Assume R and D are normally distributed random variables and (L) is a type I asymptotic random variable. As per Galamobos et al. (1982), the corresponding coefficient of variation (CoV) associated with these variables are as follows:

$$VR = 0.11$$

$$VD = 0.10$$

$$VL = 0.25$$

Assume $L_n/D_n = 1.0$, which is the live load to dead load ratio.

$$\frac{\mu_L / 1.15}{\mu_D / 1.05} = 1.0 \quad \mu_L = 1.095\mu_D$$

whereas the BS requirements become

$$\varphi\frac{R'}{0.95} = 1.4\frac{D'}{1.05} + 1.6\frac{L'}{1.15}$$

The British standard uses a partial safety factor for concrete and another for steel, which is different from ACI318, which uses one reduction factor multiplied by resistance. For simplicity, the reduction factor (φ) is equal to 0.54, which corresponds to the steel percentage between area of steel to concrete area (ρ) being equal to 1%.

The cumulative and density probability asymptotic type I random variables will be as follows:

$$F_L(1) = \exp\left(-e^{-\alpha_L(1-u_L)}\right)$$

$$f_L(1) = \alpha_L \exp\left(-\alpha_L(1 - u_L) - e^{-\alpha_L(1-u_L)}\right)$$

The parameters that describe the asymptotic live load α_L and u_L will be calculated as follows:

$$\alpha = \frac{\pi}{\sqrt{6}} \cdot \frac{1}{\sigma_L} = \frac{4.686}{D'}$$

$u = L' - (0.577/\alpha) = 0.9721\ D'$

In this case, that is, involving nonnormal distributions even though the performance function is linear, the mean values and standard deviations are unknown as these are now functions of respective failure point values. An iterative solution using Eqs. (2.5) and (2.6) is therefore necessary.

For the first iteration assume

$$l^* = \mu_L = 1.095\,\mu_D$$

For the type I extremal distribution of L

$$F_L(l^*) = 0.57$$

$$f_L(l^*) = 1.5/D'$$

and thus, using Eqs. (2.5) and (2.6), transform from nonnormal random variables to equivalent normal variables.

$$\sigma_L^N = 0.26\mu_D$$

$$\mu_L^N = 1.05\mu_D$$

Knowing the value of the mean and standard deviation of the normally equivalent normally distributed variables, the safety index can be calculated using Eq. (2.17); it is found to be:

$$\beta = \frac{\mu_R^N - \mu_D^N + \mu_L^N}{\sqrt{(\sigma_R^N)^2 + \sigma_D^2 + (\sigma_L^N)^2}} = 4.18 \tag{2.19}$$

and the failure point is

$$D^* = \mu_D - \beta\sigma_D$$

$$R^* = \mu_R - \beta\sigma_R$$

$$L^* = \mu_L^N - \alpha_L^*\beta\sigma_L^N \tag{2.20}$$

where

$$\alpha_L^* = \frac{-\sigma_L^N}{\sqrt{(\sigma_R^N)^2 + \sigma_D^2 + (\sigma_L^N)^2}}$$

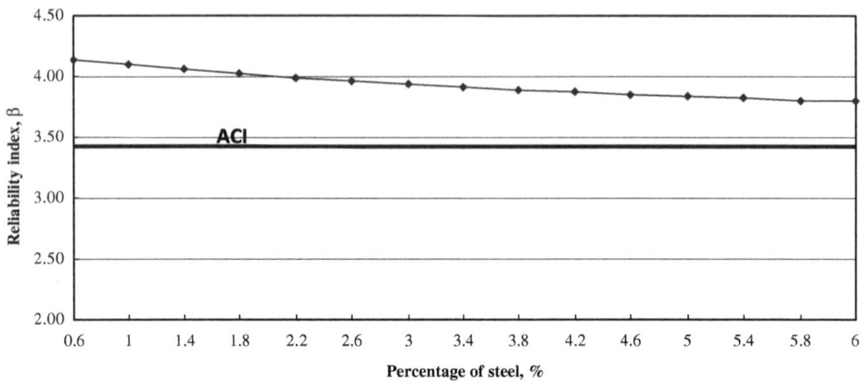

Fig. 2.5 Reliability index for BS and ACI for Different Percentages of Longitudinal Steel.

The value of L^* calculated from Eq. (2.20) is used in another iteration to improve the value of the safety index β.

After some iteration, the reliability index $\beta = 4.1$, which corresponds to the BS requirements for columns using $L_n/D_n = 1.0$ and $\rho = 1\%$.

From Fig. 2.5, one can obtain the reliability index of BS 8110 in a design reinforced concrete column, which corresponds to a different percentage for longitudinal steel bars.

Moreover, it is noticed that the reliability index has a higher value than that of the ACI code: i.e., a constant value of 3.4 (Galamobos et al., 1982) for all percentages of steel bars.

One can observe from Fig. 2.5 that the BS reliability change based on percentage of steel for ACI is the same value regardless of the percentage of the longitudinal steel, and in general the BS has a higher reliability than the ACI code.

2.2.2.4 Monte Carlo simulation technique

A great deal of research in the area of reliability uses the Monte Carlo simulation technique. The name of this simulation came from the casino games of Monte Carlo, Monaco, originating from the research work of Neumann and Ulam in 1949.

In 1967, Moses and Kinser compared the simulation with other methods of reliability analysis and concluded that the Monte Carlo simulation was the more flexible method of computing reliability.

In 1984 Rashedi (1984) also used this simulation method to calculate the mean value and standard deviation of the failure mode for brittle and ductile frames and consequently a structural system reliability index.

Monte Carlo simulation was carried out using software like M-Star, which was used by Marek et al. (1993) to understand the load and resistance factor design (LRFD). In 1995 Nikolaos also used Monte Carlo simulation to study the reliability of reinforced concrete members strengthened with carbon-fiber reinforcement. In 1997 Frangopol also used Monte Carlo simulation to study reinforced concrete girder reliability under corrosion attack. In 2000 El-Reedy used the Monte Carlo simulation technique with the program Crystal Ball to define the reliability index for a reinforced concrete column under corrosion attack and eccentricity.

Monte Carlo simulation is known as a simple random sampling method or statistical trial method that creates realizations based on randomly generated sampling sets for uncertain variables. The Monte Carlo method consists of digital generation of random variables and functions, statistical analysis of trial outputs, and variable reduction techniques. These are discussed briefly in this section.

The Monte Carlo simulation process is as follows:

(1) choose the best distribution type to present the random variable

(2) then generate a random sampling set from the selected probability distribution

(3) after that, use the generated sampling set to do the simulation.

In general, the simulation of each mode of failure is by a limit state equation. The variables in this equation will be represented by the best representative probabilistic distribution, with average, and standard deviation values.

The simulation will have two parameters: a variable and uncertainty. For example, the height of the men in a country is a variable, as it follows a normal distribution.

The distribution type of the random variable is chosen among the different probability distributions provided by the program. Fig. 2.6 presents a sketch that shows how the software works by using the Monte Carlo simulation technique, as the variable input data will be a probabilistic distribution; after running the simulation, the outputs will be statistical data and can be represented by graphs.

As the simulation starts, the modeling uses the limit state equation and its links to the main variables, which are represented by the random values. The flow chart for the Monte Carlo simulation technique is illustrated in Fig. 2.7.

At the end of the simulation, the statistical data of the variable, and also probability distribution, will be obtained, as this variable represents the limit state equation, as if you consider $Z=R/L$ so if the outcomes are less than 1 it

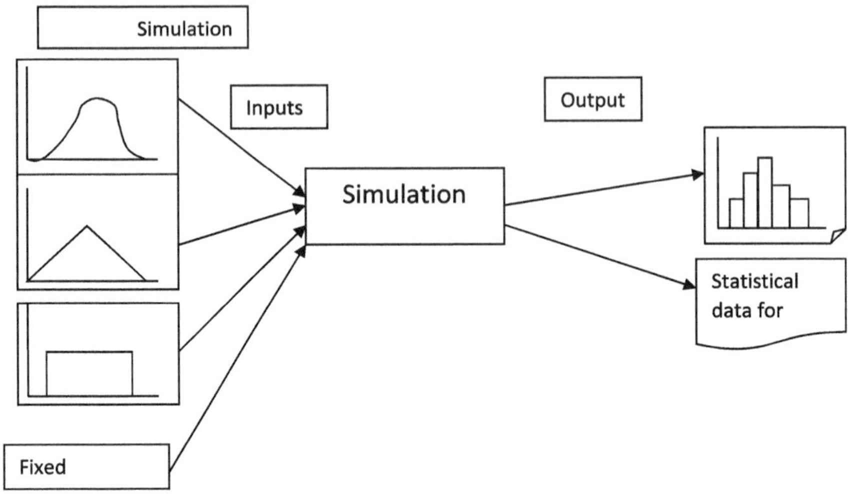

Fig. 2.6 Monte Carlo simulation.

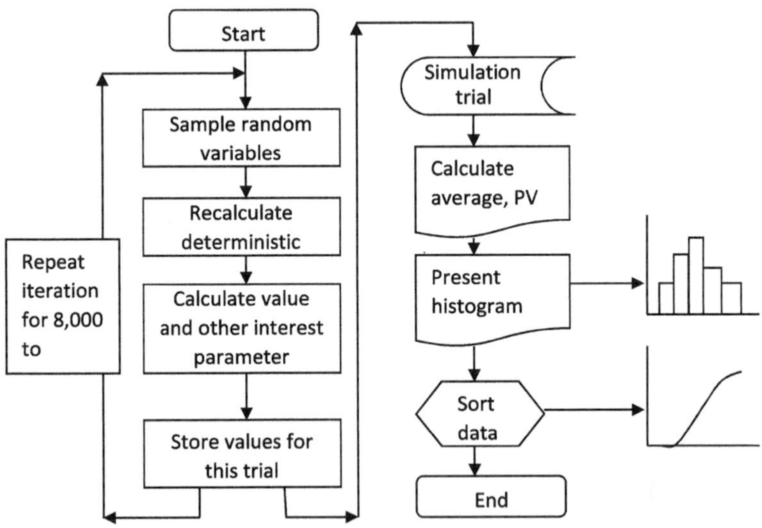

Fig. 2.7 Flow chart for Monte-Carlo simulation.

will represent a failure, and in the case of $Z = R - \mathrm{L}$ for any outcomes of Z less than 0 it means a failure.

As shown in Fig. 2.8, the input data for all the variable parameters is selected by choosing the probability, average value, and standard deviation or coefficient of variation. After running the simulation, the output will be a probabilities distribution curve.

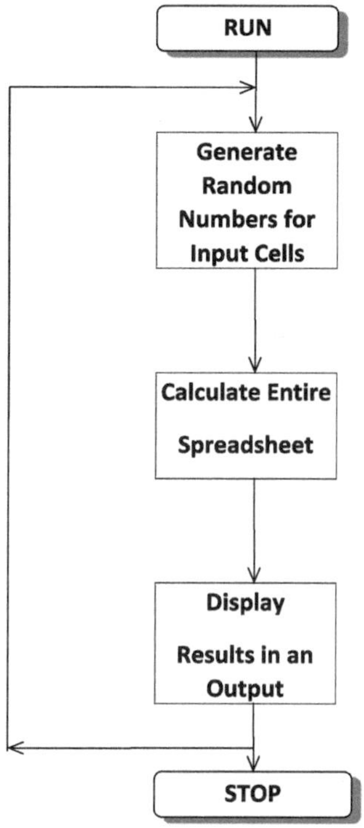

Fig. 2.8 Monte-Carlo simulation consequence.

In 1940 Von Neumann and Metropolis created random numbers using the middle-square method, which produces pseudorandom four-digit numbers:

(1) The software will start with four digits, called the seed number in Crystal Ball software; you can contact us to send you a video of using this simulation. Assume a seed value of 5555;
(2) Then square this seed number (5555)∘2 = 30,858,025;
(3) Then the software extracts the center four digits: 8580
(4) Then use the remaining digits, 3025, as the random number, and multiply by 10^{-4} to be (0.3025), less than 1 to represent the cumulative probability distribution value.
(5) Then use this number again, 3025, as the seed for the next trial. Go to step 2.
(6) As seen in Step No. 4, the generated random number usually returns a value between 0 and 1.

Therefore, any obtained random variable will be less than 1, so this number is considered the value of the cumulative curve as we previously defined its probability distribution function, based on the fact that the software will define the value of that variable that provides a cumulative value equal to its random number, as presented in Fig. 2.9, and this action will be repeated for all the defined variables by the software.

The generation method used is the *inverse transform method*. Let $F_x(x_i)$ be the CDF of the random variable x_i. By definition, the numerical value of $F_x(x_i)$ is a value in the interval [0,1]. Assuming that u_i is the generated uniformly distributed random number (0–1) u_i, the inverse transform method is used to equate ν_i to $F_x(x_i)$ according to the following equation and as presented in Fig. 2.9:

$$F_x(x_i) = u_i \text{ or } x_i = F_x^{-1}(u_i)$$

This method will be used for variables for which their cumulative distribution function has been defined by direct observation, or if there is an analytic expression for the inverse cumulative function.

Fig. 2.9 presents a method for the inverse transform technique.

For an asymptotic distribution TYPE I for the largest value is:

$$F_x(x) = \exp\left[-e^{-\alpha(x-\beta)}\right] \tag{2.21}$$

where β is the most probable value of X and α is the shape parameter. At a given probability value $F_x(x) = u$, we can obtain x from the following equation:

$$x = \beta - \frac{1}{\alpha} \ln\left(\ln \frac{1}{u}\right) \tag{2.22}$$

With an exponential distribution for the CDF,

$$F_x(x) = 1 - e^{-\lambda x} \tag{2.23}$$

The inverse function:

$$x = F_x^{-1}(u) = -\frac{1}{\lambda}(1-u) \tag{2.24}$$

The inverse transform method is effective if the inverse of the CDF can be expressed analytically. However, there are many probability distributions, such as normal and lognormal distributions, for which CDF cannot be inverted analytically. In this case other methods will be used, such as the composition method and the method of functions of random variables.

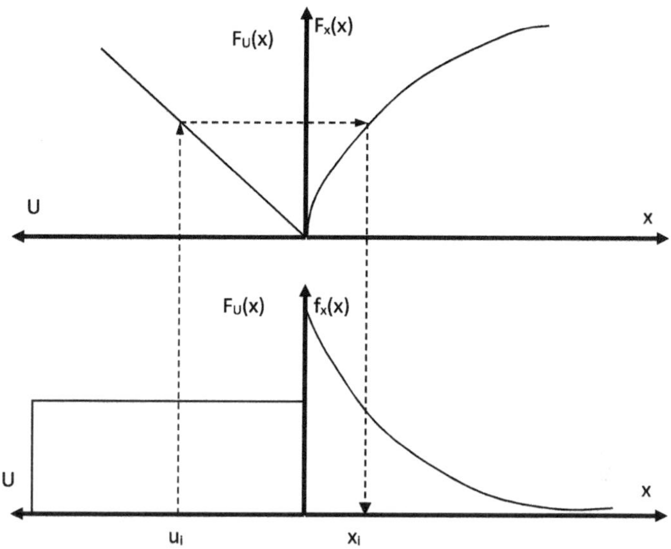

Fig. 2.9 Inverse CDF method for probability distribution.

The last methodis for a normal distribution with mean value μ and standard deviation σ, if there is a pair of independent, uniformly distributed random numbers u_1, u_2.

$$x_1 = \mu + \sigma\sqrt{-2\,\ln u_1}\cos 2\pi u_2 \tag{2.25}$$

$$x_1 = \mu + \sigma\sqrt{-2\,\ln u_1}\sin 2\pi u_2 \tag{2.26}$$

With beta distributions, the PDF of the standard beta distribution with shape parameters q and r is given by:

$$f_x(x) = \frac{1}{B(q, r)} x^{q-1}(1-x)^{r-1} \tag{2.27}$$

where $B(q, r)$ is the beta function. A value from this distribution may be generated with

$$x = \frac{u_1^{1/q}}{u_1^{1/q} - u_2^{1/r}} \tag{2.28}$$

where u_1 and u_2 are two random values from the standard uniform distribution and provided that

$u_1^{1/q} + u_2^{1/r} \leq 1$; the basis for the preceding procedure is as follows (John, 1964).

In the case of a gamma distribution, the PDF of the gamma distribution is

$$f_x(x) = \frac{v(vx)^{k-1} e^{-vx}}{\Gamma(k)} \quad x \geq 0 \tag{2.29}$$

with parameters v, k.

The inverse equation will be as follows:

$$x = \frac{1}{v}(-\ln u_3)\frac{u_1^{1/k}}{u_1^{1/k} + u_2^{1/(1-k)}} \tag{2.30}$$

where u_1, u_2, u_3 are values generated from the standard uniform distribution and provided that

$$u_1^{1/k} + u_2^{1/(1-k)} \leq 1$$

To have more accuracy by applying the Monte Carlo simulation in the case of large numbers of random variables, a large number of sampling sets is required. Thus there are limitations to obtaining high accuracy for large-scale problems, which require tremendous computational time and effort. To improve the rate of convergence, several modifications can be introduced to Monte Carlo simulation using quasi-random points.

Estimate the reliability index β of the example in Table 2.1 using the Monte Carlo simulation method with the same limit-state function, mean values, standard deviations, and distributions of the random variables.

As shown in Table 2.1, the Monte Carlo simulation results converge to three digits after 200,000 runs (β 2.85, P_f=0.0021).

Therefore the table shows that there is an error depending on the sample size.

For any structure system, there are many variables and uncertainties, for example, the theory of design, since for a concrete structure the design theory is based on the structure being monothetic and any section is the same for all the members. These two assumptions cannot be applied, since it is impossible to pour the concrete for the foundation, column, beam, and slabs for all

Table 2.1 Relation between the number of sample trials and the probability of failure.

No. of simulation	1000	10,000	50,000	100,000	200,000	500,000	1,000,000	2,000,000
β	3.0902	2.9290	2.8228	2.8175	2.8574	2.8642	2.8507	2.8499
P_f	0.001	0.0017	0.00238	0.00242	0.00214	0.00209	0.002181	0.002187

floors at the same time, and we cannot guarantee that every section on any member will have exactly the same properties. There are also many other variables that can be illustrated, as in the following sections.

2.2.3 Load variables

In European code, the term loads is changed to actions, meaning those actions that affect the structures along their lifetime: the dead load, the live load, and the lateral loads such as wind, waves, and seismic load. All these loads are variables and represented by their probabilistic distribution based on the statistical research on these types of loads.

Based on (Corotis and Chalk, 1980; Corotis and Doshi, 1980), and Hasofer (1968), the accurate evaluation of gravity loads is a very important variable in formulating a design base for safe and economical structures. Moreover, load characteristics are probably the most important input parameters in a reliability-based design.

Lind et al. (1972) recommended that in obtaining the reliability-based design codes we should take care to obtain reliable load data in a form suitable for the estimation of the key statistical parameters. Much research was done in 1970 to define live load and most studies revealed that the live load follows a lognormal distribution. The researchers mentioned that the live load consists of two parts: the sustained load and the extraordinary load. The sustained load includes furnishings and people normally found in the building or structure. The extraordinary load is usually associated with special events that may lead to a higher number of people.

In general, the lateral loads affecting most structures are seismic or wind loads, and adding wave loads for offshore structures. There are many researchers like Holmes et al. (1985) and Thoft-Christensen and Murotsu (1986) who mentioned that the Gumbel distribution is often used to model the maximum wind speed, which is the main variable in the wind load equation. Tzavellis and Shinozuka (1988) determined earthquake ground acceleration using a stationary and Gaussian random process with zero mean.

For the seismic intensity statistical variation, Travelis and Shinozuka used an extreme value distribution of type II, and to model earthquake occurrences, they used the Poisson distribution.

At the end of 1970 and the start of 1980, many researchers like Ellingwood (1979a,b) suggested consideration of a basic load combination, which is the dead load to live load combination, noting that the ratio between live load to dead load is usually in the range from 0.25 to 1.50. In 1977

Ellingwood mentioned that in 92% of all possible load combinations, the mean live-to-dead load ratios range from 0.25 to 1.0. Mirza et al. (1980) assumed that the probability of having the ratio of nominal live-to-dead loads ranging from 0.25 to 1.25 is 80%; on the other hand, in 1980 Ellingwood et al. presented a probability of 95% of having the ratio of nominal live-to-dead loads ranging from 0.25 to 1.5.

2.2.4 Structure resistance variables

In general, the resistance of a concrete structure depends on the concrete strength, steel yield strength, area of the longitudinal steel bars, and cross-sectional area of the structure member.

2.2.5 Concrete strength variations

There are many variations in concrete strength. There is a variation between the concrete strength in a structure and that in a control specimen for many reasons. For example, the control specimen will be compacted in three layer by special rode specifies its dimensions and shape in the standard. In addition, the concrete samples for both in cube or cylinder model as per the project specification will be submerged in water for 28 days; however, the curing process onsite will be 7 days, with a maximum of only 14 days. Also, in a structure there are steel bars but not in the control specimen, so there is a void and also the thickness and dimension of the member is different, and so on.

MacGregor (1976) maintained that high-strength concrete is more affected by this than low-strength concrete. He simulated the concrete strength by a normal distribution. From the available data, he concluded that the coefficient of variation of concrete test cylinders should be considered as the values 0.10, 0.15, and 0.20, to represent excellent, average, and poor quality control, respectively.

Ramakrishnan and Balaguru (1987) studied 51 datasets of concrete compressive strength collected from different sites. They found in their study that, on average the targeted concrete strength (f_{cr}) was conservative if it was obtained by applying two equations. The first equation stated that the probability was 1% that the mean of three consecutive tests was smaller than the specified strength. The second equation stated that the probability was 1% that an individual test result would fall below the specified strength by more than 3.45 MPa (500 psi).

$$f_{cr} > f_c' + 1.34\,\sigma_{fc}\ \text{MPa} \tag{2.31}$$

$$f_{cr} > f_c' + 2.33\sigma_{fc} - 3.45\ \text{MPa} \tag{2.32}$$

where

σ_{fc} = the standard deviation of concrete cylinder strength

f_c' = the specified concrete compressive strength

Arafa (1997) presented the results of 955 random concrete samples and revealed that ready-mixed concrete is well modeled by a normal distribution, whereas site-mixed concrete is well represented by a lognormal distribution and has a lower ratio between mean and nominal value of concrete strength.

In 1984 research was carried out by MacGregor et al. on the statistical resistance of reinforced and prestressed concrete members and in this study the tensile and compressive concrete strength followed a normal distribution.

The codes and standards provide guidelines to control the variation in the concrete strength by providing acceptance and refusal limits; for example, the British standard BS 5328:part 4:1990 stipulates use of a cube for measuring the concrete strength. The British standard design equation depends on using a characteristic strength, which is defined as the value of strength below which 5% of all possible test results are expected to fall; the margin between the characteristic strength and the mean strength is selected to verify this probability.

According to the British standard, these two criteria should be followed on site:

(1) The average value of any four consecutive test results exceeds the specified characteristic strength by 3 MPa.

(2) No test results fall below the specified characteristic strength by more than 3 MPa.

Similar requirements are prescribed for the flexure test: the values in (1) and (2) criteria are then 0.3 MPa.

On the other hand, ACI 318 applies the following criteria:

(1) No single test strength or the average of the strengths of two cylinders from a batch shall be more than 3.5 MPa (500 psi) below the specified compressive strength f_c, i.e., 21 MPa (3000 psi) for specified 24.6 MPa (3500 psi) concrete.

(2) The average of any three consecutive test strengths must equal or exceed the specified compressive strength, f_c.

These criteria in the standards and codes control the variation of the concrete specimen based on the fact that there is another variation between the concrete specimen and the concrete strength inside the members; most codes and standards control it by construction quality control limits and recommendations on the compaction and curing processes.

Fig. 2.10 illustrates that the quality control on-site is the main action to reduce the difference between the concrete strength in situ and the standard cube or cylinder specimens. This figure obviously shows that, as a general rule, when the concrete grade is high it will reduce the differences between the concrete on-site and that obtained from the specimen for all types of quality on the site. This figure also shows that with excellent quality control on-site, the difference between the concrete on-site and the specimen is narrow and in general the concrete coefficient of variation is about 15%; on the other hand, the coefficient of variation with poor quality control can reach 40% in situ and 35% for a concrete specimen that is very poor.

For reliability analysis applications, it is important to take the time to select the value of the concrete coefficient of variation that represents the actual condition of the site quality control. If a wrong value is selected, the output will give an inaccurate value of probability of failure for the structure.

Ellingwood et al. (1982) estimated the coefficient of variation (CoV) to be 0.207 in the average concrete control specimen. Freudenthal et al. (1966)

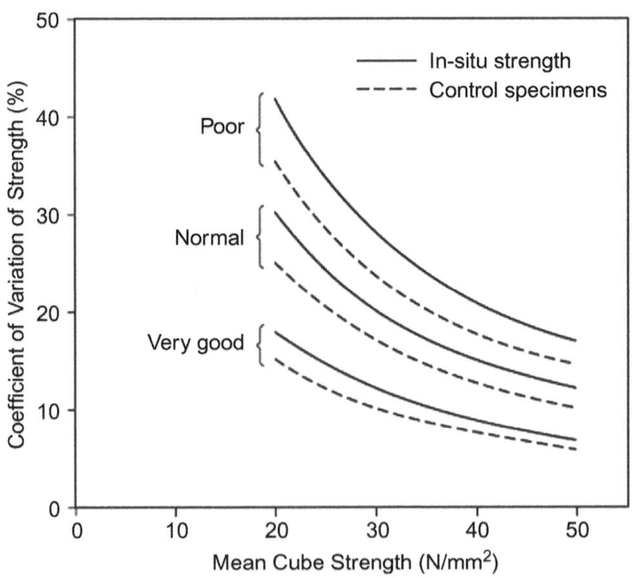

Fig. 2.10 Relation between coefficient of variation for control specimens and in situ with different concrete grade and quality grade.

stated that with poor quality control the concrete strength (f_c) can be represented by a lognormal distribution.

Mirza (1987) stated that the CoV of the in-situ concrete compressive strength for grades 35 and 20 MPa can be considered equal to 0.15 and 0.18, respectively. In 1997 Arafa showed that the coefficient of variation CoV of the strength of concrete mixed on-site had a higher value, from 0.3 to 0.4, and could be represented by a lognormal distribution, confirming the findings by Freudenthal et al. that poor quality control can be represented by this probability distribution.

Table 2.2 gives the overall standard deviation, based on ACI 214-77, for concrete in laboratory trial mixes and in the field or with concrete strength of 35 MPa.

A study by El-Reedy and Hassan (2005) focused on four levels of project quality control with different concrete grades. The probabilistic model that represented the concrete compressive strength was verified by using the chi-square and Kolmogorov-Smirnov (K-S) goodness-of-fit test methods, with 5% confidence level. These models were developed to predict the performance of structural elements and to assess the reliability level. This research showed that knowing the contractor organization and the level of its quality management system can be very useful in defining the level of reliability, as it is directly affected by the variation of the concrete on the site.

For a high-quality management system the safety factors can possibly be reduced; on the other hand, for poor management systems, as with some small projects, the safety factors must be increased in order for the management system to maintain the same reliability level. For ready mix, lower concrete grades of 20 MPa, and higher concrete grades of around 40 and 45 MPa, the concrete compressive strength is well represented by a normal

Table 2.2 Classification of standard concrete based on ACI 214-77.

	Overall standard deviation, MPa	
Standard of control	**In the field**	**Laboratory trial mixes**
Excellent	<3	<1.5
Very good	3–3.5	1.5
Good	3.5–4	1.5–2
Fair	4–5	2–2.5
Poor	>5	>2.5

distribution. For concrete grades of 30 and 35 MPa, the concrete compressive strength is well represented by a lognormal distribution.

The main objective of the experimental program was to collect concrete sample records from different projects in Egypt that varied between high, medium, and low project sizes. Different project costs depend on the size of the project; the costs of a larger organizational size for contractors, consultants, and owners for large projects are higher than those for small projects. For this phase of study, all of these projects used ready-mix concrete; the organization focused on good contract documents between the contractor and concrete manufacturer, and the owner and consultant auditing procedure of the concrete manufacturer mixing site, as quality assurance procedures.

The probabilistic models for the previous data of about 500 concrete samples comprise standard cubes (150 × 150 × 150 mm), which after 24 h were transferred to the laboratory where they were cured for 28 days before compression testing.

The data analyses for the different contractors were performed using Crystal Ball software and chi-square and K-S values to select the most suitable probability distribution that conforms with the data.

The results of the two tests with some hypothesized probability distributions are shown in Table 2.3; from this table one can find that the least value for the two tests is for a hypothesized lognormal distribution.

Note that project (A) is a larger project than (B), and so on. All the projects in Table 2.3 used ready-mix concrete; the table shows the project, the required concrete compressive strength, and the related bias factor; however, the bias factor is the ratio between the mean compressive strength and the nominal value f_c. The coefficient of variation is illustrated in the table for different contractors and concrete grades.

Table 2.3 Statistical parameters for different projects.

Project	f_c	Bias factor	CoV
(A)	22.5	1.03	0.12
(A)	35.0	1.03	0.20
(B)	40.0	1.11	0.15
(C)	45.0	1.07	0.17
(D)	30.0	1.05	0.22

Table 2.4 Proposed distribution model for different concrete grades.

Project	f_c	Distribution	χ^2	K-S
(A)	22.5	Normal	3	0.16
(A)	35.0	Lognormal	13	0.18
(B)	40.0	Normal	0.31	0.81
(C)	45.0	Normal	10.92	0.07
(D)	30.0	Lognormal	9	0.1

The tests were performed by comparing the available data with the different probability distributions and the best probability distributions to represent the data are presented in Table 2.4.

For lower strength of around 22.5 MPa and for higher grades above 40 MPa, the best model is a normal distribution; a lower grade and high-quality control provide a small value for the coefficient of variation, and also with higher concrete grades the coefficient of variation is low.

For concrete grades of 30 and 35 MPa for different projects, the concrete compressive strength is well represented by a lognormal distribution with a coefficient of variation of 0.2 and 0.22, respectively.

With low concrete grades, it is easy to achieve this strength, so the CoV is low and the bias factor is around 1.03. Also, for higher concrete grades over 40 MPa, some additives can be used to achieve the higher concrete strength so the bias factor is around 1.11 to 1.07 with CoV from 0.15 to 0.17 which is reasonable.

On the other hand, 30 to 35 MPa values are between the lower and higher grades, so there is high variation to reach the required grade; thus the bias factors are 1.03 and 1.05, respectively, with higher CoV, revealing that for any type of projects, it is necessary to increase the quality control procedure in the case of concrete grades 30 and 35 MPa being required.

In any case, the coefficient of variation of ready-mix is lower than that for mixing on-site. This is due to a lack of high quality control and this type of mix is usually performed on a small project with lower cost.

This study states that for lower grade and higher concrete grades around 40 and 45 MPa the concrete compressive strength is well represented by a normal distribution. For concrete grades of 30 and 35 MPa, the concrete compressive strength is well represented by a lognormal distribution and also, for grades 30 and 35 MPa, more efficient quality control procedures are required.

2.2.6 Steel yield strength variable

For the steel yield strength and ultimate strength, many statistical research projects have been carried out. The results of this research reveal that the shape of the distribution can be variously described as being normal, lognormal, or extremal.

As per Alpsten (19,710 studies), the results of almost 20,000 mill tests on structural steel plates and shapes with a nominal yield strength of 220 MPa fit the lognormal distribution very well. On the other hand, Allen (1970) recommended the use of a normal distribution for high-strength structural steel and for reinforcing bars; his study in 1972 found for steel grade 60 an average mill test yield strength of 66 ksi and the coefficient of variation of the yield strength was assumed to be 0.07.

Castello and Chu (1969) recommended in their study to use a beta probability distribution to represent the steel yield strength with a mean value equal to 47.7 ksi and a coefficient of variation equal to 12.4% for grade 40 reinforcement.

Bljuger (1987) investigated the ratio between mean and characteristic steel yield strength, which is called bias factor, and suggested that it equal 1.15. However, a study by Benjamin and Lind (1969) stated that, from experience, the bias factor, which is the ratio of mean to nominal steel yield strength, is equal to 1.05.

2.2.7 The nonbasic variables in design

Many studies have discussed the other variables that affect the limit state equation rather than the variation in load and resistance. In general, these variables are the results of uncertainty in the theoretical equations, structural analysis, and unknown events.

Galambos and Ravindra (1973) stated that the effect of load and resistance variables for combined dead and live loads can be represented by the following formula:

$$S = E(C_D \cdot A \cdot D + C_L \cdot B \cdot L) \tag{2.33}$$

The terms in Eq. (2.33) are defined as follows: D and L are random variables for dead and live load intensities, respectively; C_D and C_L are the deterministic coefficients that reflect the influence that transforms the load intensities to a straining action, such as moment, shear, and axial force.

A and B are random variables representing the uncertainties in the transformation of loads into load effects; and E is a random variable representing the uncertainties in structural analysis theory.

On the other hand, there are uncertainties in concrete design equations, as a result of using the rectangular stress block, the limiting strain, and also due to neglect of strain hardening; therefore the strength calculated from the code equation is different from the actual concrete strength.

Mattock et al. (1961) found that in columns the ratios between test results of calculated strength ranged from 0.97 to 1.00, with coefficient of variation CoV ranging from 0.064 to 0.074, including possible in-test variations. According to this study, in columns the average ratio between the actual strength and the calculated design strength is 0.98, with CoV equal to 0.05.

MacGregor (1976) stated that the rate of loading has a strong influence on concrete and steel bar strength. With extremely slowly applied loads or very high sustained loads, the compression strength of concrete decreases by about 75% to 80% of the short-time strength, says Rusch (1960).

In a mill test on reinforcing bars, the load is applied at a very high rate. However, during normal operation the loading rate is much lower under dead and live loads. A study by Allen (1972) revealed that the average steel test yield strength is about 27.5 N/mm^2 (4 ksi) higher than the mean static yield strength.

2.3 Concrete cross-section variation

The concrete dimensions represent another major factor of uncertainty that affects structure reliability. The defined cross-sectional dimensions in the design of a structure member will be different from the actual dimensions on the site due to many construction issues. There are always deviations from the design: for example, if the column is designed at 500 mm × 500 mm, it may actually be 499 mm × 499 mm or 501 mm × 501 mm. In real life it is impossible to match the design exactly, and this variation must also be considered in the reliability analysis.

According to Fiorato (1973), the concrete dimensions selected by the structural engineer during the design, which are the column width and overall depth, are variables due to imperfections in concrete manufacturing. Therefore, Fiorato proposed that the average dimension will be assumed to be 1.27 mm (0.05 in.) greater than the design values, with standard deviation of 7.62 mm (0.3 in.).

Therefore, to control the variations in dimensions, every code defines the allowable deviations on concrete dimensions to be the acceptance and refusal limits, since these variations directly affect the structure reliability and the reliability index. El-Reedy (2012) states that this reliability index influences the values of load and resistance design factors.

For example, Tables 2.5 and 2.6 present the allowable tolerances in the dimensions in the ACI code and British standard, and these tolerance values should be followed by the quality control and site engineer to accept or refuse the work. With high-rise buildings, there is another specification for acceptance and refusal, as the designer must state the tolerance limits in the specification of the project clearly in the special structures.

Table 2.5 Allowable dimensions tolerance in ACI code.

Item	Description	Tolerance (mm)
1	Maximum tolerance for columns, beams, and walls dimensions	
	In any span or every 6 m in any direction	+13
	Whole building dimensions	*+25*
2	Vertical alignment for column and walls surface and line of surface intersection	
	Every 3 m height	6
	Whole building height (max. 30 m)	25
3	Surface of corner column and vertical expansion joint	
	Every 6 m height	6
	Whole building height (max. 30 m)	13
4	Columns and walls execute by sliding form	
	Every 1.5 height	3
	Every 15 m height	25
	Whole building height (max. 180)	75
5	Allowable tolerance for slab and beam bottom level	
	Every 3 m in horizontal distance	+6
	Every span or every 6 m horizontal distance	±10
	Whole building length and width	±19
6	Allowable tolerance in points of level that define slab and inclined beam leveling	
	Every span or every 6 m horizontal distance	±10
	Whole building length and width	±19

Table 2.5 Allowable dimensions tolerance in ACI code—cont'd

Item	Description	Tolerance (mm)
7	Allowable tolerance for columns, beam, slab, tie beam and walls	
	Dimensions until 304 mm	+10 or −6 mm
	More than 304 mm	+13 or −10
8	Reinforced concrete foundation	
	Horizontal dimensions	−13 or +50
	Dimensions between axes	±50
	Eccentricity of column to the foundation	2% of the foundation length in deviation direction and not more than 50 mm
	Foundation thickness	Without maximum or −2%
	Foundation top level	+13 or −50
9	Stairs	
	Height for one rise	±3
	Horizontal distance for one rise	±6
	Height for one flight or group of flights for one story	±3
	Horizontal dimensions for one flight or group of flights for one story	±6

Table 2.6 Allowable tolerance in dimensions based on BS5606.

Item	Describe	Tolerance (mm)
1	Whole building dimension	
	Ground floor slab	±28
	Length or width up to 40 m	±26
2	Position on plan in relation to nearest reference	
	Foundation	±50
	Walls	±16
	All elements above foundation	±12
3	Allowable tolerance in beam depth	
	External beam (perimeter up to 600 mm)	±13
	Over 600 mm	±20
	External beam (perimeter up to 600 mm)	±12
	Over 600 mm	±16
	Level variation from target plan	±22
4	Stairs	
	Vertical height of any flight between landing	±15
	Difference in rise of any consecutive steps	±6
	Difference in level between treads	±4
	Per meter width of stair	±5
	Stairs finished vertical height of any flight between landing	±10
5	Foundation level	±16

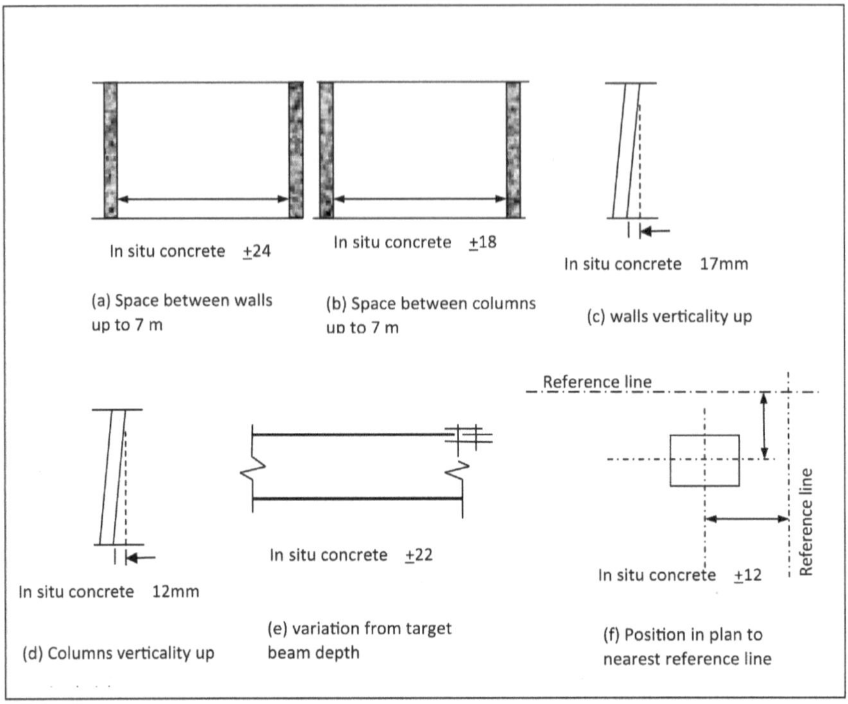

Fig. 2.11 Variation in reinforced concrete member after construction from its target.

Fig. 2.11 presents the allowable limits in the British standard of horizontal distance between columns and walls and the allowable limit for column and wall inclination and the variation of the position of the column to the nearest reference.

2.3.1 Uncertainty in steel bars during construction

The concrete cover and also the concrete depth have variations as shown in Table 2.7; the tolerance of the steel bar placement as per ACI is illustrated in Table 2.8.

Table 2.7 Allowable tolerance in depth, *d*, and concrete cover.

Effective depth	Allowable in effective depth	Tolerance in concrete cover
Effective depth (d) ≤ 8 in. (200 mm)	±0.375 in.	−0.375 in.
Effective depth (d) > 8 in. (200 mm)	+0.5 in.	−0.5 in.

Table 2.8 Tolerance in steel bars based on ACI.

Item	Member type	Allowable tolerance in inches
1	Distance between bars	
	Min. distance between bars in a beam	−0.25
	Slabs and walls in case of equal space	±2
	Stirrups	±1
2	Bending location and ends for longitudinal bars	
	Continuous beam and slabs	+2
	Ends of bars in beam and external slabs	+0.5
3	Reduce bar splice	−1.5
4	Reduce splice length inside concrete	
	For bar diameters 10–35 mm	−1
	For bar diameters 44–56 mm	−2

ACI code states that the distance between bars higher than 25 mm (1 in.) in any case must be higher than three-fourths the maximum nominal aggregate size.

Based on Brikeland and Westhoff (1972), there are usually variations in dimensions; Fig. 2.12 presents an example of dimensional variations for a beam during construction. This variation can significantly affect the size and hence the strength of concrete members, considering that the beam depth value may vary also, and its value has a higher influence on the member capacity.

Fig. 2.13 shows the variation that can occur in the columns, which can cause eccentricity.

Fiorato (1973) studied the measurements for about 6000 cast-in-place slabs with thicknesses ranging from 125 to 200 mm (5–8 in.); it was found

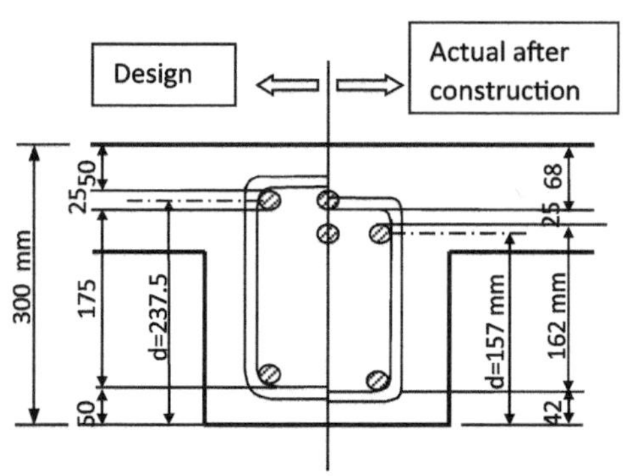

Fig. 2.12 Deviation of beam geometry from design to construction stage.

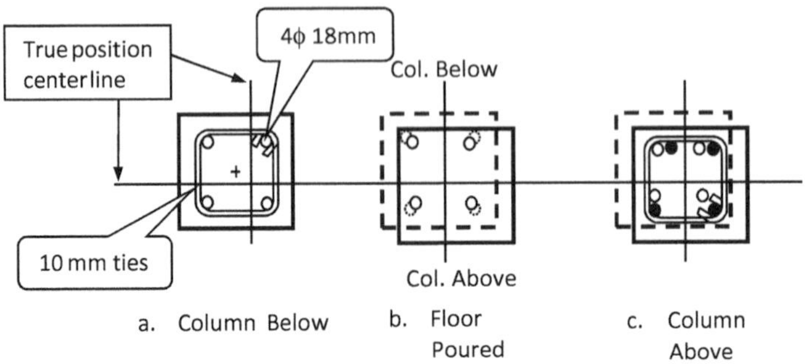

Fig. 2.13 Deviation in the column alignment during construction.

that the average thickness was around 1.25 mm (0.05 in.) more than the designed thickness, with a standard deviation equal to 7.5 mm (0.3 in.). It was also found that the effective depth, *d*, in the positive moment region on average was 6 mm (0.25 in.) less than the specified depth, with a standard deviation equal to 0.3 in. However, a negative moment region, *d*, averaged about 18 mm (0.75 in.) less than specified, with standard deviation equal to 12.5 mm (0.5 in.). With precast slabs, the average error in overall thickness and positive moment *d* was approximately 0 with standard deviation of about half those for cast-in-place members, as presented by Fiorato. Based on this, in general the reliability of precast slab is higher.

2.3.2 Variability in the steel size

There is a variation of the factory tolerance in the bar diameter during its production. It was found that for bars with diameter more 10 mm, as per CSA and ASTM specifications, up to 6% underweight is allowed on any individual bar. Areas measured on 102 specimens from five sizes are compared to their nominal areas in Fig. 2.14. By analysis of this data, it is found that mean cross-sectional area is 0.975 times the nominal value.

According to Lind et al. (1972), the cross-sectional area of the steel bars that was selected by the designer is higher than the required area an average of 1.02 times and CoV is equal to 5%. According to Allen (1972), the rolling tolerances are such that the mean area of a given reinforcing bar or group of bars is assumed to be 0.98 times the nominal area chosen by the designer, with CoV equal to 3%. The deviation between the required area of the steel and the actual is illustrated in Fig. 2.15.

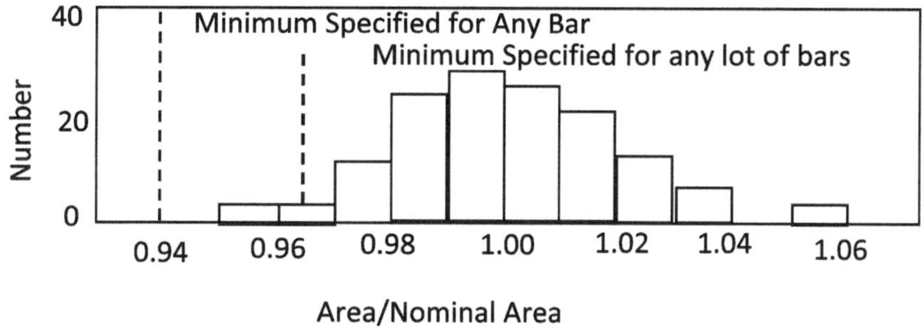

Fig. 2.14 Ratio of actual bar area to nominal area.

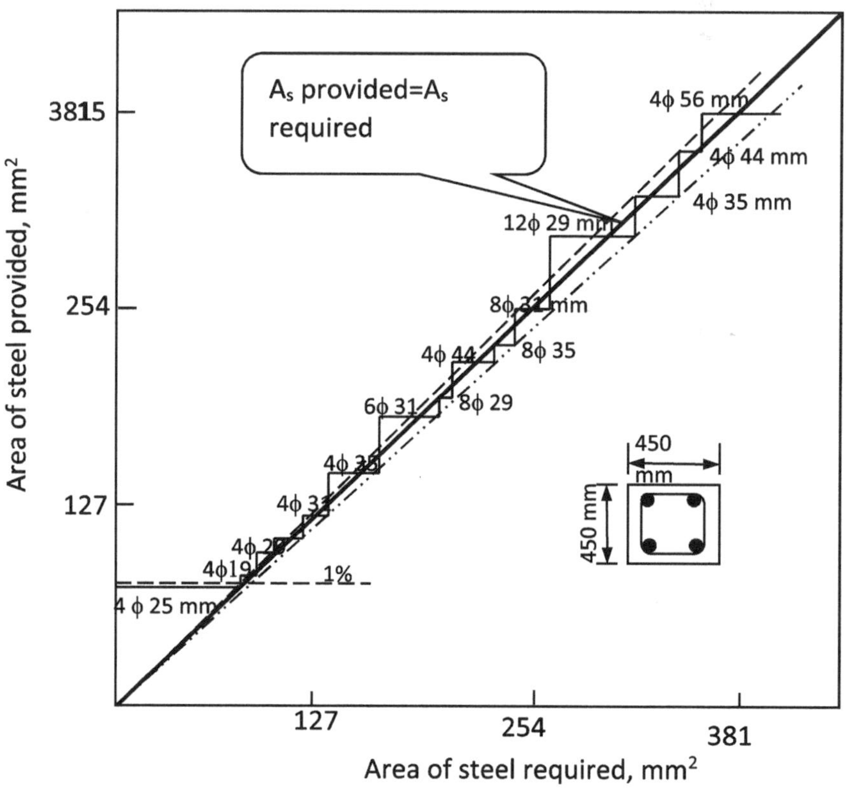

Fig. 2.15 Effect of selection of discrete bar sizes on choice of reinforcement in a tied column.

A study by MacGregor (1976) stated that the mean value of the steel bar area was equal to $1.02 \times 0.98 = 1.00$ times that chosen by the designer and the CoV of the steel bar area was equal to 6%.

Fiorato (1973) and Tso and Zelman (1970) studied the variation in reinforced concrete column dimensions by performing a measure of the width and thickness of 299 columns ranging from 300 mm (12 in.) to 750 mm (30 in.) in width from eight buildings. This data showed a mean error of +1.5 mm (+0.06 in.) and a standard deviation of 7 mm (0.28 in.).

2.3.3 Reliability approach in EC2 code

The European code is considered to be the first set of codes that considered a structural reliability approach in its equations and its philosophy of application. The European code is used for concrete and steel in EC2 and EC3, respectively. ISO codes for offshore structures for topside and underwater take the same approach.

The European code depends on using partial safety factors in the design. These partial safety factors are used for the basic variables, which are loads, resistances, and geometrical properties; through the use of partial factors and ψ, the factors are given design values, and verification is made to ensure that no relevant limit state has been exceeded.

In the European code (EC) the values of the partial factors and the ψ factors are determined in either of two ways:

(1) By calibration, based on a huge database of residential buildings.

(2) Based on the statistical evaluation of experimental data and field observations. This should be carried out within the framework of a probabilistic reliability theory.

When using the first method, either on its own or in combination with the second method, ultimate limit state partial factors for different materials and loads should be calibrated such that the reliability levels for representative structures are as close as possible to the EC target reliability index.

Fig. 2.16 shows the process in the Eurocode EC2 for defining and calibrating the partial safety factors that are used in the design equations; from this, one can see that the European code relies on the reliability theory in design.

The probabilistic calibration procedures for partial safety factors are divided into two main levels:

- full probabilistic methods (Level III), and

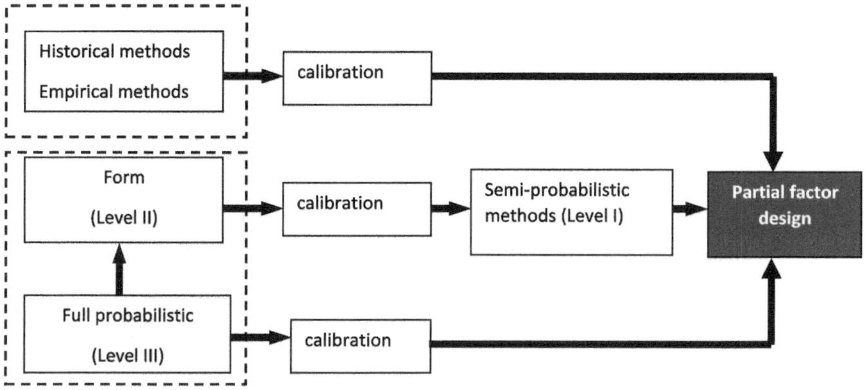

Fig. 2.16 Overview of reliability methods.

– first-order reliability methods (FORM) (Level II), as discussed previously and as seen in Eq. (2.1).

Full probabilistic methods (Level III) in principle give correct answers to the reliability problem as stated. The level III is used only to calibrate the design codes, as there is always a shortage of available statistics data. The level II method is accurate enough for structural building, but there are some approximations that should be considered.

For Level II and Level III methods, the measure of reliability should be determined as the successful probability Ps, which is equal to $(1 - P_f)$, where P_f is the failure probability for the specific failure mode, within an appropriate reference period. If the structure probability of failure is higher than 1×10^{-5} the structure is considered unsafe with respect to the code or it must be considered whether to violate the code, as most codes assume the probability of failure using this value.

In some cases, the probability of failure and its corresponding reliability index are considered only notional values that do not necessarily represent the actual failure rates, but are used as operational values for code calibration purposes and comparison of reliability levels of structures.

The Eurocodes have been primarily based on method a, as illustrated in Fig. 2.16. In this code the main target of using the reliability analysis method is to calibrate the limits state equation.

Table 2.9, presents the relation between the probability of failure, P_f, and reliability index β.

Table 2.9 Relation between β and P_f.

P_f	β
1×10^{-1}	1.28
1×10^{-2}	2.32
1×10^{-3}	3.09
1×10^{-4}	3.72
1×10^{-5}	4.27
1×10^{-6}	4.75
1×10^{-7}	5.20

Table 2.10 Target reliability index β for class RC2 structural members.

	Target reliability index	
Limit state	**1 year**	**50 years**
Ultimate	4.7	3.8
Fatigue		1.5–3.8[a]
Serviceability	2.9	1.5

[a]Depends on the degree of inspectability, reparability, and damage tolerance.

The European code EC2 provides target values for the reliability index β for different design conditions, as shown in Table 2.10 with a lifetime period of 1 and 50 years.

For the EC2 it is recommended to use lognormal or Weibull distributions for material and structural resistance parameters and model uncertainties; but in case of self-weight the normal distribution is the best probability distribution to represent it.

When the main uncertainty comes from loads that have a statistically independent maximum value in each year, the values of β for a different reference period can be calculated using the following formula:

$$\phi(\beta_n) = [\phi(\beta_1)]^n \tag{2.34}$$

where

β_n is the reliability index for a reference period of n years, β_1 is the reliability index for 1 year.

The actual probability of failure is significantly affected by human error, which is not considered in partial factor design.

2.3.3.1 Design values calibration

Reliability analysis is used to verify the partial safety factor of the code, as reliability verification is shown in Fig. 2.17, since the design values are required to be defined for all the basic variables. A design is sufficient if the limit states are not reached when the design values are introduced into the analysis models. In symbolic notation this is expressed as:

$$R_d > L_d$$

This is a practical way to ensure that the reliability index, β, is equal to or larger than the target value.

L_d and R_d can be expressed in partly symbolic form as:

$$L_d = L\{F_{d1}, F_{d2}, \ldots a_{d1}, a_{d2}, \ldots \theta_1, \theta_2\} \tag{2.35}$$

$$R_d = R\{M_{d1}, M_{d2}, \ldots a_{d1}, a_{d2}, \ldots \theta_1, \theta_2\} \tag{2.36}$$

where L is the load effect, R is the resistance, F is a specific load effect, M represents a material property, a is a geometrical property, and θ is a model uncertainty.

For particular limit states such as fatigue load, a more general formulation may be necessary to express the limit state.

The design values shall be based on the basic variable values at the first-order reliability method (FORM) design point, which was discussed earlier as the point on the failure surface ($Z=0$) closest to the average point in the space of normalized variables, as presented in Fig. 2.17.

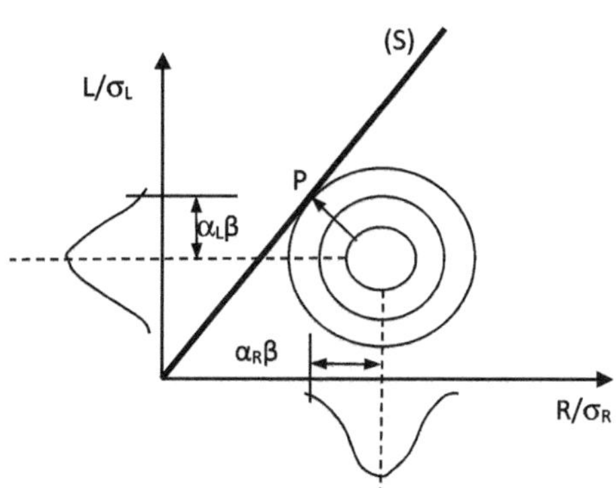

Fig. 2.17 Design point and reliability index according to the first-order reliability method (FORM).

The design values of load effects L_d and member strength R_d should be obtained such that the probability of having a more unfavorable value is as follows:

$$P(L > L_d) = \phi(+\alpha_L \beta) \quad (2.37)$$

$$P(R > R_d) = \phi(-\alpha_R \beta) \quad (2.38)$$

where β is the target reliability index, and α_L and α_R with $|\alpha| \leq 1.0$ are the values of the FORM sensitivity factors. As per EC considerations, the values of α will be negative for unfavorable loads and load effects, and positive for resistances.

The European code states that the values of α_L and α_R will be equal to -0.7 and 0.8, respectively, provided that.

$$0.16 < \sigma_L / \sigma_R < 7.6 \quad (2.39)$$

where σ_L and σ_R are the standard deviation of the load effect and resistance, respectively.

$$P(L > L_d) = \phi(-0.7\beta) \quad (2.40)$$

$$P(R \leq R_d) = \phi(0.8\beta) \quad (2.41)$$

If the condition in Eq. (2.36) is not satisfied, $\alpha = \pm 1.0$ should be used for the variable with the larger standard deviation, and $\alpha = \pm 0.4$ for the variable with the smaller standard deviation.

When the action model contains several basic variables, expression (2.40) will be used for the leading variable only. Therefore the design values are defined as follows:

$$P(E > E_d) = \phi(-0.4 \times 0.7 \times \beta) = \phi(-0.28\beta) \quad (2.42)$$

For $\beta = 3.8$, the values are defined by Eq. (2.42). The expressions provided in Table 2.11 should be used for deriving the design values of variables with a given probability distribution.

Table 2.11 Design values for various distribution functions in Eurocode.

Probability distribution	Design values
Normal	$\mu - \alpha\beta\sigma$
Lognormal	$\mu \exp(-\alpha\beta V)$ for $V = \sigma/\mu < 0.2$
Gumbel	$u - (1/a)\ln\{-\ln\Phi(-\alpha\beta)\}$
	Where $u = \mu - (0.577/a); a = \pi/\sigma(6)^{0.5}$

In these expressions, μ, σ and V are, respectively, the average value, the standard deviation, and the coefficient of variation of a given variable. Load variables should be based on the same reference period as for β.

2.3.3.2 Reliability verification formats in Eurocodes

As per EN 1990 to EN 1999, the design values of the basic variables M_d and F_d are usually not introduced directly into the partial safety factor design equations. They are introduced in terms of their representative values M_{rep} and F_{rep}, which may be:

- characteristic values, which are *values* of the design parameter with a prescribed or intended probability of being exceeded, such as for loads, concrete strength, and geometrical properties;
- nominal values, which are treated as characteristic values for material properties and as design values for geometrical properties.

The representative values M_{rep} and L_{rep} should be divided and/or multiplied by the appropriate partial factors to obtain the design values M_d and L_d.

Design values of actions L, material properties M, and geometrical properties a are given in expressions (2.43), (2.44), and (2.45), respectively.

The load design value L_d can be expressed in general terms as:

$$L_d = \gamma_f \ L_{rep} \tag{2.43}$$

with:

$$L_{rep} = \psi L_k \tag{2.44}$$

where

L_k is the characteristic value of the load;

L_{rep} is the relevant representative value of the load;

γ_f is a partial factor for the load that takes into account the possibility of unfavorable deviations of the load values from the representative values. Note that ψ is either 1.00 or ψ_0, ψ_1, ψ_2.

The design value M_d of a material or product property can be expressed in general terms as:

$$M_d = \eta \frac{M_k}{\gamma_m} \tag{2.45}$$

where

M_k is the characteristic value of the material and the section properties;

η is the mean value of the conversion factor, taking into consideration the effects of volume, its scale effects, moisture, temperature, and any other relevant parameters;

γ_m is the partial factor for the material, which takes into consideration the possibility of deviation of a material or section dimensions from their characteristic value; and the random part of the conversion factor η.

Design values of geometrical data such as dimensions of members that are used to assess load effects and/or member strength may be represented by nominal values:

$$a_d = a_{nom} \tag{2.46}$$

Considering that the effects of deviations in geometrical data—for example, inaccuracy in the load application or supports location—have significant influence on the structure reliability, such as the column second-order effects, the design values of geometrical data will be calculated as follows:

$$a_d = a_{nom} \pm \Delta a \tag{2.47}$$

where Δa takes into account the probability of unfavorable deviations from the nominal values, i.e., the cumulative effect in the case of the simultaneous occurrence of several geometrical deviations.

For that, the upper value for design resistance is used, and Eq. (2.45) takes the form:

$$M_d = \eta \gamma_{fM} M_{k,\,\mathrm{sup}} \tag{2.48}$$

where γ_{fM} is an appropriate factor greater than 1. Note that Eq. (2.45) may be used for capacity design.

As per EC, the design values for model uncertainties will be considered in the design equations by using the partial safety factors γ_{sd} and γ_{Rd} applied on the total model, such that:

$$L_d = \gamma_{SD} L\left\{\gamma_{gi} G_{kj} \gamma_p P; \gamma_{q1} Q_{k1}, a_d \ldots\right\} \tag{2.49}$$

$$R_d = R\{\eta M_k/\gamma_m; a_d \ldots\}/\gamma_{Rd} \tag{2.50}$$

The coefficient ψ accounts for reductions in the design loads variable and is applied as ψ_0, ψ_1, or ψ_2 to simultaneously occurring, accompanying variable loads.

The following simplifications may be made to Eqs. (2.49) and (2.50) when required.

(a) On the loading side (for a single action or where linearity of action effects exists):

$$L_d = L\{\gamma_{L,i} L_{rep,i}, a_d\} \tag{2.51}$$

On the member strength or resistance side, the general format is given in Eq. (2.52), and further simplifications may be given in the relevant Eurocode material. The simplifications should only be made if the level of reliability is not reduced.

$$R_d = \frac{1}{\gamma_{Rd}} R\{M_{d,i}; a_d\} = \frac{1}{\gamma_{Rd}} R\left\{ n_i \frac{M_{k,i}}{\gamma_{m,i}}; a_d \right\} i \geq 1 \tag{2.52}$$

Nonlinear resistance and loads models and multivariable load or resistance models are commonly encountered in Eurocodes. In such instances, the previous relations become more complex.

Fig. 2.18 shows the different partial factors available as per EN 1990 and also obviously presents the relation between individual partial factors in Eurocodes.

The following equation is used by EC to obtain the ψ_0 factors for two variable loads. The assumptions for this equation are as follows:

- the combination of two loads that are independent of each other;
- the basic period (T_1 or T_2) for each load is constant, where T_1 is the greater basic period;
- the values of the load within the basic periods are constant;
- the intensities of a load within basic periods are uncorrelated;

The distribution functions refer to the maximum within the reference period T. These distribution functions are total functions that consider the probability that an action value is zero during certain periods.

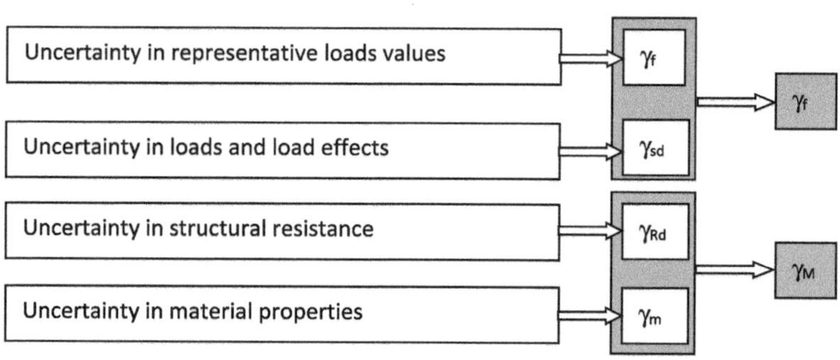

Fig. 2.18 Relation between partial safety factor and different uncertainties in EC2.

As per the EC, the expression for ψ_0 for the case of the variable loads and the general formula is as follows:

$$\psi_0 = \frac{F_s^{-1}\left\{\phi(0.4\beta')^{N_1}\right\}}{F_s^{-1}\left\{\phi(0.7\beta)^{N_1}\right\}} \tag{2.53}$$

where

$$\beta' = -\phi^{-1}\{\phi(-0.7\beta)/N_1\} \tag{2.54}$$

As an approximation for very large N_1 values, this formula will be used:

$$\psi_0 = \frac{F_s^{-1}\{\exp[-N_1\phi(0.4\beta')]\}}{F_s^{-1}\{\phi(0.7\beta)\}} \tag{2.55}$$

with

$$\beta' = -\phi^{-1}\{\phi(-0.7\beta)/N_1\} \tag{2.56}$$

For a normal approximation, apply this formula:

$$\psi_0 = \frac{1 + (0.28\beta - 0.7\ \ln N_1)V}{1 + 0.7\beta V} \tag{2.57}$$

By using a Gumbel approximation for this formula:

$$\psi_0 = \frac{1 - 0.78V[0.58 + \ln(-\ln\phi(0.28\beta)) + \ln N_1]}{1 - 0.78V[0.58 + \ln(-\ln\phi(0.7\beta))]} \tag{2.58}$$

where

$F_s(\cdot)$ is the probability distribution function of the extreme value of the accompanying load in the reference period T; T_1 is the greater of the basic periods for loads to be combined; N_1 is the ratio T/T_1 approximated to the nearest integer; V is the coefficient of variation of the accompanying action for the reference period.

Consequences classes

Based on the European code it presents the relation between the reliability index and the structure risk assessment, which is mainly a function of the consequence or impact on people, economics, or the social or surrounding environment in the case of a structure failure.

For reliability differentiation, consequences classes (CC) may be established by considering the structural consequences of failure, as given in Table 2.12.

Table 2.12 EC2 consequences classes.

Reliability classes	Consequences class	Description	Buildings types	Minimum value of β	
				1 year reference period	50 years reference period
RC3	CC3	**High** consequence for loss of human life, *or* economic, social or environmental consequences **very great**	Grandstands, public buildings where consequences of failure are high (e.g., a concert hall)	5.2	4.3
RC2	CC2	**Medium** consequence for loss of human life, economic, social or environmental consequences **considerable**	Residential and office buildings, public buildings where consequences of failure are medium (e.g., an office building)	4.7	3.8
RC1	CC1	**Low** consequence for loss of human life, *and* economic, social, or environmental consequences **small or negligible**	Agricultural buildings where people do not normally enter (e.g., storage buildings), greenhouses	4.2	3.3

The consequences classification criteria are important in terms of consequences of failure of the structure or structural member concerned. EC permits decisions to be made during the design to have some specific members of the structure to be designed in a higher or lower consequences class than for the entire structure. At the present time, according to the EC requirements, the reliability is related to the structural members of the construction works.

The reliability classes are defined by the reliability index (β) concept. Three reliability classes, RC1, RC2, and RC3, can be associated with the three consequences classes, which are CC1, CC2, and CC3, as shown in Table 2.12. The recommended minimum values for the reliability index associated with the reliability classes are shown in Table 2.12.

In general, doing the design using EN 1990 with the partial factors obtained from EN 1991 to EN 1999 leads us to have a structure with a reliability index β of more than 3.8 for a 50-year lifetime period, which is mentioned in the EC code by reference period.

Reliability classes for members of the structure above RC3 are not considered further, since these structures each require individual consideration.

The EC code states that to obtain reliability differentiation between classes of load, partial factor γ_L factors are to be used in fundamental combinations for design situations. For example, for the same design supervision and execution inspection levels, a multiplication factor K_{LI}, as shown in Table 2.13, will be applied to the partial factors based on the reliability class.

In particular, for class RC3, K_{LI} factors apply only to unfavorable loads. Reliability differentiation may also be applied through the partial factors of resistance γ_M. However, this is not normally used as an exception in the case of fatigue verification.

The level of quality control for the design and execution will be associated with the classes of γ_L. A three-level system for control during design and execution has been assumed in the code as presented in Table 2.14. Design supervision levels and inspection levels associated with the reliability classes are suggested.

Table 2.13 K_{LI} factor for loads.

	Reliability class		
K_{LI} factor for loads	**RC1**	**RC2**	**RC3**
K_{LI}	0.91	1.0	1.1

Table 2.14 Design supervision level.

Design supervision level	Structure type	Characteristics	Minimum recommended requirements for checking of calculations, drawings, and specifications
Level 3	Relating to RC3	Extended supervision	Third party checking: Checking performed by an organization different from that which has prepared the design
Level 2	Relating to RC2	Normal supervision	Checking by different persons than those originally responsible and in accordance with the procedure of the organization
Level 1	Relating to RC1	Normal supervision	Self-checking: Checking performed by the person who has prepared the design

There can be cases, such as lighting poles and other similar noncritical structures, where, for reasons of economy, the structure might be in RC1 but be subjected to higher corresponding design supervision and inspection levels.

2.3.3.3 Engineering design quality in EC

Using the asset management approach, one of the main barriers that prevents the failure of any structure is good design that is matched well with the project code and standards, with up-to-date technical practice. The European code focuses on the engineering phase and provides a level of inspection for the design calculations and the drawings, to avoid any error or oversight and to guarantee that the design complies with the project specifications.

Table 2.14 illustrates the required engineering design supervision level, which is related to the structure reliability category. There are three levels of engineering design check, which correspond to three categories of structure reliability that is related to the consequences class as presented in Table 2.12; every level of design check strategy will affect the structure reliability.

2.3.3.4 Execution quality in EC

As discussed earlier, there is a relation between the structure reliability and the quality control on the site. The EC provides a guide published by EN

Table 2.15 Inspection levels.

Inspection levels	Structure type	Characteristics	Requirements
Level 3	Relating to RC3	Extended inspection	Third-party inspection
Level 2	Relating to RC2	Normal inspection	Inspection in accordance with the procedures of the organization
Level 1	Relating to RC1	Normal inspection	Self-inspection

1992 to EN 1996 and EN 1999 that defines different levels of quality inspection on the site, as shown in Table 2.15.

Table 2.15 presents the inspection levels, which depend on the structure reliability class and define the quality requirements for each level of inspection.

2.4 Pipeline reliability analysis

For pipeline integrity and reliability, every major oil and gas company has its own criteria. In general, there are three levels of safety based on different conditions:

Low level: In the case of failure there is no human injury and only minor environmental damage and economic consequences.

Normal level: When a failure occurs, there will be a risk of human injury and significant environmental and economic consequences.

High level: When a failure occurs, there shall be a risk of human injury, and higher environmental and economic consequences.

The Sotberg et al. (1997) studies proposed target reliability levels. The target reliability level needs to be evaluated considering the implied safety level in the existing rules and codes; these levels are as shown in Table 2.16.

2.4.1 Limit state equation for the pipeline

It is very important to calculate reliability by defining the limit state equation. The limit state equation for the pipeline is as follows:

If only internal pressure is considered, the partial safety factor approach leads to the design format as follows:

Table 2.16 Reliability target levels as per Sotberg et al. (1997).

	Safety classes		
Limit states	**Low**	**Normal**	**High**
SLS	10^{-1} to 10^{-2}	10^{-1} to 10^{-2}	10^{-1} to 10^{-2}
ULS	10^{-1} to 10^{-2}	10^{-1} to 10^{-2}	10^{-1} to 10^{-2}
FLS	10^{-3}	10^{-4}	10^{-5}
ALS	10^{-4}	10^{-5}	10^{-6}

$$P_C \geq \gamma P_L$$

where

P_C: characteristic strength of the pipe according to a criterion

P_L: characteristic load (internal pressure)

Bai and Bai (2005) presented a new design equation for dented pipes with cracks in operation with respect to fracture; the criterion can be formulated as

$$P_L = \leq \frac{12t}{\gamma D} \cdot \frac{2\sigma_p}{\pi} \cos^{-1}\left(\exp\left(-\frac{\pi K_{mat}^2}{Y^2 8a\sigma_p^2}\right)\right) \tag{2.59}$$

where

K_{mat} was provided by Rolfe and Novak (1970):

$$K_{mat} = \sigma_Y \sqrt{\frac{0.6459C_v}{\sigma_Y} - 0.25} \tag{2.60}$$

with K_{mat} in MPa $(\text{mm})^{0.5}$, C_v in mm N, and σ_y in MPa.

Y is the geometry function and can be obtained from the following formula:

$$Y = \frac{F}{\sqrt{Q}}\left(1 - 1.8\left(\frac{D_d}{D}\right) + 5.1H\left(\frac{D_d}{t}\right)\right) \tag{2.61}$$

where D_d is the denting depth.

Q is the geometry correction factor, which is a function of the dent depth a and its half-length c

$$Q = 1 + 1.464\left(\frac{a}{c}\right)^{1.65} \quad \text{for } a/c \leq 1 \tag{2.62}$$

H can be obtained from the following formula:

$$H = H_1 + (H_2 - H_1)\sin^P \phi \tag{2.63}$$

where

$$P = 0.2(a/c) + 0.6(a/t) \tag{2.64}$$

$$H_1 = 1 - 0.34(a/t) - 0.11\left(a^2/c \cdot t\right) \tag{2.65}$$

$$H_2 = 1 + G_1(a/t) + G_2(a/t)^2 \tag{2.66}$$

And Factor F can be obtained from the following formulas:

$$F = f_\phi \cdot g \cdot f_w\left(M_1 + M_2(a/t)^2 + M_3(a/t)^4\right) \tag{2.67}$$

where

$M_1 = 1.13 - 0.09(a/c)$

$M_2 = 0.89/(0.2 + (a/c)) - 0.54$

$M_3 = 0.5 - (1/(0.65 + (a/c))) + 14(1 - a/c)^{24}$

$g = 1 + (0.1 + 0.35(a/t)^2)(1 - \sin\phi)^2$

ϕ is parametric angle of the elliptical crack

$$f_\phi = \left((a/c)^2 \cos^2\phi + \sin^2\phi\right)^{0.25} \tag{2.68}$$

f_w is a finite width correction factor calculated from the following formula:

$$f_w = \left(\sec\left((c/D)(a/t)^{0.5}\right)\right)^{0.5} \tag{2.69}$$

The limit state function can be formed based on failure criteria for the specified case. Bursting of a pipe will happen at an uncontrolled tearing point when the equivalent stress exceeds the flow stress.

The bursting failure will lead to pipe rupture. The limit state function based on a new fracture criterion can be formulated as:

$$g(Z) = \frac{4t_c\sigma_{fc}}{\pi D_c} X_M X_1 X_f M_s^{-1} \cos^{-1}\left(\exp\left(\frac{-\pi M_s^2 K_{mat}^2}{8\sigma_{fc}^2 Y^2 a X_y^2 X_f^2}\right)\right) - X_p \cdot P_d \tag{2.70}$$

where P_d is the design pressure that can be estimated from the new design, parameters M_s. and K_{mat} are given by Eqs. (2.70) and (2.60), respectively, by introducing uncertainties into the corresponding random variables, and the subscript c indicates the characteristic values of corresponding variables.

$$M_s = (M_T.t - a)/(M_T(t - a)) \tag{2.71}$$

where M_T is obtained from the following formula, developed by Kiefher and Vieth (1989):

$$M_T = \left(1 + 0.6275x^2 - 0.003375x^4\right)^{0.5} \text{ for } x \leq 7.07$$

$$M_T = 0.032x^2 + 3.3 \text{ for } x > 7.07$$

where $x = L/(Dt)^{0.5}$

L is the crack total length;

D is the pipe nominal outside diameter.

The collapse stress is calculated from the following formula, from ASME31G for rectangular defects in a pipe:

$$\sigma_p = \alpha.\sigma_y \frac{t - a}{t - aM_T^{-1}} \tag{2.72}$$

where σ_y is the pipe yield strength and the parameter α value is around 1.25; this value will decrease with the increase of yield strength.

The uncertainty variables (X) in the limit states equation are as shown in Table 2.17.

Bai and Bai (2005) presented a case of a pipeline with diameter equal to 1066.8 mm with materials API5L60 with pressure 1.9 MPa; the relation between the reliability index and dent depth with probability of failure is as shown in Fig. 2.19.

Fig. 2.20 illustrates the relation between reliability index and probability of failure with dent depth to thickness ratio and Fig. 2.21 shows the effect of crack to thickness ratio on the probability of failure and reliability index.

Table 2.17 Basic probabilistic parameters descriptions.

Random variable	Distribution	Mean	CoV
Wall-thickness factor, X_t	Normal	1.04	0.02
Flow stress factor, X_f	Normal	1.14	0.06
Flow stress model, X_m	Normal	0.92	0.11
Max pressure factor, X_p	Gumbel	1.05	0.02
Crack length factor, X_L	Normal	1.00	0.1
Crack depth, a	Exponential	0.1	1.0
Dent depth factor, X_D	Normal	0.9	0.05
Y function factor, X_Y	Log-normal	1.00	0.1
Charpy energy, C_v	Log-normal	63.0	0.1
Young's modulus, E	Normal	210	0.03

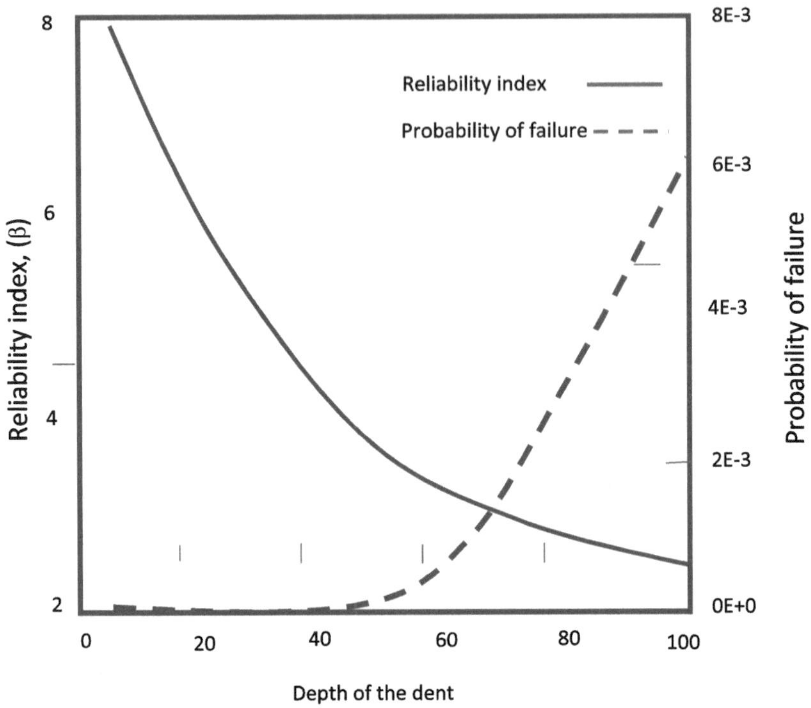

Fig. 2.19 Relation between reliability index, probability of failure and dent depth.

2.4.2 Reliability model

Corrosion in a pipeline becomes unacceptable when the corrosion depth exceeds the allowable corrosion depth. This can be described by the limit state function, as presented by Edwards et al. (1996):

$$g(X) = dc - C_R\, t \tag{2.73}$$

where

X contains all the basic uncertainty variables;
dc is maximum allowable corrosion depth;
C_R is the corrosion rate;
t is the duration of wet service.

2.4.3 Determination of corrosion rate

Corrosion rates (C_R) can be based on empirical equations such as shell models (deWaard and Lotz, 1993; deWaard et al., 1995), or on experience/measured data.

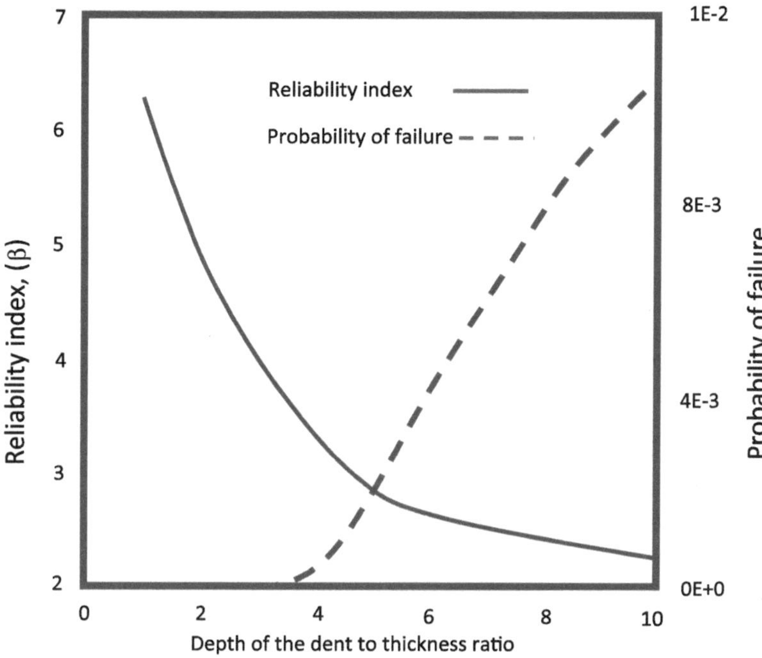

Fig. 2.20 Reliability index due to effect dent depth to thickness ratio (Dd/*t*).

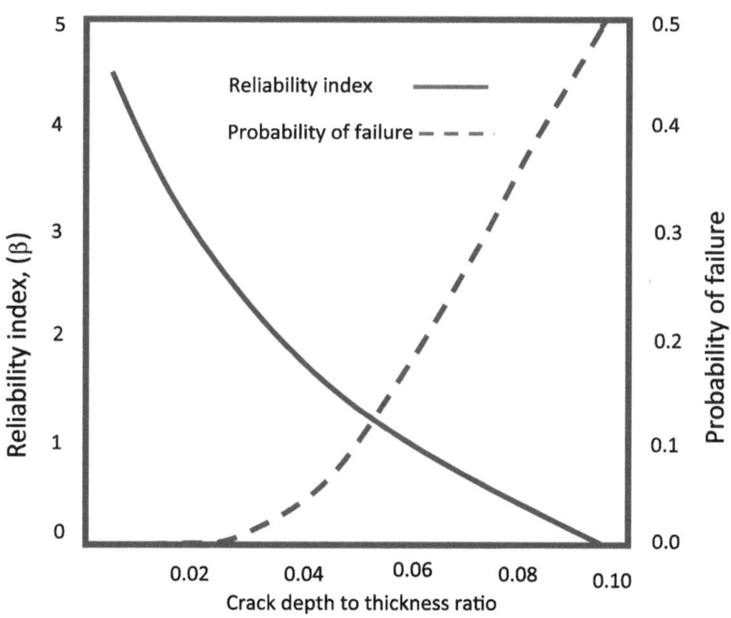

Fig. 2.21 Relation between pipeline reliability and crack depth to thickness ratio.

In a pipeline design, a combination of the preceding and sound engineering judgment from a qualified corrosion engineer is necessary. In the following examples, a deterministic corrosion rate (based on deWaard et al. (1995)) is multiplied by a probabilistic variable, Xm, called the corrosion rate uncertainty factor. This factor is developed based on operation experiences and engineering judgment.

Sharifi et al. (2019) discussed the reliability of pipelines under lateral buckling failure. For this study they applied FORM, SORM, and Monte-Carlo simulation. For the lateral buckling the effect of the fluid temperature is also considered and the buckling mode shape is presented as per Bai and Bai (2014). For this study, the mean of Young's modulus is 210×109 and its CoV is equal to 0.05; however, the normal distribution presents the pipeline wall thickness and its diameter with a mean value of 0.024 and 0.6156, respectively, with CoV of 0.05.

The effect of corrosion on the pipeline thickness is considered to be 0.4 mm/year as per the Association of Corrosion Engineers (NACE) if there is no field test data, based on Vanayi and Eslami (2015). Their study revealed that the pipeline design based on DnV can withstand 30 years without lateral buckling failure.

2.5 Offshore structure reliability

Most of the owner companies are focused on the integrity of their facilities and specifically of their platforms, which are the major assets. The structure integrity represents the structure reliability along its lifetime, which requires calculation of its probability of failure, and having a criteria for the reliability index or the probability of failure based on the consequences in the case of failure.

Many research and development projects have been carried out to define the structural probability of failure, which is the reverse of the structural reliability. An offshore structure reliability analysis (OSRA) is often required in the reassessment of existing structures; this is particularly the case for fixed platforms, which experience wave-in-deck loading at high return periods. An OSRA allows the risk of platform failure to be quantified and the effects of any measures to mitigate the consequences to be understood. A key input into this analysis is metocean loading in extreme conditions. Another is the probability of those conditions occurring.

There are two main methods used to define the structure reliability and its probability of failure: the first-order reliability method (FORM) and second-order reliability method (SORM), or simulation using the Monte-Carlo technique. Most researchers use second moment methods, as they can calculate the variability of load and resistance.

In a DnV report (1995) the second moment method was used, but the report described four limit state equations: ultimate limit state (ULS), which represents danger of failure such as an inelastic member, free drifting or sinking of the structure in the case of a tension leg platform (TLP); fatigue limit state (FLS), which represents failure due to cyclic loading as in the case of fatigue in the joint; collapse limit state (CLS), which represents a complete collapse from abnormal effects; and also the serviceability limit state (SLS), which represents the failure to perform its function under normal conditions.

Another method, which has been used in recent years, is that proposed by Ersdal (2005). In this research, a simple probabilistic model was proposed that included a relation between the reserve strength ratio (RSR) and the annual failure rate for existing structures located in the North Sea in shallow or intermediate water depth.

For this study, wave load was calculated by using the annual maximum wave height with probability of exceedance equal to 10^{-2}. It was assumed that the annual maximum wave height is represented by a Gumbel distribution.

The ultimate capacity was assumed equal to the calculated wave loading multiplied by the RSR, using the following formula:

$$\mathrm{UCR} = \mathrm{RSR} \times \left(\mathrm{C1} \times \mathrm{H100}^{\mathrm{C3}}\right)$$

where

RSR is the Reserve Strength Ratio (RSR), C1 H100 $^{\mathrm{C3}}$ is assumed to be the design loading. C1 and C3 are the load coefficients that must be extracted from a curve fitted to the results of load calculation.

Monte-Carlo simulation was used to obtain the probability of failure based on the total number of simulations and the number of samples falling into the failure domain.

From this study, the relationship between RSR versus annual probability of failure was also established by this formula:

$$P_f = 10^{-0.0886 - 1.9976\,\mathrm{RSR}}$$

This study did not consider the corrosion effect and pile failure as one of the failure modes. Therefore, there are some limitations to this study.

Some research in 2006 and 2007 focused on calculation of the probability of failure, such as studies of seasonal storms in the Gulf of Mexico. In these studies, the performance of fixed steel jacket platforms in Katrina and Rita were assessed.

However, some research studies did the simulation but only considered two variables, so the benefit of the simulation was not obtained. By using simulation, we can present the variability of each element, which is very complicated, and the factors, which are also very complicated to do for offshore structures. Therefore, most research in this area defines the variability of the loads and represents the wave by its statistical parameters. Some other studies considered resistance and load as probabilistic and all other factors as deterministic.

In the ideal case, which would provide more accuracy but needs further research and development, all the variables would be considered in the simulation, and there would be integration between the simulations by using Monte Carlo, such as Crystal Ball software or B-risk, and offshore structure analysis using SACS software. In this case it would be required to run the SACS software for at least 10,000 times, as discussed in the previous sections. This approach needs some software development to connect Monte Carlo with the offshore structure analysis package, as it is very complicated to represent the platform by equations. This approach will be costly but it is more representative of the actual case for calculating the offshore structure probability of failure.

The consequences are the main factors in risk assessment as well as defining the target reliability as per HSE report, 1999. This report states that the fatalities induced by structural failure of offshore platforms occur primarily when the support of the deck fails or the platform capsizes, i.e., system failure occurs.

Failure of individual components (members, joints) commonly does not lead to fatalities. Clearly, the risk of fatalities would depend upon whether platforms will be evacuated before or during the accidental scenario or not. For instance, the likelihood of fatality caused by storm overload in the Gulf of Mexico (GOM) may be less than in the North Sea (NS), because most platforms in GOM can be evacuated in the face of a storm (hurricane), while this is not the case in the NS.

In structural risk assessment, the probability of fatalities, pollution, and property loss is estimated, and the target level should in principle be referred to each of these consequences. The focus has been on fatalities and the reference value is the annual death rate in society, which is on the order of

$P_d = 10^{-4}$ (Flint, 1976; Jordaan and Maes, 1991). Paté-Cornell (1993) suggests 10^{-3} to 10^{-4} to be an upper bound of an acceptable fatality rate.

2.6 FPSO integrity

The integrity of FPSOs worldwide is a challenge for each operator, as there are many mature FPSOs, so it is required to obtain efficient integrity management. This is more critical for FPSOs because some of them were tankers converted to an FPSO. For any tanker, its inspection and assessment are related to its class. Most of the classes like DnV, ABS, and others work by time-based inspection, which can be applied easily to vessels as they can go to the dry dock for repairs and complete rehabilitation. However, for the FPSO this is complicated, as it means a total shutdown for long periods. Mansour et al. (1996) presents the risk and reliability technology for ships, with demands from the owners for more flexibility than the classification services are providing. From a commercial point of view, the tankers that are converted to FPSOs are being operated in service beyond their service life. In most cases the standard survey cycles do not align with the operating practice and also the oil and gas industry requires more efficient and rational ways to maintaining machinery and structures. Basu and Lee (2006) present a flexible approach to risk-based methods for the hull structure inspection. Lee et al. (2006) presented four approaches that are used in ABS and implemented software for a technique to do risk-based inspection for FPSO, as follows:

(1) Screening assessment, pre- and postsurvey scope assessment. The method identifies the most critical compartments and structures within a compartment. The likelihood of strength failure is calculated deterministically and expressed as the ratio of class renewal thickness over the as-gauged thickness.

(2) Degradation assessment, which is a deterministic approach using safe hull analysis. This method considers the corrosion rates that may change over the FPSO's lifetime.

(3) Structure reliability analysis level I, in which the reliability will be calculated based on using safe hull assessment as per American Bureau of Shipping guidance for safe hull-dynamic loading (2001).

(4) Structure reliability analysis level II, which is the reliability calculation based on using dynamic load analysis and spectral fatigue analysis as per American Bureau of Shipping guidance for spectral fatigue analysis (2002).

For these four methods, the consequences are based on present criticality classes for the particular component of interest. For example, the criticality of the hull girder in failure is higher than the primary and secondary member, and so on. As per Lee et al. (2006), a crude measure of consequences is the proportion of the entire structure that participates in the failure.

Generally, a small localized failure will have lesser consequences than if a large stiffened panel failed. (Of course, the possibility of progressive failure must be guarded against.) Applying this kind of thinking to a generic tanker/ship-shaped FPSO structure, it appears reasonable to consider the structure in three broad categories (as is often done in ship design). Three levels (primary, secondary, and tertiary) of structure components are identified. The primary level of structure includes the hull girder. The secondary level of structure is composed primarily of large stiffened panels. The tertiary level includes plates, stiffeners, and small stiffened panels.

There is a relation between class inspection rules and the four methods, but for the reliability level II the class rules have little influence, so this method provides justification for the variation from the class rules. This does not necessarily mean that the inspection frequencies and scopes differ more from the class requirements, but rather the basis for the inspections is driven more directly from the structural analysis and structural reliability analysis results.

For the over 150 years of knowledge and experience within the class rules, and in order to justify an alternative to the class guidelines, a significant foundation of structural analysis and structural reliability analysis must be developed.

References

Ahmed, M.A., 1990. Reliability Analysis of Ductile Plan Structures (Ph.D. thesis). Faculty of Engineering, Cairo University, Giza.

Allen, D.E., 1970. Probabilistic study of reinforced concrete in bending. Proc. ACI 67, 989–993.

Allen, D.E., 1972. Statistical Study of The Mechanical Properties of Reinforcing Bars. Building Research Note No. 85, National Research Council, Ottawa.

American Bureau of Shipping, 2001. Guidance Notes on Safehull-Dynamic Loading Approach for Floating Production, Storage and Offloading (FPSO) Systems.

American Bureau of Shipping, 2002. Guidance Notes on Spectral-Based Fatigue Analysis for Floating Production, Storage and Offloading (FPSO) Systems.

Ang, A.H., Cornell, C.A., 1974. Reliability based structural safety and design. J. Struct. Div. 100, 1755–1769. ASCE, No. ST9, Proc. Paper 10777, Sept.

Ang, A.H., Tang, W.H., 1984. Probability Concepts in Engineering Planning and Design Volume II Decision, Risk and Reliability. John Wiley & Sons, Inc., New York.

Arafa, A.M., 1997. Statistics for concrete and steel quality in Saudi Arabia. Mag. Concr. Res. 49 (180), 185–193.
Bai, Y., Bai, Q., 2005. Subsea Pipelines and Risers. el-sevier.
Bai, Q., Bai, Y., 2014. Lateral buckling and pipeline walking. In: Bai, Q., Bai, Y. (Eds.), Subsea Pipeline Design, Analysis, and Installation. Gulf Professional Publishing, Boston, pp. 221–253 (Chapter 10).
Basu, R., Lee, A.K., 2006. A Flexible Approach to the Application of Risk-Based Methods to the Inspection of Hull Structure. WMTC2006-CS023, London,.
Benjamin, J., Lind, N.C., 1969. A probabilistic basis for a deterministic code. J. Amer. Conc. Inst. 66 (11), 857–865.
Bljuger, F., 1987. Probabilistic analysis of reinforced concrete columns and walls in buckling. ACI Struct. J. 84 (2), 124–131.
Brikeland, P.W., Westhoff, L.J., 1972. Dimensional tolerance-concrete, state-of-art reportb5, Tech. Comm 9, Proc. of Int. Conf. on planning and design of tall buildings, Lehigh Univ., Vol.1b, pp. 845–849.
Castello, J.F., Chu, K., 1969. Failure probabilities of reinforced concrete beams. ASCE J. Str. Div. 95 (ST10), 2281–2340.
Cornell, C.A., 1967. Bounds on the reliability of structural systems. J. Struct. Div. 93 (ST1), 171–200.
Corotis, R.B., Chalk, P.L., 1980. Probability models for design live loads. J. Struct. Div. ASCE 106 (ST10), 2017–2033.
Corotis, R.B., Doshi, V.A., 1980. Probability models for live load survey results. J. Struct. Div. ASCE 106 (ST10), 2017–2033.
Dagher, H.m., Kulendran, S., Peyrot, A., 1993. System reliability concepts in design of transmission lines. J. Struct. Eng. ASCE 119 (1), 223–340.
deWaard, C., Lotz, U., 1993. Prediction of CO2 Corrosion of Carbon Steel. CORROSION/93, paper no. 69.
deWaard, C., Lotz, U., Dugstad, A., 1995. Influence of Liquid Flow Velocity on CO2 Corrosion: A Semi-Empirical Model. CORROSION/95, paper no. 128.
Ditlevesn, O., 1979. Narrow reliability bounds for structural system. J. Struct. Mech. 7 (4), 453–472.
DnV, 1995. Guideline for offshore structure reliability analysis. DnV Technical report, 95-3203, Det Norsek veritas 1995.
Det Norske Veritas (USA) Inc. (DNV), 1996. Guidleine for Offshore Structural: Reliability Analysis, Document Number 95-3203, 5 September 1996.
Edwards, J.D., Sydberger, T., Mork, K.J., 1996. Reliability based design of CO_2-corrosion control. In: CORROSION'96, Paper No. 29. NACE International.
El-Reedy, M.A., 2012. Reinforced Concrete Structural Reliability. CRC Press.
El-Reedy, M.A., Ahmed, M.A., 1998. Reliability-Based Tubular Joint of Offshore Structure Based Inspection Strategy, Offshore Mediterranean Conference. OMC.
El-reedy, M.A., Hassan, M.A., 2005. Concrete quality control related to management performance. In: 11th International Colloquium on Structural and Geotechnical Engineering, May, E05MAT14 -1-18.
El-Reedy, M.A., Ahmed, M.A., Khalil, A.B., 2000. Reliability Analysis of Reinforced Concrete Columns. PhD diss., Faculty of Engineering, Cairo University, Giza.
Ellingwood, B., 1979a. Reliability of current reinforced concrete design. J. Struct. Div. ASCE 105 (ST4), 1377–1388.
Ellingwood, B., 1979b. Reliability based criteria for reinforced concrete designs. J. Struct. Div. ASCE 105 (ST4), 713–727.
Ellingwood, B., MacGregor, J.G., Galambos, T.V., Cornell, C.A., 1982. Probability based load criteria-load factors and load combinations. J. Struct. Div. 108, 978–997. ASCE, No. ST5, Proc. Paper 17068.

Ersdal, E., 2005. Assessment of Existing Offshore Structures for Life Extension (PhD thesis). University of Stavenger. Department of Mechanical and Structural Engineering and Material Science, Stavanger.
Fiorato, A.E., 1973. Geometric Imperfections in Concrete Structures. Document D5:1973, Statens Institute for Byggnads for Skning, Stockholm. 219 pp.
Flint, A.K., 1976. Design objectives for offshore structures in relation to social criteria. In: Proc. BOSS Conf. Trondheim: Tapir.
Frangopol, D.M., 1985. Sensitivity of reliability-based optimum design. J. Struct. Div. 111 (8), 1703–1721.
Frangopol, D.M., Curley, J.P., 1987. Effects of damage and redundancy on structural reliability. ASCE J. Struct. Eng. 113 (7), 1533–1549.
Frangopol, D.M., Lin, K.Y., Allen, C., 1997. Reliability of reinforced concrete girders under corrosion attack. J. Struct. Div. 123 (3), 286–297.
Freudenthal, A.M., Garrelts, J.M., Shinozuka, M., 1966. The analysis of structural safety. J. Struct. Div. 92 (ST1), 267–325.
Galambos, T.V., Ravindra, M.K., 1973. Tentative Load and Resistance Factor Design Criteria for Steel Buildings, Research Report No. 18, Civil and Environmental Eng. Dept., Washington University, St. Louis.
Galamobos, T.V., Ellingwood, B., MacGregor, Cornell, C.A., 1982. Probability based load criteria: assessment of current design practice. J. Struct. Div. 108, 959–977. ASCE, No. ST5, Proc. 17067.
Garson, R.C., 1980. Failure mode correlation in weakest-link systems. J. Struct. Div. 106 (St8), 1797–1810.
Hasofer, A.M., 1968. Statistical model for live floor loads. J. Struc. Div. ASCE 94 (ST10), 2183–2196.
Hasofer, A.M., Lind, N.C., 1974. Exact and invariant second-moment code format. J. Eng. Mech. 100 (EM1), 111–121.
Holmes, J.D., Pham, L., Leicester, R.H., 1985. Wind load estimation and safety index. In: Proc. 4th Int 1. Conf. On Struct. Safety and Reliability, Japan, May 1985, pp. 131–139.
HSE report, 1999. Target levels for reliability-based assessment of offshore structures during design and operation, Report 060/1999.
John, M.D., 1964. Erzeugung von Betaverteilten und Gammaverteilten Zufallazahlen. Metrika 8, 5–15.
Jordaan, I.J., Maes, M.A., 1991. Rationale for load specifications and load factors in the new CSA code for fixed offshore structures. Can. J. Civil Eng. 18 (3), 454–464.
Kiefher, J.F., Vieth, P.H., 1989. A modified criterion for evaluation the remaining strength of corroded pipe. RSTRENG, Project PR 3-805, Pipeline Research Committee, American Gas Association.
Lee, A.K., Serratella, C., Wang, G., Basu, R., Spong, R., 2006. Flexible Approaches to Risk-Based Inspection of FPSOs. OTC 18364, Energo Engineering, Inc.
Lind, N.C., Ravindra, M.K., Sin, W.C., 1972. Some Example of Code Calibration. University of Waterloo Solid Mechanics, Waterloo, ON. Report No. 17.
MacGregor, J.G., 1976. Safety and Limit States Design for Reinforced Concrete. Canada by The University of Alberta.
Mansour, A., Wirsching, P., Ayyub, G., White, B., 1996. Probability-Based Ship Design Implementation of Design Guidelines of Ships: A Demonstration. Ship Structure Committee Report SSC-392, Washington, DC.
Marek, P., Gustar, M., Tikalsky, P.J., 1993. Monte Carlo simulation – tool for better understanding of LRFD. J. Struct. Eng. 119 (5), 1586–1599.
Mattock, A.H., Kriz, L.B., Hongnestad, E., 1961. Rectangular concrete stress distribution in ultimate strength design. Proc. ACI 57, 875–928.
Melchers, R.E., 1983. Reliability of parallel structural systems. J. Struct. Eng. 109 (11), 2651–2665.

Mirza, S.A., 1987. Probability-based strength criterion for reinforced concrete slender columns. ACI Struct. J. 84 (6), 459–466.
Mirza, S.A., Kikuchi, D.K., MacGregor, J.G., 1980. Flexural strength reduction factor for bonded prestressed concrete beams. J. ACI 77 (4), 237–246.
Mori, Y., Ellingwood, B.R., 1994. Maintaining reliability of concrete structures I: role of inspection/repair. J. Struct. Eng. 120 (3), 824–845.
Moses, F., 1982. System reliability development in structural engineering. J. Urban Plan. Dev. Div. ASCE. 93 (ST5), 147–164.
Paté-Cornell, M.E., 1993. Risk management for existing energy facilities: a global approach to numerical safety goals, in ageing of energy production and distribution systems. Appl. Mech. Rev. 46 (5).
Plevris, N., Triantafillou, T.C., Veneziano, D., 1995. Reliability of RC members strengthened with CFRP laminates. J. Struct. Div., ASCE 121 (7), 1037–1044.
Rackwitz, R., Fiessler, B., 1978. Structural reliability under combined random load sequences. Comput. Struct. 9, 489–494.
Ramachandran, K., Baker, M.J., 1985. New reliability bounds for series systems. In: Proc. 4th Intl. Conf. On Structural Safety and Reliability, Japan, pp. 1157–1169.
Ramakrishnan, V., Balaguru, P.N., 1987. Computer use for statistical analysis of concrete test data. J. ACI, 136.
Rashedi, M.R., 1984. Studies on Reliability of Structural Systems, Ph.D. Thesis. Case Western Reserve University, Cleveland, OH.
Renyi, A., 1973. Letters on Probability. Wayne State University Press, Detroit.
Rolfe, S.T., Novak, S.R., 1970. Slow-bend KIC testing of medium hightoughness steel. Review of development in plane strain fracture toughness testing. ASTM Spec. Tech. Publ. 463, 124–159.
Rusch, H., 1960. Researches toward a general flexural theory for structural concrete. Proc. ACI 57, 1–28.
Schueller, G.I., Stix, R., 1987. A critical appraisal of methods to determine failure probabilities. Struct. Saf. 4, 293–309.
Sharifi, S.M., Taheri, A., Pool, M.B., 2019. Assessment of offshore pipeline reliability against lateral buckling. Int. J. Maritime Technol. 12, 41–48.
Shinozuka, M., 1983. Basic analysis of structural safety. J. Struct. Div. 109 (3), 721–740.
Simiu, E., Filliben, J., 1975. Statistical Analysis of Extreme Winds. National Bureau of Standards Technical Note No. 868, National bureau of standards, Washington, DC.
Sotberg, T., Moan, T., Bruschi, R., Jiao, G., Mork, K.J., 1997. The SUPERB project: recommended target safety levels for limit state based design of offshore Pipelines. In: Proc. of OMAE'97.
SriVidya, A., Ranganathan, R., 1995. Reliability based optimal design of reinforced concrete frames. Comput. Struct. 57 (4), 651–661.
Surahman, K., Rojiani, K., 1983. Reliability based optimum design of concrete structures. J. Struct. Eng. 110 (3), 671–672.
Thoft-Christensen, P., Murotsu, Y., 1986. Application of Structural Systems Reliability Theory. Springer-Verlag, New York.
Tso, W.K., Zelman, L.M., 1970. Concrete Strength Variations in Actual Structures. Report EMO TR 200-3, Canada Emergency Measures Organization, pp. 981–988. Also in Proc. ACI, Vol. 67, Dec. 1970.
Tzavellis, C., Shinozuka, M., 1988. Seismic reliability of rigid frame structures. J. Eng. Mech., ASCE 114 (11), 1953–1970.
Vanayi, H., Eslami, A., 2015. A review of inspection and corrosion rate determination methods, in oil and gas pipelines. In: 6th Iranian Pipe and Pipeline Conference, Iran. Beheshti International Conference Center, Tehran.

CHAPTER 3

Inspection techniques

3.1 Introduction

The main activity in the integrity management process is inspection. Many companies are marketing specialized services for the inspection of offshore structures, subsea pipelines, and even onshore structures and associated facilities. Research and development in this sector is ongoing, attempting to define better, easier inspections that do not affect operations.

The target and responsibility of the integrity engineer is to determine the best method of inspection that provides the required data with reasonable accuracy, to assist in the next step of the assessment. The integrity engineer is also responsible for defining the inspection scope of work. Periodic inspections should be defined based on the integrity strategy. As discussed in the previous chapter, time-based inspections have commonly been done in the past, but the newer, optimized trend for inspections is to use risk-based inspections, specifically for offshore structures and FPSOs, as the cost of inspection is more higher than that in case of onshore structure, so for offshore structure it will be more economic to go through using risk based inspection approach. However, some companies are still using time-based inspections, including for the topside of offshore structures.

Table 3.1 summarizes the code that covers the inspection scope and interval.

In general, the main element of the inspection is the visual inspection; realistically, about 80% of the inspection depends on the visual inspection, along with the competency of the inspector and the consultant engineer. The successful evaluation and assessment of the structure depends largely on the experience of the engineer who does the assessment and inspection. Some countries provide a structure consultant engineer certification to evaluate and repair the structure for those who have over 25 years experience rather than the design of a new structure as it needs only 15 years experience to be certified as a consultant engineer for design.

Sometimes a junior engineer or third party is sent to do the inspection, part of which is the visual inspection. They should be well trained and familiar with the inspection process and what aspects must be focused on,

Asset Integrity Management for Offshore and Onshore Structures
https://doi.org/10.1016/B978-0-12-824540-8.00003-X

Table 3.1 Different facilities' inspection intervals and related codes.

Facilities type	Codes	Inspection interval
Structural inspection	API RP2A	Topside 1 year Jacket 5 years
Pressure-relieving devices	API 510 and 576	Interval less than 5 years
Storage tank	API 650	
Inspection and maintenance of storage tank	API 653	
Piping	API 570	Piping inspection based on corrosion rate
Pressure vessel with pressure above 15 psig	API 572	15 years or half the remaining life
Pressure relief device testing and maintenance	API 576	–
Piping, tubing, valves (non-control valve), and fitting depends on the corrosion rate and remaining life	API 574	Every 5 years or half the remaining life

including methods for taking photographs and the minimum acceptable resolution. All of the photos and small measurements can be very important in decision making during the assessment or in deciding on the next level of inspection, using tools that are explained in the next sections. If the inspection is inadequate and the photos don't represent actual conditions, much time and effort will be wasted, specifically for offshore structures.

The photos taken should represent the structure system and the level of deficiency of the structure member; thus it is preferred that the inspector have a structural engineering background and experience when inspecting structure members. For example, in Fig. 3.1, the inspector placed a pen to define and highlight the crack width when the tools to measure the crack width were not available.

3.2 Offshore structures

The subsea inspection is very important in evaluating an existing structure. The scope of work of the inspection is defined in API SIMS as a guideline, but every company with a fleet of platforms has their own standards for inspection.

Fig. 3.1 Crack width assessment.

The main scope for a subsea inspection, which is usually performed by a remote operating vehicle (ROV) with debris removal by divers, includes the following items:

1. Visual inspection
2. Define the condition of the anodes
3. Measure the CP
4. Flooded member detection
5. Magnetic particle inspection (MPI) for the joints
6. Measure the member thickness
7. Define the scour depth if it exists and define the local and global scour.

The techniques of the underwater inspection for the offshore structure will be as described in the following sections.

3.2.1 Flooded member inspection

The main goal of the flooded member inspection is to determine whether the subsea tubular structure member is full of water, which means that there are cracks in one of its two ends. Gamma ray or ultrasonic methods, using an ROV, can be employed for this. Gamma ray is most commonly used for flooded member detection (FMD). It requires a gamma source from one point and the detector located on the other end, and both are fixed on a yoke attached to the ROV. The ROV is operated from above the sea surface to be positioned over the required member and the detector reading is counted. This reading is then compared with an expected count that is calculated beforehand, considering the water absorption characteristics between the yoke and the member, the thickness of the steel member, and the air within the member. If the member is flooded, then the count reading will be lower than expected due to the absorption of the gamma radiation by the water. This member is then flagged for a close visual inspection of its two ends to determine the crack location and dimensions.

When ultrasonic testing (UT) is used for FMD, the surface is first cleaned in order to have good acoustic contact and an ultrasound probe is fixed to the outside of the member (Fig. 3.2).

Water is a better transmitter of ultrasound than air, and a flooded member is detected by observing echoes that are received by the probe and viewed on a screen. For an air-filled member, the echo will be that of the back wall. For a water-filled member, the ultrasound is transmitted through the water and a second echo will be generated at the interface between the water and the back face of the tube. This back-face echo is an indication that the member is flooded. Fig. 3.3 shows a typical UT trace for a flooded member.

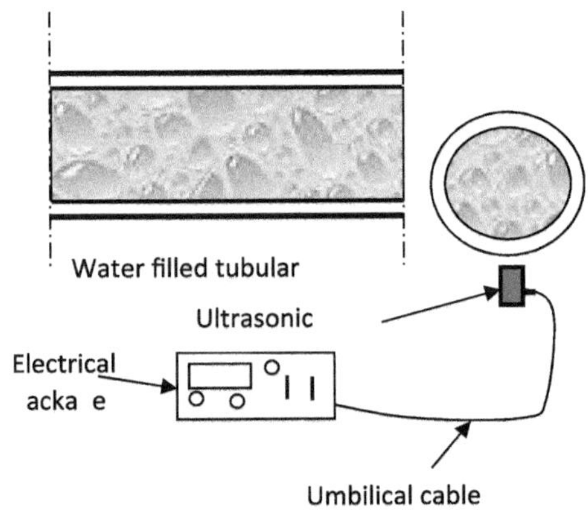

Fig. 3.2 Equipment arrangement for UT FMD.

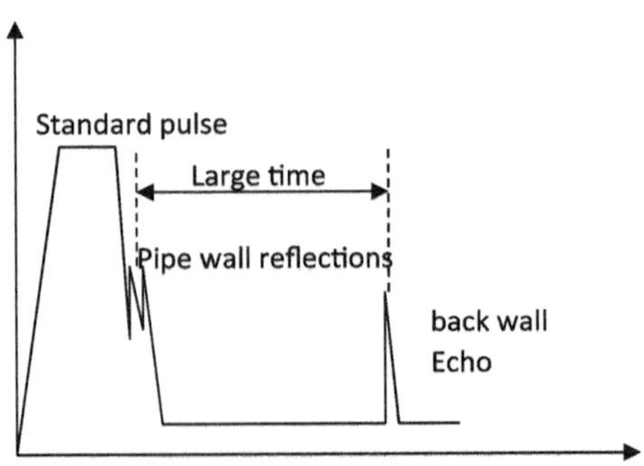

Fig. 3.3 FMD UT signal reading.

For gamma FMD, the maximum distance between source and detector determines the maximum member diameter that can be inspected. The distance between the two is limited not only by the size of the yoke, which is dependent on the payload capability of the host vehicle, but also by the strength of the radioactive element relative to the thickness of the steel walls of the members requiring inspection. The size of a gamma source normally varies according to member diameter and wall thickness. In general, as large a source as possible is used to give meaningful results, but without saturating the detector so that a drop cannot be observed.

The host vehicle should be of sufficient size to allow access and to carry the FMD equipment, and it should be able to operate in the environmental conditions present.

The range of member diameters typically surveyed by gamma FMD methods is from 12 inches (0.3 m) to 2.6 m. FMD is currently performed on members with a wall thickness in the range of 8–63.5 mm.

The yokes can be adjusted to any position for taking readings on inclined members. Fig. 3.4 shows that the positions in which readings may be taken on the member can also be varied. Typically, readings are taken at the "6 o'clock–12 o'clock" or the "3 o'clock–9 o'clock" positions. The yoke can be adjusted to make the necessary readings possible.

The majority of flooded or cracked members occur as a result of unforeseen circumstances, such as the presence of a gross defect caused by poor fabrication, poor design practice, or accidental damage. Poor fabrication or design errors manifest themselves in the form of unexpected fatigue cracking, in which joints crack at a time well before their design lifetime. Because the occurrence of this damage is not expected, the inspection strategy should enable detection of unforeseen damage.

Reference to design fatigue lives is not a suitable means for justifying the use of FMD. The traditional S-N fatigue design process does not allow for large defects or poor design, assuming welded joints to be "nominally perfect." Operationally, however, it should be assumed that a large defect or design error could potentially exist in every member and could lead to through-thickness failure of the member and subsequent flooding, and could potentially develop to a size that could lead to final severance.

The most appropriate means of assessing an individual member for the effect of a crack of a size allowing flooding is through a fracture mechanics assessment. This assessment will determine, for the size of defect, whether failure by fracture or collapse will occur under the prescribed conditions.

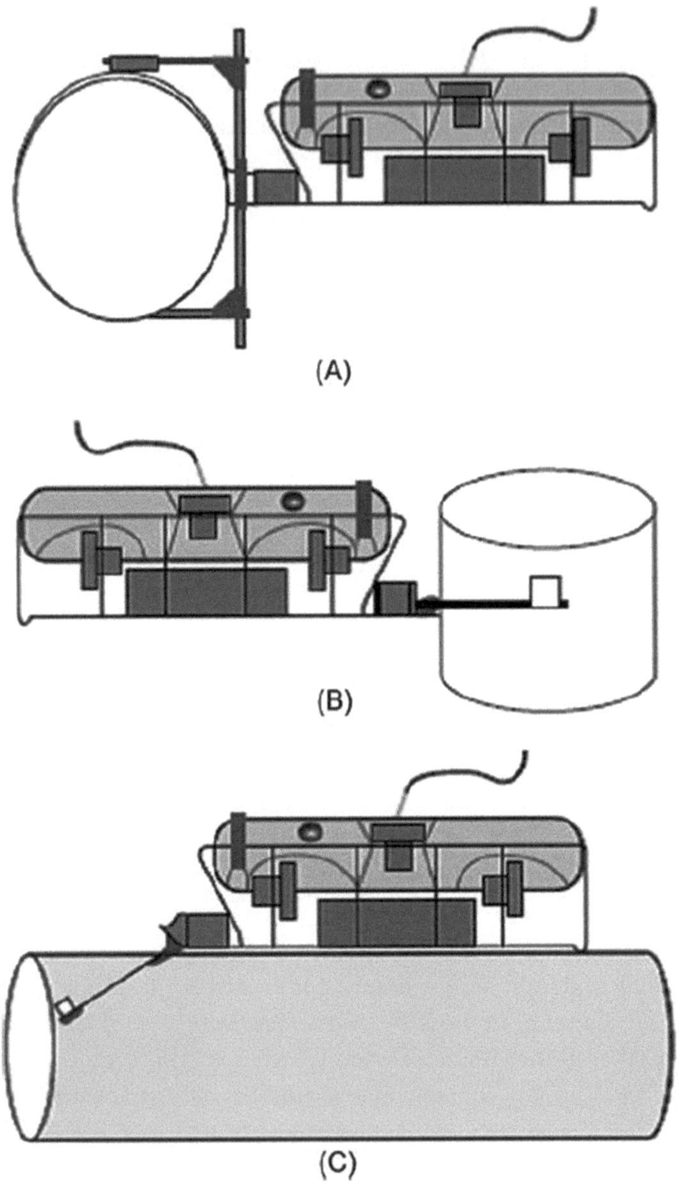

Fig. 3.4 (A) Horizontal member standard survey mode for FMD. (B) Vertical member survey mode for FMD. (C) Horizontal member survey mode for FMD.

It should first be determined that the member can sustain through-thickness damage of a size sufficient to cause leakage without failing completely under the anticipated loading conditions. If this cannot be demonstrated, then FMD is not a suitable method for ensuring that the member's integrity is maintained. If it is determined that through-wall damage is possible but that complete severance is not, then it is necessary to demonstrate that FMD can detect the flooded condition before final severance or some other degree of damage is detected over a particular inspection interval. Given the extreme variability associated with all factors influencing fracture and collapse performance, a probabilistic fracture mechanics assessment should be undertaken to establish the probability of failure of an individual weld and, ultimately, the probability of failure of a complete member.

An alternative method for predicting failure frequencies is the use of historical data. There are many studies of a fracture mechanics approach to calculation of the probability of a member's leaking and the probability of complete severance. The studies consider member failure rates obtained from historical review. It should also be noted that other analytical approaches, many of them based on fracture mechanics, may be employed to determine the probability of failure of a member.

It is worth mentioning that the FMD technique is expensive but it is worthwhile in the case of higher water depth with many tubular joints and with a company having a fleet of offshore platforms, so the cost will not be very high. However, in the case of a small operating company with one or two platforms, it is more beneficial to do a closed visual inspection for every tubular joint by using the magnetic particle inspection method (MPI). Using this technique gives full confidence concerning the tubular joint condition. Fig. 3.5 presents the MPI for a tubular joint.

3.2.2 Magnetic particle test

This test is used as part of the close visual inspection and is considered to replace the FMD in the case of shallow depth and fewer joints. Since this method is more accurate but higher cost than FMD, it is a better choice for an inspection plan for one or two platforms, as FMD would be more expensive. FMD can be better used in an inspection program for a fleet of platforms.

The benefit of MPI is that it determines whether or not there are cracks in the joint and can define the length of the crack, which is of benefit later during the assessment process. This inspection tool is shown in Fig. 3.5.

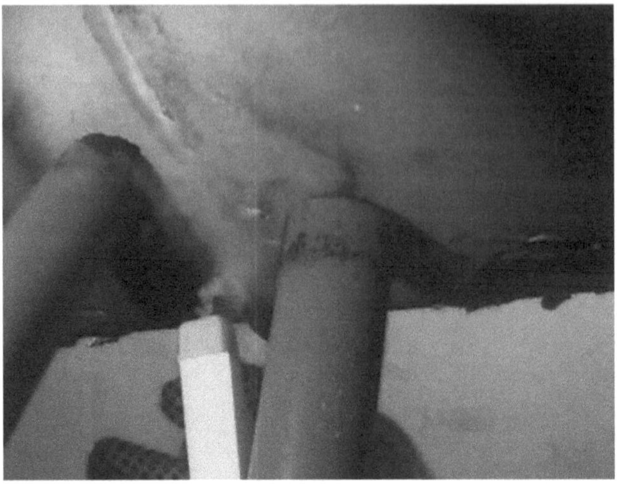

Fig. 3.5 MPI technique for subsea joints.

It uses a two-probe connection to an electrical supply and a special iron powder with a light color that fills the groove of the cracked joint due to the magnetization from the electrical circuit.

3.2.3 The inspection report

The inspection report is a very important document and will remain in the archive system for the platform, as it will be a reference for any future assessment, if any modifications are needed, and can be of assistance during the next inspection plan.

The format of the report should be set by the company based on their integrity management system. I prefer not to rely on the inspection company format, as it differs from one inspection company to another. This concept applies for subsea or topside inspection and also for onshore structures, even concrete or steel.

The report format should be as follows:

- Title page for the project
- Report Title/Author/Company/Date, etc.
- Contents
- Executive Summary (highlight here any defects posing a risk)
- Inspection company's name
- Dates of survey
- Inspections completed and inspections omitted and reasons for omission
- Anomalies found

- Conclusions
- Outstanding actions
- Inspection results by area
- Written summary
- Inspection datasheets
- Drawings
- Anomaly sheets (including NDT reports where available)
- Photographs/Video stills (should be defined by date and time)
- Operations analysis
- Summary of time breakdown (chart)
- Summary of work breakdown (chart)
- Daily reports
- Inspection team details
- Copies of up-to-date CVs
- Film/video logs
- Summary of all film/video logs

3.2.4 Topside inspection

The topside inspection is usually a visual inspection, with UT used in some cases to measure the remaining thickness after corrosion. On the topside, UT is not easy to deploy, as it needs a scaffolding or abseiler for the height to carry out measurements.

Currently, many companies use time-based inspections for the topside, as the cost of inspection is not extreme, but an expert is required to do the visual inspection, perform the required test, and issue the inspection report in a professional way to support the desk work for the assessment.

The main element to measure is the wall thickness, to define the thickness loss of the member. This is carried out by the UT. In addition, a thickness gauge that uses the same concept as UT, but provides only the thickness measurement, is shown in Fig. 3.6.

3.2.5 Offshore structure inspection plan based on ISO

The inspection plan will be defined after the exposure level is determined, as shown in Tables 3.2 and 3.3.

It is important to highlight that the timing of the first periodic level I inspection shall be determined from the date platform installation was completed. However, the timing for the first periodic level II and level III inspections shall be determined from the date of the baseline inspection.

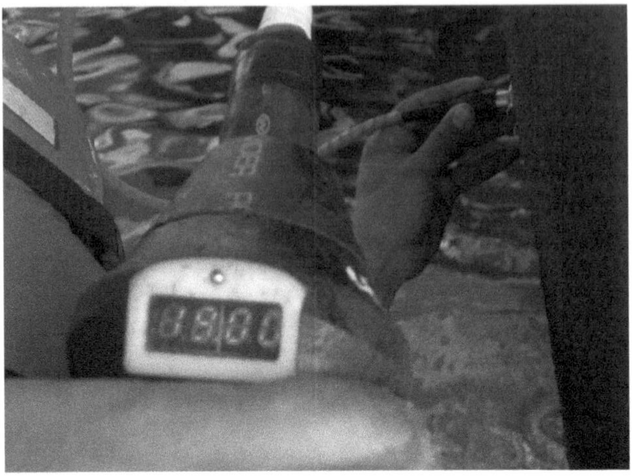

Fig. 3.6 Thickness measurement.

Table 3.2 Determination of exposure level.

Life-safety category	Consequence category C1 high consequence	C2 medium consequence	C3 low consequence
S1 Manned non-evacuated	L1	L1	L1
S2 Manned evacuated	L1	L2	L2
S3 Unmanned	L1	L2	L3

Table 3.3 Maximum inspection intervals for default periodic inspection program.

Exposure level	Level I inspection	Level II inspection	Level III inspection	Level IV inspection
L1	Annual	3 years	5 years	Determined from level III inspection results
L2	Annual	5 years	10 years	Determined from level III inspection results
L3	Annual	5 years	Not required	Not required

The ISO identifies the level of survey as presented in Table 3.4.

Table 3.4 ISO identified level of survey.

Level I periodic inspections	A level I inspection shall consist of: • the effectiveness of the corrosion protection system • deterioration of coating systems • excessive corrosion • bent, missing, or damaged above-water members • indications of obvious overloading, design deficiencies, and any use that is inconsistent with the structure's original purpose • general inspection of all structural members in the splash zone and above water • concentrating on the more critical areas such as topsides, legs, girders, trusses, etc.
Level II periodic inspections	A level II inspection shall consist of a general underwater visual inspection to detect the presence of any or all of the following: • excessive corrosion • accidental or environmental overloading • scour or sea floor instability • damage detectable in a visual swim-around survey • design or construction deficiencies • presence of debris • excessive marine growth The inspection shall include the measurement of cathodic potentials of pre-selected critical areas
Level III periodic inspections	Detection of significant structural damage during a level II inspection shall initiate a level III inspection. The level III inspection, if required, should be conducted as soon as conditions permit A level III inspection shall consist of: – underwater close visual inspection (CVI) of pre-selected areas – flooded member detection (FMD) can provide an acceptable alternative to CVI of pre-selected areas – close visual inspection of pre-selected areas for corrosion monitoring shall be included as part of the level III inspection
Level IV periodic inspections	A level IV inspection consists of: • underwater NDT of areas pre-selected from results of a level III inspection or from known or suspected damage • detailed inspection and measurement

Continued

Table 3.4 ISO identified level of survey—cont'd

	• a level III and/or level IV inspection of fatigue sensitive joints of damaged areas • focus on areas susceptible to cracking can be necessary to determine if damage has occurred • monitoring fatigue sensitive joints, and/or reported crack-like indications, may be an acceptable alternative to analytical verification

A level I inspection shall consist of a below-water verification of the performance of the cathodic protection system (for example, a drop cell), and the above-sea level inspection will be through visual survey. This inspection shall include a general examination of all structural members in the splash zone and above water, concentrating on the condition of the more critical areas such as topsides legs, girders, trusses, etc. If above-water damage is detected, non-destructive testing shall be used when visual inspection cannot fully determine the extent of damage. If the level I inspection indicates that underwater damage is possible, a level II inspection should be conducted as soon as conditions permit.

A level II periodic inspection shall consist of a general underwater visual inspection. The inspection shall include the measurement of cathodic potentials of pre-selected critical areas. Detection of significant structural damage during a level II inspection shall initiate a level III inspection. The level III inspection, if required, should be conducted as soon as conditions permit.

A level III periodic inspection shall consist of an underwater close visual inspection (CVI) of pre-selected areas or areas of known or suspected damage. Such areas shall be cleaned to remove marine growth, to allow a thorough inspection.

The selection of areas that are required to inspect will be according to an engineering structure assessment for areas particularly susceptible to structural damage and to areas where repeatable inspections are desirable in order to monitor integrity over time.

The technique of flooded member detection (FMD) can be used successfully instead of CVI of pre-selected areas. Engineering judgment should be used to determine optimum use of FMD or close visual inspection of joints. It is worth mentioning that inspection level III in API recommends the close visual inspection of pre-selected areas for corrosion monitoring.

Detection of significant structural damage during a level III inspection shall initiate a level IV inspection, where visual inspection alone cannot determine the extent of damage. The level IV inspection, if required, should be conducted as soon as conditions permit.

The target of the FMD technique is to determine the condition of the tubular bracing members of steel jacket structures that directly affect the structural integrity. As previously described, the method relies on the detection of through-wall cracks by the leakage of water into the internal, air-filled, space and the successful detection of that water. The detection is made through the use of gamma radiation or ultrasonic techniques, both of which are applied externally to a tubular member, usually by means of a remotely operated vehicle (ROV) in the case of gamma FMD, or by diver or potentially an ROV with ultrasonic FMD.

A level IV inspection shall consist of underwater non-destructive testing (NDT) of areas pre-selected from results of a level III inspection or from known or suspected damage. A level IV inspection shall include detailed inspection and measurement of damaged areas.

A level III and/or level IV inspection of fatigue sensitive joints and/or areas susceptible to cracking can be necessary to determine if damage has occurred. Monitoring fatigue sensitive joints and/or reported crack-like indications may be an acceptable alternative to analytical verification.

An example of a survey scope of work (SOW) in brief is presented in Table 3.5, which serves as a guideline in preparing the inspection scope of work.

3.3 Piping and pipeline inspection

The time frequency for piping inspection as proposed by API571 is shown in Table 3.6.

Piping and pipeline inspection codes are related to API 570, revision 2009. This code specifies risk-based inspection to define the inspection frequency and some of the parameters considered for probability of failure (but I prefer to call it a likelihood of failure).

The likelihood-of-failure factors that should be considered are:

a. the construction material quality and its selection for the fluid it carries;
b. equipment design conditions, relative to operating conditions;
c. the codes and standard used in the design and their revisions;
d. the corrosion monitoring program efficiency;
e. the past inspection and maintenance program quality control and quality assurance;
f. both the pressure retaining and structural requirements; and
g. operating conditions both past and projected.

Table 3.5 Inspection survey SOW overview.

Survey	Generic scope of work	Recommendation
General visual	Underwater general visual inspection of full structure with particular reference to members, joints and appurtenance connections, to detect the presence of excessive corrosion, accidental or environmental overloading, fatigue damage, design or construction issues	All damage and all other anomalies found to be reported in accordance with owner or operating company specification On attached plan (first level above waterline) confirm or add location of jacket legs, conductor guides, conductors, j-tubes, casings, risers, boat landings, etc.
Marine growth	Measurements of compressed marine growth at different locations that should be defined based on the water depth and the region of this platform	See schematic for Marine Growth measurement locations. Visually check for areas of ripped or missing marine growth. Document and report, immediately, any locations of missing or ripped marine growth to the Company Representative
Scour	Visually inspect the seafloor up to 7.5 m (25 ft.) out for possible erosion and scouring or instability Check if the piles are exposed and whether the mudmat and bottom braces are on the bottom Take readings of the bottom depths	If the pile is exposed, note if there is any corrosion or damage and check for mudmats Note any bottom irregularities and bottom depth
Cathodic protection	Readings to be taken on legs on outboard truss chord, and legs at the mean water level and halfway between each horizontal framing plan reading	See schematic for CP reading locations Measure all potentials in millivolts with reference to an Ag/AgCl half-cell
Anodes	Visual examination to locate, count, grade, and confirm the physical condition as loose, missing, or damaged of all anodes if Original or Retrofit. Potential readings and detailed dimensional survey to be taken on two anodes	Grade 4 anodes should be reported as anomalous in accordance with the owner or operator specification Anodes identified as loose, damaged, or missing should be recorded as anomalous

Table 3.5 Inspection survey SOW overview—cont'd

Survey	Generic scope of work	Recommendation
Debris	Visual inspection to record the quantity, type, size and location of debris hanging on or in the jacket, in contact with the bottom elevation, on the seafloor up to 10.0 m (30 ft.) out or in contact with any risers, pipelines or conductors	Report the amount of debris inside, on, and around the platform. Note whether the debris is metallic, hazardous, obstructing or contains items that could have caused damage. Include a map of debris on the seabed. Remove, subject to confirmation from the Company
Risers	• Visually inspect the risers • Take CP readings between riser clamps. Record riser coating type and depth of termination • Inspect the pipelines to the point of entry into the seafloor or 10 m (30 ft.) from platform, whichever is closer	Document and report immediately to the Company Representative any soil disturbance or pipeline distortion within 10 m (30 ft.) of the platform or any hydrocarbon leaks from any source
Caissons	Visual inspection of casings, casing clamps, and intakes. Take CP readings on clamps and casing mid-sections General visual survey of the conductors with particular attention given to examine for conductor movement	Note total number of conductor slots, number of conductors driven, and if any are undrilled Intakes shall be cleaned of marine growth if necessary
Flooded members	Flooded member detection for all the member or a specific member in the platform	Provide a Schematic for FMD locations If flooding is detected, the Company Representative may request an additional survey to establish the cause

Table 3.6 Frequency of piping inspection based on API570.

Type of piping class	Visual external inspection	Thickness measurement by external NDT
Class 1: Highest potential of resulting in an immediate emergency if a leak occurs: flammable services prone to brittle fracture/ rapidly vaporizing streams/H_2S streams	5 years	Lesser of 5 years or half the remaining life
Class 2: Services not in other classes: slowly vaporizing streams/fuel gas/natural gas	5 years	Lesser of 10 years or half the remaining life
Class 3: Services that are flammable but do not significantly vaporize when they leak and are not in high activity areas: hydrocarbons operating below the flash point/distillate and product lines to and from storage	10 years	Lesser of 10 years or half the remaining life

The consequences factors, or in other words the factors that impact humans, are health, safety, and environment (HSE) and financial losses. According to risk-based inspection (RBI API) 580, the consequences mainly depend on the type of fluid inside the pipe and also the type of potential release in the case of a leak or failure, such as explosion, fire, toxic exposure, or others.

With a pipeline the corrosion may be internal or external, so it is necessary to focus on the inspection types based on each pipe type. On the other hand, this code provides time-based inspection, but API 570 deals with four classes of pipes. Class 1 is for services with the highest potential of resulting in an immediate emergency if a leak were to occur. The Class 2 classification includes the majority of unit process piping and selected off-site piping. Typical examples of these services include but are not necessarily limited to those containing the following:

a. on-site hydrocarbons that will slowly vaporize during release such as those operating below the flash point;
b. hydrogen, fuel gas, and natural gas;
c. on-site strong acids and caustics.

Class 3 services include those that are flammable but do not significantly vaporize when they leak and are not located in high-activity areas such as tank farm piping. Class 4 services contain essentially non-flammable and non-toxic fluids, such as water and nitrogen, and the inspection intervals are similar to those shown in Table 3.7.

Table 3.7 Recommended maximum inspection intervals.

Class level	Thickness measurement	External visual inspection
Class 1: Requires an immediate emergency in case of leak	5 years	5 years
Class 2: Most of the unit process piping and selected off-site piping	10 years	5 years
Class 3: Containing fluids that are flammable but do not significantly vaporize when they leak and are not located in high-activity areas such as tank farm piping	10 years	10 years
Class 4: Containing fluids that are essentially non-flammable and non-toxic, for example, water and nitrogen	optional	optional
Injection points for potential corrosive injection/mix points can be defined by RBI as per API580	3 years	By class

3.4 Inspection methods

For subsea pipelines, the inspection is similar to that for offshore structures, using ROV for the whole pipeline and defining any defects or leaks or any other issues. This general visual inspection is done by a continuous longitudinal profile along the pipeline, recording any anomalies such as scour, suspension, damage, lost weight coat, damage to the protective coating, metal loss or damage, and providing a description and location of any debris that contacts the pipeline and preferably also recording that within 2 m of the pipeline.

The cathodic protection reading is also measured to define its condition and to guarantee the isolation between the pipelines and the structures. It is required to test the effectiveness of electrical isolation of the pipelines and risers. For this test the percentage of depletion in the anode is also defined to record its condition. The anode potential or current (amps) between the anodes and pipeline and the anode spacing are also defined.

Regarding internal corrosion for onshore and offshore pipelines, the pipe thickness can be measured by using an "intelligent pig," also called "smart pigging," which involves an electronic device that moves inside the pipe with the fluid and measures the piping thickness for the whole piping length.

Another method used for onshore pipelines is the guided wave, which employs a special tool clamped around the pipe to transmit the guided waves. The insulation can be removed from this section only and the wave spreads longitudinally to detect any metal loss on the piping thickness (Fig. 3.7).

3.5 Tank inspections

There are special guidelines for tank inspections in API653.

The API considers two strategies regarding inspection interval: time-based maintenance and risk-based inspection. The time-based maintenance as presented in the API is given in Table 3.8. As per API 650, the interval from initial service date until the first internal inspection shall not exceed

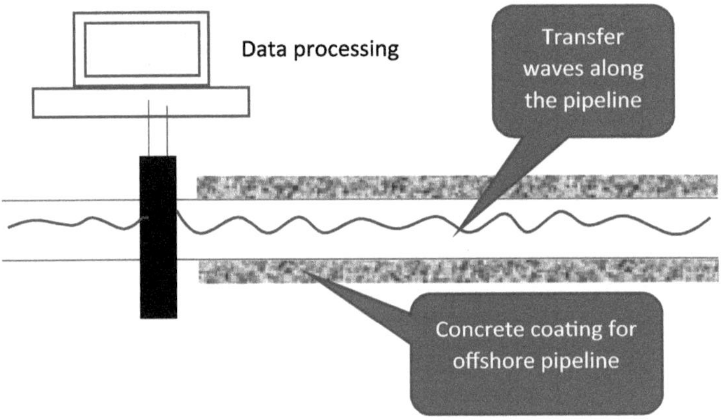

Fig. 3.7 UT scan along pipeline.

Table 3.8 Steel storage tanks inspection interval guide.

Condition of the tank bottom	Add to initial interval inspection
Fiberglass reinforced lining of the tank bottom	5 years
There is an internal thin film coating as per API RP 652	2 years
There is a cathodic protection of the tank bottom on the soil side	5 years
Bottom corrosion allowance more than 0.150 in.	Actual corrosion allowance-0.15 in./corrosion rate
In the case of stainless steel bottom designed as per API 650 with a qualified corrosion specialist, which presents very low risk of crack or corrosion failure	10 years

10 years, unless a tank has one or more prediction method or tools for leak detection, corrosion mitigation. The guidelines in Table 3.8 for inspection intervals are to reduce the risk of failure or occurrence of any leaks, as safeguards for the tank.

Risk-based inspection, which matches the integrity management approach of API 653 in revision 2014, provides the likelihood and consequences factors that should be considered. The likelihood factors are:

1. The age of the bottom plates and its original thickness and weld type.
2. Analysis methods used to determine the product-side, soil-side, and external corrosion rates for both shell and bottom and the accuracy of the methods used.
3. The last inspection data and the inspection history.
4. The soil report data including the soil resistivity, soil bearing capacity, and settlement calculation or recommendation.
5. The quality of the design and the tank pad and the reputation of the designer office.
6. The quality control during construction and the record of the contractor for similar tanks in the same year.
7. The system for water drainage.
8. The cathodic protection (CP) system design and its efficiency.
9. The maintenance history for the CP.
10. The internal corrosion rates due based on the existing product service.
11. The operating temperatures, in particular if they match with the design.
12. The condition and age of the internal coating or lining and its type.
13. The quality of the maintenance applied for previous repairs and alterations and whether it matches with the API.
14. The codes and standards used in design, construction, and even in repair and alteration for the tank bottom.
15. The quality and materials used in construction.
16. The previous inspection data including the scope of inspection and the method and tools of inspection and the inspection company history, with a record of any functional failure such as floating roof seals and drain.
17. The last inspection survey for settlement.
18. The quality control and assurance system during construction of the tank and foundation and slope of the bottom.
19. The documentation and records covering the tank method of installation.

The following factors affect the consequences and can impact on business in the case of failure; these factors should be reviewed by the integrity team as they are considered as guidelines.

1. The tank volume and its product type.
2. Tank bottom with a release prevention barrier (RPB) details (single, double, RPB, internal reinforced linings).
3. Mode of storage tank failure, such as slow leak, tank bottom rupture, tank shell brittle fracture, or other.
4. The environmental condition around the tank area, for example, wetlands, surface water, ground water, drinking water aquifers, and bedrock.
5. The efficiency of the leak detection system.
6. The mobility of the product in the environment, including releases to soil, product viscosity, and soil permeability.
7. The estimated cost to remediate potential contamination.
8. The expected cost for tank cleaning and repair.
9. The expected cost in the case of the tank total loss or reuse.
10. The impact of failure on safety, health, and environment.
11. The design of dike and its capabilities such as its condition and volume.
12. The effect on company reputation in the case of tank failure.

3.5.1 Settlement survey

The critical settlement is the differential settlement that affects the tilt of the tank, which increases the liquid head. The settlement measurement will be done after performing the hydrostatic test. It is preferred to be done at the first time of inspection and as it is known from soil mechanics that for the sandy soil the settlement will happen after construction and settlement will have occurred after hydrostatic testing; for clay soil settlement can take from 8 to 10 years, but the greatest amount of settlement will occur after the hydrostatic testing.

The inspection is to be done through measuring points circumferentially around the tank. The maximum distance between the measured points should be less than 32 ft., so there will be around 12 points around the tank and 6 diameters inside the tank, and the distance of the diameters between the measurement points should be no more than 10 ft. According to API 563 the minimum number of measuring points around the tank is tank diameter/10 with the same constraint of the distance between points not being more than 32 ft. (Figs. 3.8 and 3.9).

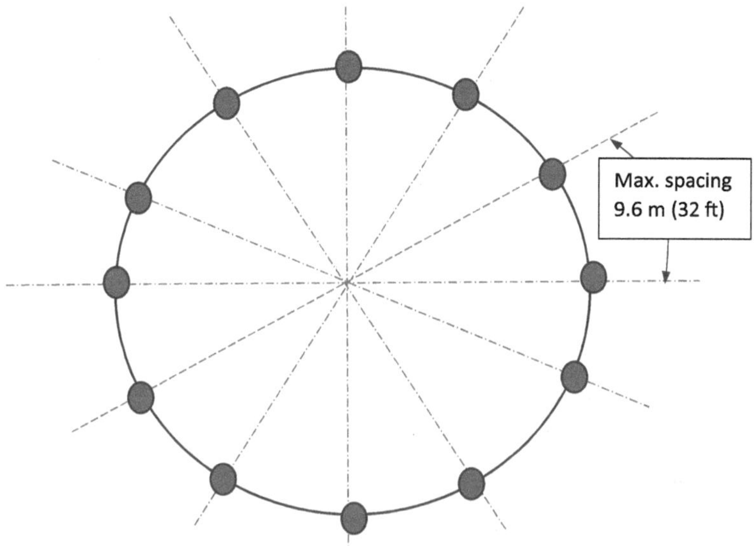

Fig. 3.8 The settlement measurement around the shell.

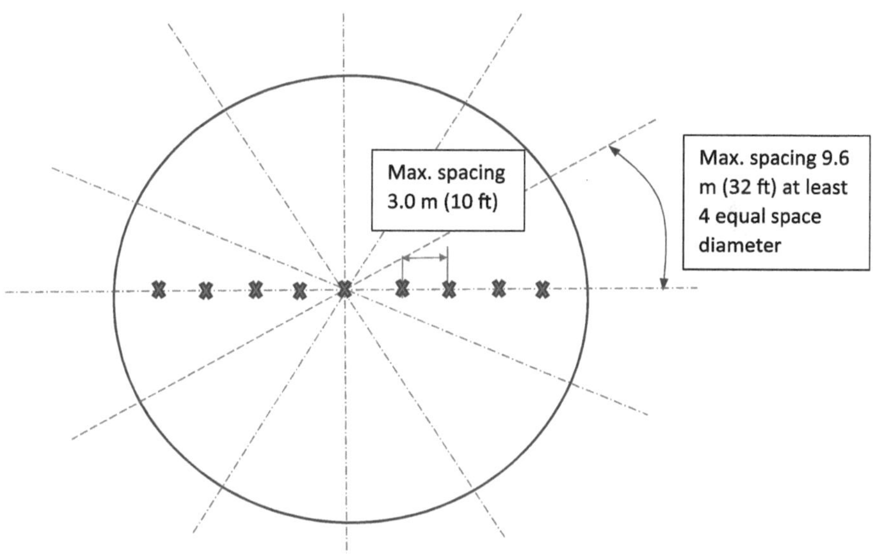

Fig. 3.9 Internal bottom tank settlement measurement.

3.6 Onshore structures

The onshore structures include both steel and concrete structures. The steel structures, such as pipe racks, have a higher risk in case of failure and also in case of workshop or warehouse and similars which are traditional structures in oil and gas plant. Concrete structures are administration buildings, rooms, control rooms, concrete tanks, or dikes, and the foundations underneath all the facilities on the site.

3.6.1 Steel structures

Steel structures must receive regular inspections, usually a time-based inspection, often every 5 years. The integrity management system, as discussed later, will define the time interval between inspections based on the criticality of the structure.

The main inspection tool is the visual inspection, so the condition of the structures will be visually inspected and compared with the original design to determine if there are any modifications. All the inspections findings and photographs should be reflected in an inspection report having the agreed format, as described in Section 3.2.3.

If corrosion has occurred, it is required to measure the current thickness and redo the assessment; another important finding is the condition of the bolts of the joints, specifically in areas with high corrosive effects, such as near the sea. It is mandatory to record the condition of the bolt and whether there are any cracks in the welded connection. This needs special consideration in areas with high vibration.

3.6.2 Concrete structures

As discussed, there are many concrete structures and foundations on a site. The main challenge in concrete structures is the strength, which varies depending on the condition of the steel bars from a corrosion point of view, and also the strength of the concrete when no data is available, in the case of mature structures. This inspection is also required if the load increases, for any change of mode of operations or any modifications needed.

There are many inspection methods used to define concrete strength, so it is important to select a suitable technique that matches the evaluation requirements for accuracy and the condition of the building, its structure, and also the cost.

3.6.2.1 Concrete strength NDT

To define the concrete strength for an existing structure there are many types of NDT (Table 3.9); the best method possible for the particular case should be selected that does not interrupt operations unduly.

Table 3.9 NDT to define material properties and evaluate the existing condition as per ACI 228.2.

Property	Test methods		Remarks
	Primary test	Secondary test	
Compressive strength	Cores for compression testing as per ASTM C 42 and C 39	Penetration resistance ASTM C 803	Penetration test is not traditional to use
Relative compressive strength	Rebound number (ASTM C 805) Ultrasonic pulse velocity (ASTMC 597)	–	Rebound number influenced by near surface properties; ultrasonic pulse velocity gives average result through thickness
Tensile strength	Splitting-tensile strength of core (ASTMC496)	In-place pull off test (ACI 503R; BS 1881; Part 207)	Assess tensile strength of concrete
Local or global strength and behavior	Load test, deflection, or strain measurements	Acceleration, strain, and displacement measurements	Ascertain acceptability without repair or strengthening; determine accurate load rating
Static modulus of elasticity	Cores compression test (ASTMC469)	–	–
Dynamic modulus of elasticity	Resonant frequency testing of sawed specimens (ASTM C 215)	Ultrasonic pulse velocity (ASTM C 597); impact echo; spectral analysis of surface waves (SASW)	Requires knowledge of density and Poisson's ratio (except ASTM C 215); dynamic elastic modulus is typically greater than the static elastic modulus
Shrinkage/expansion	Length change of drilled or sawed specimens (ASTM C 341)	–	Measure of incremental potential length change
Resistance to chloride penetration	Ponding test (90-day) (AASHTO-T-259)	Electrical indication of concrete's ability to resist chloride ion penetration (ASTM C 1202)	Establishes relative susceptibility of concrete to chloride ion intrusion; assess effectiveness of chemical sealers, membranes, and overlays

Continued

Table 3.9 NDT to define material properties and evaluate the existing condition as per ACI 228.2—cont'd

Property	Test methods		Remarks
	Primary test	**Secondary test**	
Air content, cement content; and aggregate properties (scaling, alkali-aggregate reactivity, freezing-and-thawing susceptibility)	Petrographic test of concrete samples removed from structure (ASTM C 856, ASTM C 457); Cement content (ASTM C 1084)	Petrographic test of aggregates (ASTM C 294, ASTM C 295)	Provide the cause(s) of distress; degree of damage; and originally cast and current concrete quality
Alkali-silica reactivity	Cornel VSHRP rapid test	–	It will be done on the site if visual assessment reveals the cause is presence of alkali-silica reactivity
Carbonation test	Phenolphthalein test	Other pH indicators (e.g., litmus paper)	In case of carbonation this test defines the carbonation depth and susceptibility of steel reinforcement to corrosion
Fire damage	Petrography; rebound number (ASTM C 805)	Ultrasonic pulse velocity	Rebound number permits demarcation of damaged concrete
Freezing-and-thawing damage	Petrography	SASW; impulse response	–
Chloride content	Chemical analysis by acid-soluble (ASTM C 1152) and water-soluble (ASTM C 1218)	Specific ion probe	Define chloride depth inside concrete and susceptibility of steel reinforcement to corrosion

Table 3.9 NDT to define material properties and evaluate the existing condition as per ACI 228.2—cont'd

Property	Test methods		Remarks
	Primary test	**Secondary test**	
Electrical resistance of concrete	AC resistance using four-probe resistance meter	SHRP surface resistance test (SHRP-S-327)	AC resistance useful for evaluating effectiveness of admixtures and cementitious additions; SHRP method useful for evaluating effectiveness of sealers
Reinforcement location	Cover meter; ground penetrating radar (GPR) (ASTM D 4748)	X-ray and y-ray radiography	Steel location and distribution; and measure the concrete cover
Corrosion potentials	Half-cell potential (ASTMC 876)	–	Identification of location of active reinforcement corrosion
Corrosion rate	Linear polarization (SHRP-S-324 and S-330)	–	Corrosion rate of embedded steel; rate influenced by environmental conditions
Locations of delamination, voids, and other hidden defects	Impact-echo; infrared thermography (ASTM D 4788); impulse-response; radiography; GPR	Sounding (ASTM D 4580); pulse-echo; SASW; intrusive drilling and borescope	Assessment of reduced structural properties; extent and location of internal damage and defects; sounding limited to shallow delamination
Concrete component thickness	Impact-echo (I-E); GPR (ASTM D 4748)	Intrusive probing	Verify thickness of concrete; provide more certainty in structural capacity calculations; I-E requires knowledge of wave speed, and GPR of dielectric constant
Reduction in steel area	UT thickness gage By direct contact with steel	Intrusive probing; radiography	Measure area reduction in steel section Provide more certainty about strength of structural member

Core test

The main trusted test to evaluate an existing concrete structure is the core test. This test is neither destructive nor non-destructive, so it is referred to in some textbooks as a semi-destructive test. This is a very common test and is usually used when it is necessary to define the concrete strength of an existing structure, such as the need to make decisions in the case of steel or concrete deterioration, the need to increase the load on a building, or in the case of fire acting on the building requiring decisions to be made. This test provides a reasonably accurate strength value that can be used in assessment and evaluation. Sometimes a cube or a cylinder yield less than an acceptable criteria from the code, so this test is used directly to resolve any conflict and reach a final decision. This test is not very expensive and it provides the best accuracy, as it presents concrete strength inside the member directly.

It is worth mentioning that a core test is performed by cutting of cylinders from the concrete member, and this can affect the integrity of the structure; thus the required number of samples should be minimized, according to the standard restrictions, as the required number will provide adequate accuracy of results without affecting the structural safety.

Specifically, with a mature structure the number of core samples and their locations should be defined carefully and all precautions should be taken to define the temporary supports and their location; this should be designed after defining the load path in case a member's capacity is reduced dramatically.

The location of the steel bars should be defined first using cover meter tools (to be discussed later) to avoid cutting the steel bars if possible. A competent engineer is required for this, as any mistake can be disastrous.

The codes and specifications provide some guidelines as to the number of cores to test, as follows, where V is the volume of the concrete member:

- $V \leq 150\,\text{m}^3$ take 3 cores
- $V > 150\,\text{m}^3$ take $(3+\ (V-150/50))$ cores

Taking this into consideration, the degree of confidence in the core test depends on the number of core test samples, which should be a minimum to maintain the structural integrity. The relation between the number of cores and confidence is shown in Table 3.10.

Some precautions must be taken regarding the core size. For example, the permitted diameter is 100 mm with a maximum aggregate size of 25 mm, and 150 mm with a maximum aggregate size that does not exceed 40 mm.

It is recommended to use a core with a diameter of 150 mm if possible, for more accurate results, as shown in Table 3.11. This table presents the

Table 3.10 Relation between core numbers and confidence in results.

Number of cores	Deviation between expected strength and actual strength (confidence level 95%)
1	±12%
2	±6%
3	±4%
4	±3%

Table 3.11 Different core size characteristics.

Test	Diameter (mm)	Length (mm)	Possible problem
One	150	150	May contain steel reinforcement
Two	150	300	May cause more cutting depth to concrete member
Third	100	100	Not allow for the maximum aggregate size 25 mm May cut with depth less than required
Last	100	200	Less accurate data

relationship between the dimensions of the sample and potential problems at each size. Therefore the inspection engineer should review this table to assist in the decision.

Some research studies have mentioned that the core test can be performed with a core with a diameter of 50 mm with a maximum aggregate size not more than 20 mm overall; it should be mentioned that it was observed that small core sizes give different results than large sizes.

It is very important, due to the criticality of this test and that it cannot be repeated, that the laboratory test should be certified and have a good reputation and the test equipment should be calibrated with a certificate of calibration from a certified company.

The core machine is equipped with a cylinder made from a special alloy mixture with diamond powder, to provide the capability to cut the concrete by rotating the body of the cylinder at a certain speed. The technician who uses this machine should be well trained and competent on it.

Fig. 3.10 illustrates a core sample device used to test the foundation under a machine in a gas plant. After the cylinder core is cut and removed, there are two common methods used. When cutting steel bars, it is preferable to fill the hole with grouting with high compressive strength, or a cylinder can be poured with the same diameter and after the core is removed,

Fig. 3.10 How to take a core.

the concrete cylinder is inserted and an epoxy added to create an adhesive bond between the old and new concrete. No matter which method is used, it should be carried out directly after removing the core, to maintain the structural safety.

Before doing the compression test, the laboratory team should examine and take a photo of every core and note the voids inside the core sample to define the void criteria. A void is considered small if it measures between 0.5 and 3 mm, average if between 3 and 6 mm, and large if more than 6 mm. It is also important to note the presence of honeycomb; also, the shape, type, color, and gradient of aggregates and the sand characteristics should be defined. The laboratory is responsible for measuring the dimensions, weight, and density of every core sample, the steel bar diameter, and the distance between the reinforcement bars.

The sample must first be prepared before performing the test, as the ends must be flat and perpendicular. The core must be trimmed to the correct length; in any case the length should be not less than 95% of the core diameter and no more than double the core diameter. In Fig. 3.11 you can see the shape of the core after removing it from the concrete member.

For flattening the end surfaces, a concrete cutting chainsaw or steel cutting disk is typically used. Then, based on the standard, the two ends of the core sample are prepared by covering them with mortar or sulfide and the sample is submerged in water at a temperature of $20 \pm 2\,^{\circ}C$ for at least 48 h before performing the test.

For testing, the sample is put in a crushing machine and the load applied gradually, at a regular and continuous range of 0.2–0.4 N/mm^2, until it reaches the maximum load, at which the sample has been crushed.

Fig. 3.11 Core sample.

The actual strength for a cylinderical core is found by determining the crushing stress obtained from the test using the following equation and considering the core's length to diameter ratio (λ).

- For a horizontal core the strength calculation will be as follows:

  $$\text{Estimated actual strength for cube} = 2.5/(1/\lambda) + 1.5 \times \text{core strength}$$

  where λ = core length/core diameter.

- For a vertical core the strength calculation will be as follows:

$$\text{Estimated actual strength for cube} = 2.3/(1/\lambda) + 1.5 \times \text{core strength} \quad (3.1)$$

Occasionally, the reinforcement steel bars are in the core perpendicular to the core axis; in this case the previous equations need to be multiplied by a correction factor, calculated as follows:

$$\text{Correction factor} = 1 + 1.5\,(s\varphi)/(\text{LD}) \tag{3.2}$$

where L is the length of the core, D is the core diameter, s is the distance from steel bar to edge of core, and φ is the diameter of the steel bar.

All efforts should be made to avoid cutting steel bars, but if this is not possible, the preceding correction factor should be used; however, this correction factor should be in the range of 5% to 10%. If the correction factor provides a higher value than 10%, then the results of the core will have a low accuracy and cannot be trusted. Another core should be taken.

It is worth mentioning that if the core is taken from an existing structure for any reason, even if it is to solve a conflict between the contractor and the owner or consultant, it will be taken after 28 days, so the concrete strength from the core will be different from that of the standard cylinders, depending on the age of the concrete when taking the core. A study was carried out by Yuan et al. (1991) and their comparison is presented in Table 3.12.

The following points should be considered when evaluating the test results from the cores:

- In this test the core will be submerged in water and this action will cause a reduction in the concrete strength up to about 15% for the strength of dry concrete.
- The equation used to obtain the expected actual concrete strength does not consider any effect due to differences in direction between the core and the standard cylinder or cube direction.
- Most codes and standards agree that the concrete will be accepted if the average strength of the cores is at least 75% of the required strength and the calculated strength for any core is not less than 65% of the required strength.
- With prestressed concrete, the concrete strength will be accepted if the average strength of the cores is at least 80% of the required strength and the calculated strength for any core is not less than 75% of the required strength.

Table 3.12 Relation between strength value from standard cylinder test and core test at different ages.

Age (days)	Standard cylinder strength (MPa)	Core strength (MPa)	fc(core)/fc (cylinder at 28)
7	66	57.9	0.72
28	80.4	58.5	0.73
56	86.0	61.2	0.76
180	97.9	70.6	0.88
365	101.3	75.4	0.94

Rebound (Schmidt) Hammer test

This test is a well-known non-destructive test that provides an estimation of concrete compressive strength. This test is usually available as it is not expensive and is simple to apply compared with other tests, but on the other hand it provides a less precise value for concrete strength. It can give an indication of the concrete strength, but the accuracy can be enhanced as we will discuss.

The rebound hammer, or Schmidt hammer, measures the hardening of the concrete surfaces by the rebound number. Curves associated with these tools represent the relation between the rebound number and the concrete strength for different locations of the application of the rebound test, whether it is applied with a side or vertical up, to measure a slab from the bottom, or vertical down, to measure the up surface of a foundation, and so on.

The rebound hammer test is shown in Figs. 3.12 and 3.13, which illustrate different types of rebound (Schmidt) hammer. There are different types of rebound hammers on the market, but all of them typically provide an impact energy of about 2.2 N/mm.

Also, there are many types of Schmidt hammer data readings available, as shown in the samples in the figures, with both analog and digital types now on the market. The digital versions can store the data in machine memory for further analysis.

The rebound hammer must be calibrated before use. Calibration tools will come with the rebound hammer package and the calibration procedure and allowable limits must be within the manufacturer's recommendations.

Since at least 15 readings are required (more is better), the area in which the test is applied should be about 300 mm × 300 mm. It is preferable to test

Fig. 3.12 Rebound hammer.

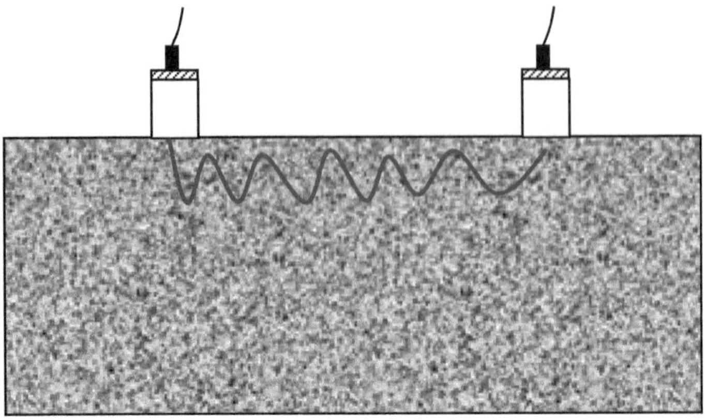

Fig. 3.13 Surface transmission.

on a surface that has no changes after casting, or a surface that has not had any smoothing done to it during the casting process. Clean the surface before doing the test.

On the testing surface, a net of perpendicular lines 20–50 mm apart in both directions should be drawn by pencil, and the intersection points will be the points to do the test on. The test point in any case must be away from the edge of the concrete by about 20 mm.

From Figs. 3.12 and 3.13, one can see that the surface should be cleaned before applying the test and the rebound hammer should be perpendicular to the surface being testing.

The following precautions must be taken during the test:

- The hammer shall be perpendicular without any inclination to the surface to which it will be applied in any conditions, because the hammer direction directly affects the rebound number reading, as it is affected by the impact of hammer weight.
- A wet surface gives a rebound hammer reading less than that of a dry surface, by up to 20%.
- The concrete member undergoing test shall not be vibrated.

The inspector will utilize the rebound hammer curves that provide the relation between concrete compressive strength and rebound number. These curves are obtained from the manufacturer. However, it is highly recommended to increase the accuracy of this test to an acceptable level by taking readings on the surface of standard concrete cubes and then crushing the concrete cubes by standard crushing machines, to obtain a calibration of the curves based on actual testing.

The tool calibration is necessary to be done from time to time, as the spring inside the rebound hammer loses some of its stiffness with age.

It has been found that the cement type affects the readings results: concrete with high alumina cement will give higher results of two times more than that of concrete with ordinary Portland cement. Concrete with sulfate-resistant cement will give a reading result of about 50% less than that for ordinary Portland cement. Also, higher cement content provides a lower reading than concrete with less cement content, but in any case the gross error is only 10%.

The number of readings must be high enough to provide reasonably accurate results. The minimum number of rebound hammer readings is 15; as stated earlier, more readings are recommended to achieve a more accurate result.

Extreme reading values should be excluded from the calculation and the average taken of the other remaining values. The concrete strength represents the compressive strength value; the standard deviation can also be calculated to give an indication of the concrete quality.

Ultrasonic test

The ultrasonic pulse velocity test is another non-destructive test. It measures the velocity of ultrasonic pulse transmission through the construction member, by measuring the time required between sending and receiving the pulse. By knowing the distance between the sender and receiver, the pulse velocity can be determined.

A calibration is done for these velocities by previously knowing the concrete strength and its mechanical characteristics for the test sample and measuring the velocity compared to it. After that the equipment can be used on any other concrete using the same procedure to identify compressive strength, dynamic and static modulus of elasticity, and Poisson ratio.

The UT equipment used should have the capability to record time for lengths ranging from 100 to 3000 mm accurately, to ±1%. The ultrasonic tool manufacturer should provide the procedures for equipment operation in different temperatures and humidity. Before starting the test, check for the availability of a power transformer sender and the receiver of natural frequency vibrations between 20 and 150 kHz, considering that the frequency appropriate for most practical applications for concrete tests is 50–60 kHz.

Figs. 3.13–3.15 illustrate the different methods of wave transmission.

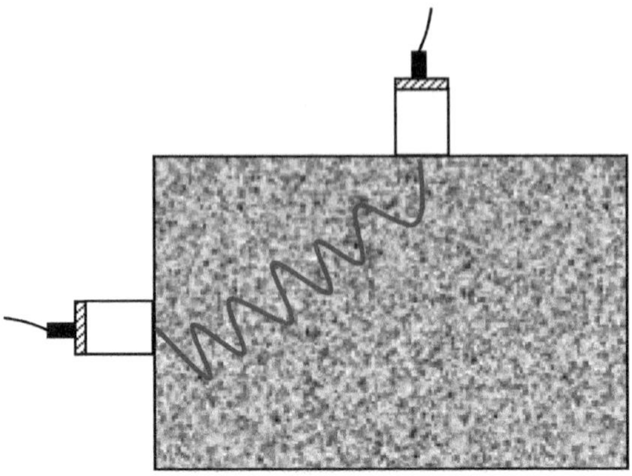

Fig. 3.14 Semi-direct transmission.

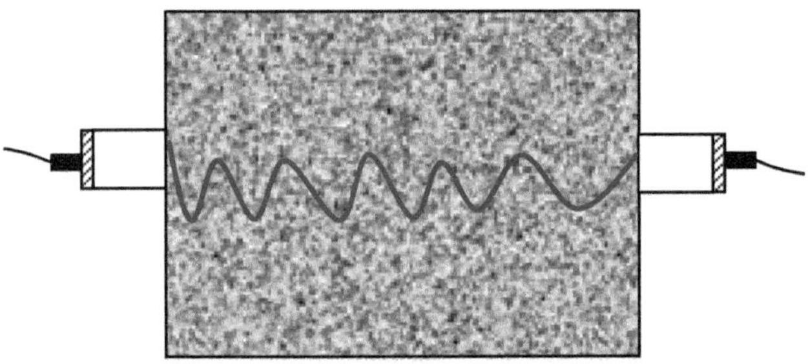

Fig. 3.15 Direct transmission.

Surface transmission is shown in Fig. 3.13, semi-direct transmission is clearly shown in Fig. 3.14, and the third method of wave transmission is direct transmission, as presented in Fig. 3.15.

The ultrasonic equipment is equipped with two rods of metal with lengths of 250 and 1000 mm. The first rod is used in the defining of zero of the measurement and the second rod is used in the calibration. In both cases, the passage time of waves through the rod is known from the manufacturer's procedure and recommendations.

Hence, connect the ends of the rod to sender and receiver and measure the time for pulse transmission and compare it with the known reading; if there is any deviation, the small rod adjusts the zero of the equipment to provide the known reading.

The long rod is used with the same procedure to define the result accuracy; in this case the difference between the two readings should be no more than ±0.5% in order to have measurements with the required accuracy.

Note that the value of wave velocity inside the steel is two times more than the velocity value inside the concrete. Therefore the reinforcement steel bars inside the concrete member will certainly affect the accuracy of the reading, as they will give a higher value for the wave impulse velocity. To avoid that influence, the location of the steel reinforcement bars must be defined previously with respect to the path of the ultrasonic pulse velocity.

To correct the reading for steel bars parallel to the path of the pulse wave, as illustrated in Figs. 3.16 and 3.17, a calculation for pulse velocity is used, as shown in the following equations, given knowledge of the wave path; time of transmission, which will be calculated by the equipment; the wave velocity in steel; and the distance between the two ends.

$$V_c = K \cdot V_m \tag{3.3}$$

$$K = \gamma + 2(a/L)\left(1 - \gamma^2\right)^{0.5} \tag{3.4}$$

$$L_s = (L - 2b) \tag{3.5}$$

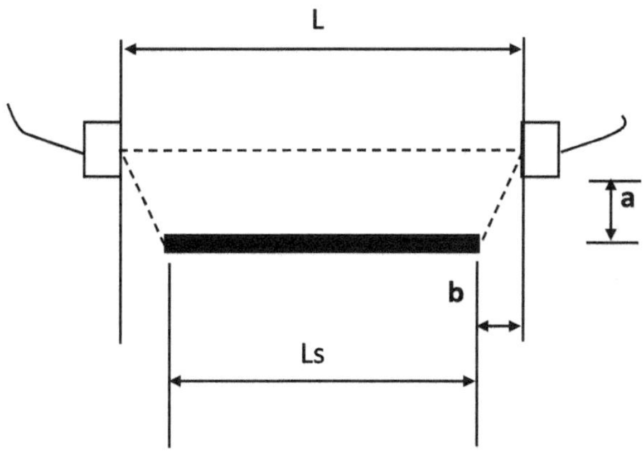

Fig. 3.16 UT wave parallel to the steel bars.

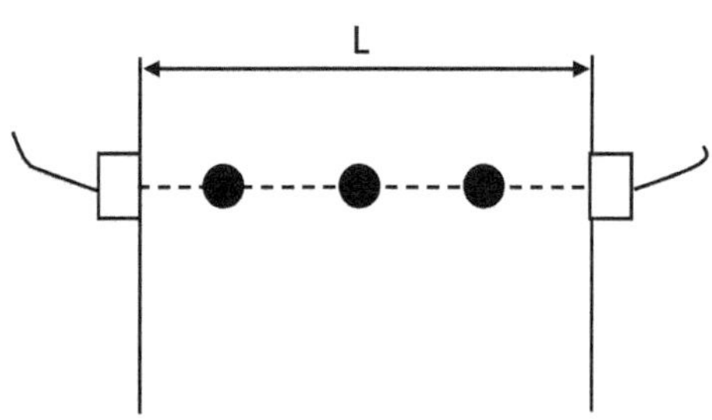

Fig. 3.17 Wave of UT perpendicular to the steel bars.

where V_m is the pulse velocity from transmission time from the equipment, and V_c is the pulse velocity inside the concrete; the value of γ varies depending on the steel bar diameter.

The effect of the steel bar can be ignored if the diameter is equal to or less than 6mm or if the distance is large between the steel bar and the end of the equipment.

If the steel reinforcement bars' axis is perpendicular to the direction of pulse transmission, as shown in Fig. 3.18, the effect of the steel bars on the reading will be less. The effect can be considered zero if a transmission source of 54 kHz is used and the steel bar diameter is less than 20mm. The previous equation can be used in the case of a lesser frequency and a diameter higher than 20mm and by changing the value of γ according to the bar diameter, using the data delivered by the equipment.

Other factors should be considered that affect the accuracy of ultrasonic testing, such as the temperature, as presented in Table 3.13, and thawing and concrete humidity.

Fig. 3.18 Concrete cover measurement

Table 3.13 Temperature effect on pulse transmission velocity.

Temperature (°C)	Percentage correction of the velocity reading (%)	
60	+5	+4
40	+2	+1.7
20	0	0
0	−0.5	−1
−4	−1.5	−7.5

The recommended precautions during the test:

1. Using the reference bar to adjust zero is very important; if it is ignored it will directly affect the accuracy of the result.
2. The concrete surface should be well leveled and smooth. If this is not possible, the impact of the surface must be taken into account.
3. Temperature affects the ultrasonic velocity measurement according to Table 3.13, and this should be considered with an increase or decrease in temperature of 30 °C.
4. When comparing the quality of concrete between various components of the same structure, similar circumstances should be taken into account in all cases in terms of the composition of concrete, moisture content, age, temperature, and type of equipment used.

The velocity value depends on the concrete density. A higher concrete density means fewer voids inside the concrete, which means high concrete strength and higher quality. Table 3.14 provides the relationship between the concrete quality and the UT velocity.

By knowing the ultrasonic transmission pulse velocity, one can obtain static and dynamic modulus of elasticity, as presented in Table 3.15.

Table 3.14 Relation between concrete quality and pulse velocity.

Pulse velocity (km/s)	Concrete quality degree
>4.5	Excellent
4.5–3.5	Good
3.5–3.0	Fair
3.0–2.0	Poor
<2.0	Very poor

Table 3.15 Relation between elastic modulus and pulse velocity.

	Elastic modulus (MPa)	
UT velocity (km/s)	Dynamic	Static
3.6	24,000	13,000
3.8	26,000	15,000
4.0	29,000	18,000
4.2	32,000	22,000
4.4	36,000	27,000
4.6	42,000	34,000
4.8	49,000	43,000
5.0	58,000	52,000

3.6.3 General test methods comparison

From the preceding discussion, the different methods for determining the hardened concrete strength have advantages and disadvantages, which are summarized in Tables 3.16 and 3.17.

Moreover, the cost of these tests varies, as well as their accuracy. A comparison of the concrete member strength and the accuracy of the measured strength to the actual is presented in Table 3.17. The table also compares test cost and speed.

Note that the obtained results from the tests will be less than the concrete strength specified in the drawings or project specifications, as this depends on the standard cube or cylinder compressive strength after 28 days. The cubes or cylinder standard specimens will be poured and compacted based on the standard and the curing process will be done by submerging the samples in water for 28 days.

For example, during the concrete structure construction the curing will not occur within 28 days and you obviously cannot submerge the concrete

Table 3.16 Comparison between test methods.

Test method	Possibility of damage	Precaution requirements
Load test	Possible member loss	– Member shall be isolated or the load transferred to the adjacent members – take safety precautions considering possible member failure
Cores	Holes to be made good	– Limitations of core size and numbers – safety precautions for critical member
Ultrasonic	None	Only requires two smooth surfaces
Rebound hammer	None	Only requires a smooth surface

Table 3.17 Accuracy comparison between test methods.

Test method	Damage to concrete	Representative to concrete	Accuracy	Speed of the test	Cost
Load test	Variable	Good	Good	Slow	High
Cores	Moderate	Moderate	Good	Slow	High
Ultrasonic	None	Good	Moderate	Fast	Low
Rebound hammer	Unlikely	Surface only	Poor	Fast	Very low

slabs or columns in water continuously for 28 days. The temperature on-site is certainly different than that in the laboratory during the test as well, and the concrete pouring and compaction methods are different for the cubes or cylinders than those in the actual member. All these variations are taken into consideration inside the partial safety factors or reductions in the design standard and codes, but when testing, the strength should be measured after complete hardening of the concrete.

3.6.3.1 *Steel in concrete corrosion measurement test*

This NDT is used to determine the reason for the corrosion, to define the best repair methods and whether corrosion has begun or not. In addition, this method can be used to determine when corrosion will start on the steel bars and the corrosion rate. All this information is mandatory for defining the integrity management system as part of risk-based inspection and taking a proactive approach. There are some special facilities, such as concrete tanks that contain seawater for firefighting or cooling systems, or even in the water flooding systems in oil plants. So, it is important to monitor these tanks so these facilities can avoid any need for their shutdown.

Chloride content penetration

Determination of the maximum chloride content inside concrete is crucial in preventing corrosion of the steel bars within a reinforced concrete structure. Table 3.18 illustrates the limits of standardization of the content of ions of dissolved chlorides in the concrete.

The ACI provides clear limits for different content of chlorides. Limits of the concrete content of dissolved chlorides ions are defined in ACI 318R-89, and chloride ion limits are stated in the ACI Committee 201, 357, 222. These are results for the components of concrete, or the total result of chlorides penetrating reinforced concrete during the life of a structure. American specifications ACI COMMITTEE 357 recommended that the water used in

Table 3.18 Maximum allowable chloride ions in concrete.

Condition around concrete	Maximum dissolved chloride ions in concrete water—% of cement weight
Reinforced concrete exposed to chloride	0.15
Dry reinforced concrete and totally protected from humidity during use.	1.0
Different structure members	0.3

the concrete mixture should not contain more than 0.07% chloride in the case of reinforced concrete, or 0.04% in the case of pre-stressed concrete.

In 1992 the European Code (ENV206) provided the limits of chlorides inside the concrete and identified these limits according to the type of application. The EC mentioned that the chloride content should be a maximum of 0.01% of the weight of cement for plain concrete, 0.04% of the weight of cement for reinforced concrete, and 0.02% of the weight of cement concrete for pre-stressed concrete.

Finally, the EU specifications prevent the use of any additives that have chloride or calcium chloride in reinforced concrete or pre-stressed concrete (Table 3.19).

Concrete cover measurements

As discussed earlier, the corrosion start time for steel bars depends on the concrete cover thickness. Any shortfall of concrete cover less than the project specification requirement can dramatically affect the structure integrity over time, reducing its lifetime. For marine structure projects or any special structures, the concrete thickness measurement is part of the quality control requirement.

The main benefit of the concrete cover thickness measurement is to explain the causes of corrosion and identify areas that have the capability to corrode faster.

It is noted that the measurement of the thickness of the concrete cover needs to be defined as axis *y*, *x* to determine the thickness of the concrete cover at every point on the structure.

Table 3.19 ACI recommendation for the maximum acceptable chloride ions.

Type of member	Dissolved* (in water)	Total**	Dissolved*** (in acid)	Total****
Pre-stressed concrete	0.06		0.06	0.08
Reinforced concrete exposed to chloride during use	0.15	0.1	0.1	0.2
Dry reinforced concrete	1.0			
Different structure members	0.3	0.15		

*(ACI 318R-89) ACI building code.
**ACI Committee 202.
***ACI Committee 357.
****ACI Committee 222.

Equipment to measure the thickness of concrete cover ranges from simple to complex. Fig. 3.18 shows the process of measurement. According to Bungey (1993) for bridges the concrete cover thickness can be obtained by radiograph, but this method is costly.

The magnetic cover method is a simple method, but the distance between steel bars can influence the accuracy of the concrete cover thickness measurement, as this method depends on the supply of electricity from a 9-V battery, while the second side of measurement for the potential voltage envelope when completed by connecting with the buried the steel bars.

Fig. 3.19 illustrates a magnetic cover test device on the concrete surface and Fig. 3.20 presents the direction of the device from which the maximum and minimum signals can be obtained.

The British standard is the only standard that cares about concrete cover measurement after construction, as per BS1881, Part 204. In 1993, Allred studied the accuracy of the cover thickness measurement when many steel

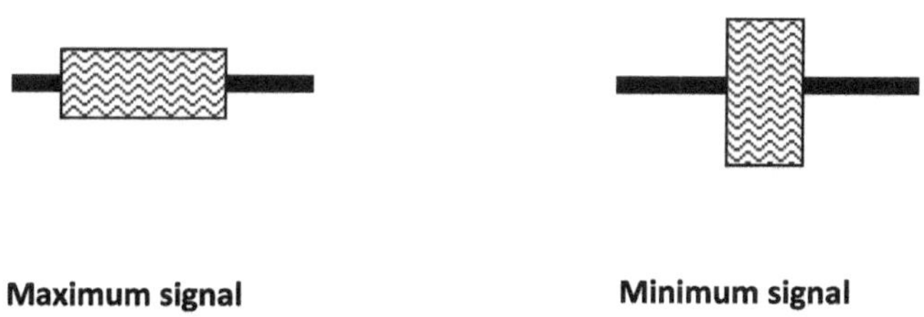

Fig. 3.19 Correct location of the device.

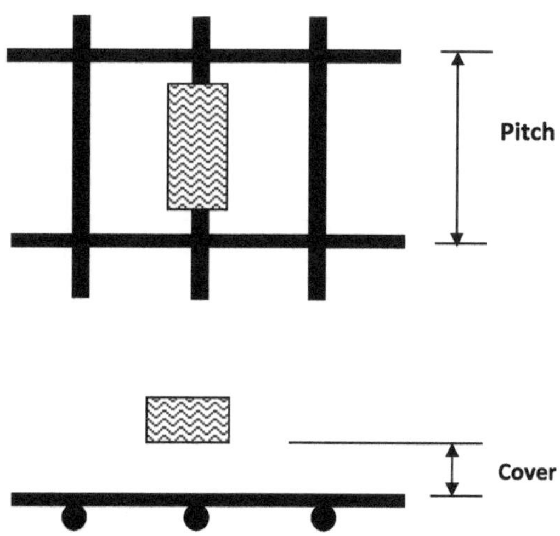

Fig. 3.20 Device location on concrete surface.

bars are close to each other, and the study revealed a benefit of using more than one measurement head to enhance the reading accuracy and mentioned that small heads have a strong influence on the accuracy of the reading.

Therefore a problem with this method is that a high percentage of steel reinforcement in the concrete section will give inaccurate data. The person who works with this tool should be competent and well trained, to be aware of anything that can affect the reading: for example, bolts, steel wires, and others.

The concrete cover is a major influence on corrosion, as is clearly obvious in Fig. 3.21, which presents the relation between the nominal concrete cover thickness and the probability of corrosion occurring. It can be found that after 40 years, with concrete cover equal to 120 mm, the probability of corrosion occurring is approximately zero. However, in the case of concrete cover thickness equal to 90 mm, the probability increases to 5%. If there is a reduction of the concrete cover to 70 mm, the probability increases to around 22%.

Carbonation depth measurement

Carbonation depth is based on the carbon dioxide propagation inside the cover of the concrete, which changes the concrete alkalinity value and lets its value decrease. The concrete alkalinity value is expressed by desired pH values around 12–13; however, with carbon dioxide propagation and reaction with cement components, the alkalinity will reduce until the pH value reaches about 8. At this point, the passive protection layer around the steel

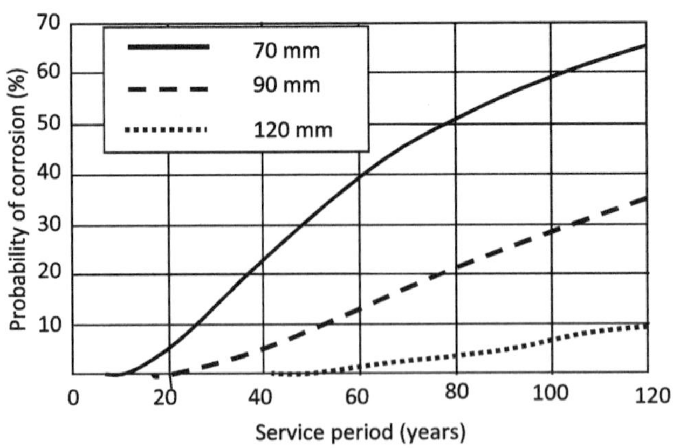

Fig. 3.21 Effect of increased nominal concrete cover on the probability of corrosion occurring.

bars will vanish and the corrosion process will start. Therefore it is very important to define the carbonation depth and determine if it is far away from or near the steel bars.

The test procedure is to spray the surface of broken concrete, using special tools, with phenolphthalein dissolved in alcohol to obtain the carbonation depth. The solution color becomes pink when it touches the surface of concrete with an alkalinity pH value of about 12–13.5, and the color will be gray or blue if the concrete loses its alkalinity and the pH value is less than 9.

The carbonation depth can also be examined by taking a core, as shown in Fig. 3.22. In actuality, carbonation depth and chloride measurements will both be done on a core in the case of mature structures, to avoid inadvertently causing any weakness in the structure.

The best solution is a phenolphthalein with alcohol and water solution of 1 mg phenolphthalein with 100 mm alcohol/water (50:50 mixing ratio) or more alcohol than water. Building Research Center in 1981 and Parrott (1987) recommended that if the concrete is completely dry, humidify the surface using water.

If the thickness of carbonation about 10–5 mm, it is going to break passive layer around the steel bar; and this is presence from 5 mm more than the depth of change color from the concrete surface.

Note that some of the aggregate or some concrete mixtures have a dark color, which makes reading phenolphthalein very difficult; therefore the surface should be cleaned during the test.

Chlorides test

Chlorides testing is described in AASHTO T260-84 1984, ASTM D1411-82 1982, and BS1881, Part 124, which depends on an analysis of the samples

Fig. 3.22 Carbonation depth measurement.

of concrete powder to obtain the chloride quantity. This test is performed by using a rotary drilling machine to drill inside the concrete to make a hole and then collecting the concrete powder output from this hole. Several separate samples will be taken at different depths. The hole depth will be varied to increase the accuracy; it is usually from 2 to 5 mm. In most cases, the first 5 mm from the concrete surface, the chlorides concentration is very high, particularly in structures exposed to sea water.

The concrete powder taken at every drilling depth will be added to a solution of acid, which is determined by the amount of chloride concentration. There are two principal ways of measuring: in situ (Quantab strips) and determining the electron ion (Specific Ion Electron), which in recent experience is more accurate, but requires high-cost equipment. To carry out experiments on chloride concentration, trained professional workers are needed.

When readings of chloride concentrations at various depths are assembled, draw the graph to present the chloride concentration with depth from the surface of the concrete into the concrete. From this figure, one can determine whether the chloride attacks the concrete from the outside, or whether it is inside, from concrete components like aggregate, water, or others. Fig. 3.23 shows the difference between the two cases, outside or inside. Note that the methods of repair and the protection methods are totally different.

It can be seen from this figure that in the case of chloride attack from outside, the first reading of the chloride's concentration is high and then it gradually decreases with the depth until reaching a constant value.

On the other hand, if the chloride is inside the concrete from a component, it can be shown from the figure that near the surface the chloride

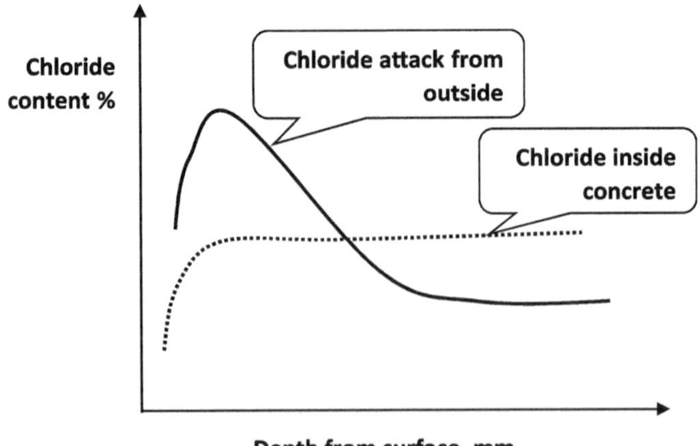

Fig. 3.23 Comparison between chloride attack from outside and within the concrete mix.

concentration is low, and it increases with depth until reaching constant values with the depth.

The effect of the chloride's concentration on the steel reinforcement must be determined, because when the chloride concentration exceeds 0.6 from hydroxyl concentration, that causes failure in the passive layer around the steel reinforcement. This ratio is approximately 0.2%–0.4% chlorides by weight of cement or 1 Ib/cubic yard concrete or 0.5% chloride by the weight of concrete.

To evaluate the probability of corrosion occurring depending on the percentage of chloride content by weight of cement or concrete, as presented in Table 3.20, we can evaluate and define the concrete structure service life against chloride attack.

Half-cell

Half-cell tools are usually found in any oil and gas field to monitor the cathodic protection system for the pipeline and the facilities; a well-trained team can assist in performing this test very professionally. This system is used to define whether the bars have lost their passive protection or not. If the reinforcement steel bars have lost passive protection due to exposure to carbonation or chloride, corrosion will start (Fig. 3.24).

The half-cell equipment complies with ASTM C876 199. In general, this equipment consists of a metal rod in solution with the same metal ions, such as a copper rod in a saturated solution of copper sulfate. This type is considered one of the most common types of equipment. Other types of equipment, such as a bar of silver in a silver chloride solution, can be linked through the voltmeter and other part of the Potentiometer will be connected to the steel bars, as in Fig. 3.25.

If the steel bars have a passive protection layer, the potential voltage will provide a small value with a range from 0 to 200 millivolts copper/copper sulfate. In the case of breakdown of the passive protection layer and some

Table 3.20 Risk of corrosion depending on the chloride content.

Chloride content (%)		
By weight of cement	**By weight of concrete**	**Risk of corrosion category**
>2.0	>0.36	Certain
1.0–2.0	0.18–0.36	Probable
0.4–1.0	0.07–0.18	Possible
<0.4	<0.07	Negligible

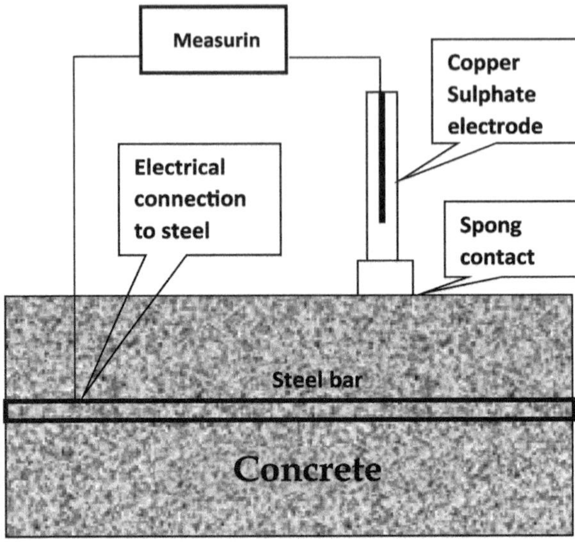

Fig. 3.24 Half-cell machine.

Fig. 3.25 Half cell test.

quantity of steel melting as a result of the movement of ions, the potential voltage value is about 350 megavolts. When the value is higher than 350 megavolts, it means that the steel has already started to corrode.

There are some recommendations to use half-cell equipment consisting of silver and a silver chloride solution or mercury and a mercury oxide solution rather than copper/copper sulfate, as the copper requires regular maintenance and thus the readings will have some errors.

To do a survey of a large surface area of concrete to determine any bars that have been corroded, silver column is put in the solution of the rod cell and take a measurement points on the surface with same pattern of space between points in two directions and then draw a contour line between these points to define the color code for each contour area to facilitate take a better decision for assessment and repair.

The half-cell measurement of the steel reinforcement bars is to determine the probability of causing corrosion to the steel bars. The ASTM C867 specification presents a method to clarify the data obtained from measuring the potential voltage. These values are shown in Table 3.21 for different equipment types and the criteria of the corrosion conditions are defined in each case. A test on a foundation is presented in Fig. 3.25 and its reading is illustrated clearly in Fig. 3.26.

Table 3.21 ASTM specification for steel corrosion for different half-cell.

Silver/ silver chloride	Copper/ copper sulfate	Standard hydrogen electrode	Calomel	Corrosion condition
>–106 mV	>–200 mV	>+116 mV	>–126 mV	Low (10% risk of corrosion)
–106 to –256 mV	–200 to –350 mV	+116 to –34 mV	–126 to –276 mV	Intermediate corrosion risk
<–256 mV	<–350 mV	<–34 mV	<–276 mV	High (<90% risk of corrosion)
<–406 mV	<–500 mv	<–184 mV	<–426 mV	Severe corrosion

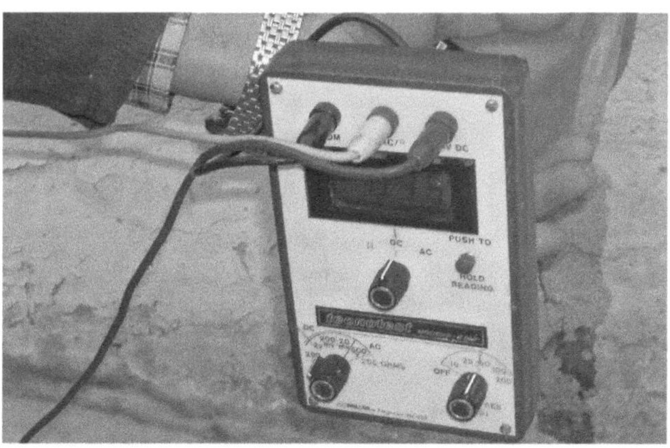

Fig. 3.26 Half-cell reading.

References

Bungey, J.H. (Ed.), 1993. Non-destructive testing in civil engineering. In: International Conference of the British Institute of Non-Destructive Testing. Liverpool University.

Parrott, L.J., 1987. Steel corrosion in concrete: How does it occur? Materials Protection 6, 19–23.

Yuan, R.L., et al., 1991. Evaluation of core strength in high-strength concrete. Concr. Int. 13 (5), 30–34.

CHAPTER 4

Assessment of structures and pipelines

4.1 Introduction

The main milestone in any integrity process is the assessment. This important task will be completed after collecting all the required data and doing the inspection, as it cannot be completed based simply on assumptions, since this would be useless from an integrity point of view. Thus all the required data should be in hand before doing the assessment. Sometimes, the engineering firm handling the work will say, "We make this assumption, based on the client's request." This can be acceptable for new construction, but not for evaluation of existing structures. However, this can involve a conflict of interest, as the engineering firm focusing on manhours, variations order and the safe design for the structure only but do not care whether it is safe for the next 20 years. On the other hand, the owner focuses on the structure being reliable during its lifetime. It is therefore recommended that an inhouse team be tasked with the integrity assessment rather than outside engineering, but a suitable project management scheme must be in place. As on the other side as it will explained in this chapter that we can accept the structure can withstand and to be unsafe from the code point of view and in this case we will say "the structure violates the code." However, some engineers or even engineering firms which are not familiar with brown field they will say that the structure is unsafe which reveal to demolish the structure decision, Or he mentions violate the code but mention some statement to let himself on the safe side.

4.2 Assessment for onshore concrete structures

Most of the reinforced concrete structures in oil and gas plants, refineries, or petrochemical sites are foundations under equipment, normal administration or accommodation buildings, or concrete tanks for different uses. The main cause of deterioration of a reinforced concrete structure is corrosion. Corrosion occurs very rapidly for fields that are near the seacoast, and is a main reason for reinforced concrete structure deterioration in residential

Asset Integrity Management for Offshore and Onshore Structures
https://doi.org/10.1016/B978-0-12-824540-8.00004-1

buildings rather than industrial ones. With bridges, the corrosion usually occurs in cold climates and is mainly due to deicing by using salt.

In practical life, an engineer who sees a sign of corrosion on a steel reinforcement typically wants to go directly to a traditional repair method. I usually say to them, "Don't jump to conclusions." The first major step is to evaluate the structure precisely, identify the environmental conditions surrounding the structure, and ask about its history.

The assessment of the structure should be done in a proper way by doing the required inspection and collecting the required information. Then the assessment will accurately match the existing conditions and, based on that, the repair scope, whether minor or complete rehabilitation, will be fit for the purpose. In addition, the prediction of the structure's remaining life will be more accurate.

As a general principle, if the diagnosis of disease is correct, the right medicine will be chosen. In concrete structures, the first action is to perform the assessment by starting to diagnose the cause of any structural defects or deterioration, whether it is from corrosion or other reasons. If it is due to corrosion, it is important to define the causes that led to the corrosion as well. The remaining lifetime calculation of the structure mainly depends on an accurate structure assessment procedure and technique.

The process of evaluating any structure consists of two phases. The first is the initial assessment of the structure. The target of this phase is to define the problem and develop a plan for detailed inspection and evaluation of the structure. The target of the second phase is to further develop a detailed assessment for the structure, by precisely defining the problem and the reasons for its occurrence. In addition, in this phase the detailed inspection of the whole structure is implemented.

In 1984, Technical Report 26 issued by The Concrete Society, as well as the American Concrete Institute, presented the basis for the inspection of a structure damaged by reinforcement corrosion in order to assess it. It was found that there are many reasons for the deterioration of reinforced concrete structures, from the presence of cracks to failure of the concrete cover. For example, some of these reasons are related to increases in stresses, plastic shrinkage, frost, cracks due to bad design, movement of the wooden form during construction, the presence of aggregate with the capability of shrinking, contamination with alkalinity, or others.

Some structures, such as foundations or other underground structures like piping culverts, tunnels, or general foundations, have surface contact directly with soils. In this case sulfate attacks it and causes the corrosion.

There are many types of cracks categorized by shape, direction, and depth. Therefore the engineers responsible for performing the structure assessment must have strong experience in this area, so that by seeing the cracks or delamination, they can specify precisely the reason for the deterioration. As a rule of thumb, the wrong diagnosis will result in the wrong repair and wrong maintenance plan, causing a waste of money and having a major impact on the structure safety.

4.3 Concrete structure probability of failure

In the past most of the textbooks stated that reinforced concrete is a durable material. However, this statement has changed over time. Many reinforced concrete structures are exposed during their lifetimes to different stresses as a result of different environmental conditions such as corrosion and expansive aggregate reactions, which attack the concrete and steel reinforcement, as presented by these researchers (Kilareski, 1980; Cady and Weyers, 1984; West and Hime, 1985; Takewaka and Matsumoto, 1980; Mori and Ellingwood, 1994a; Lin, 1995; Thoft-Christensen, 1995).

In the structural integrity management system, the main step is to do the assessment for the as-is condition of the structure. All the factors that affect the structure integrity and its probability of failure with time will be discussed in this chapter.

Therefore, the target is to discuss the effect of age on concrete strength, according to previous studies in different references. In addition to that, this chapter presents a parametric study on the reliability of reinforced concrete columns, considering the corrosion of the steel bars over time. The effects of age are discussed regarding the reliability of reinforced concrete columns, which is considered the most critical element on the structure from a redundancy point of view, considering wet and dry conditions, the effect of longitudinal steel ratio in reinforced concrete columns on the reliability, and the effect of corrosion rate on the reliability of reinforced concrete columns, respectively.

The effect of corrosion initial time on the reliability of reinforced concrete structure members is presented. The effect of corrosion on the reliability of eccentrically loaded columns is also studied and all these factors will assist in evaluating the structure in a reasonable manner with higher accuracy, considering that the column is a very critical element, due to its low redundancy, rather than the beam or slab.

4.3.1 Concrete strength with age

Many research studies have been performed to determine concrete strength after 28 days. In most cases, concrete compressive tests are made at 28 days, when the strength of concrete is considerably lower than its long-term strength. Many different recommendations are given in various codes and standards to calculate the concrete strength with age.

Although a great deal of research in this area is available, the environmental conditions surrounding each test and research study should be considered. This should be reflected in the codes and standards, but take care if using a code or standard not implemented from your country.

In 1976 Baykof and Syglof presented the variation of concrete strength with age by comparing the gain in strength of concrete specimens stored in wet and dry conditions. As shown in Fig. 4.1, they found that in dry conditions after 1 year there was no increase in concrete strength; however, the strength of specimens stored in a wet environment (at 15°C) was considerably increased with time (Table 4.1).

In 1989 a very important study was reported by George W. Washa et al., which had been carried out over 50 years. Their paper reported on concrete specimens stored in special environmental conditions to determine the concrete strength with age. The specimens were cured in water for 28 days before placement outdoors on leveled ground in an uncovered open location. Thermocouple data indicated that the outdoor compressive cylinders were subjected to about 25 cycles of freezing and thawing in each winter.

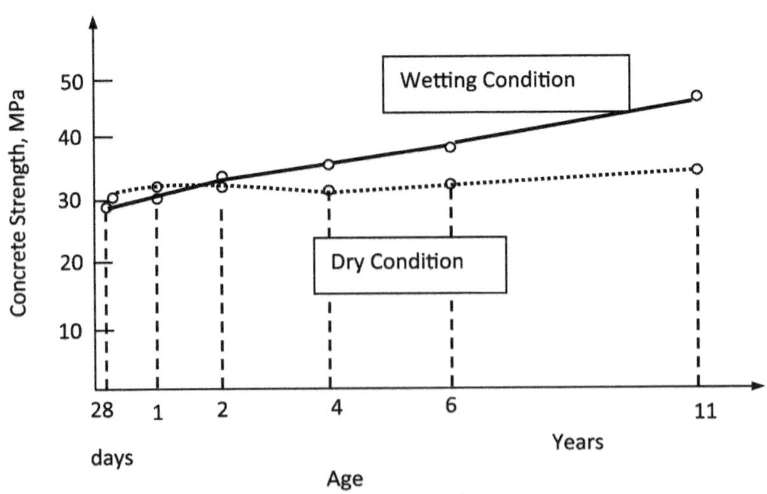

Fig. 4.1 Variation of concrete strength with time (years) from Baykof and Sigalof (1984).

Table 4.1 Factor of increase in concrete strength with age at different conditions (Baykof and Sigalof, 1984).

	Environment	
Years	**Dry**	**Wet**
1mo	1.00	1.00
1	1.25	1.20
2	1.25	1.30
4	1.25	1.40
6	1.25	1.50
10	1.25	1.63

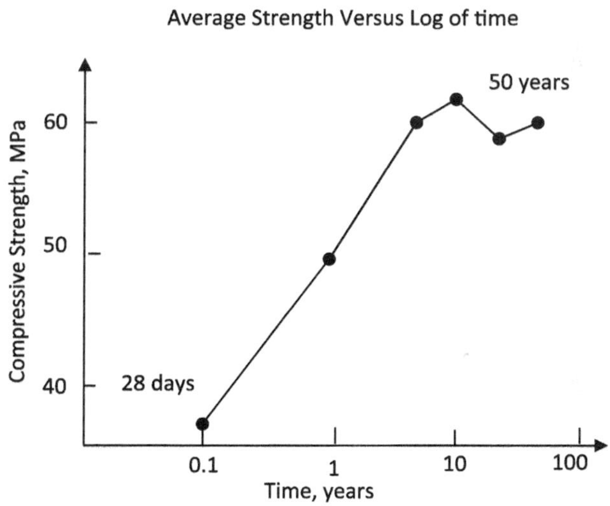

Fig. 4.2 Concrete compressive strength with time. *(Adapted from Washa, G.W., Wendt, K.F. 1975. ACI J. Proc. 72 (1), 20–28.)*

The relative humidity in the location of their laboratory normally varies from 65% to 100%, with an average of 75%. The annual precipitation, including snowfall, was about 800 mm. Air temperatures were usually in a range between 32 and 35°C.

The average concrete compressive strength with time as reported in their study is presented in Fig. 4.2.

In their study, cement with a relatively low C_2S content and a high surface area generally increased as the logarithm of age for about 10 years. After 10 years the compressive strength decreased or remained essentially the same, as shown in Fig. 4.2.

MacGregor (1983) used the same study and an equation that defined the relation between compressive strength and concrete age less than or greater than 10 years; the 28-day specified compressive strength of concrete equaled 281.5 kg/cm^2 (27.6 MPa). The corresponding mean of compressive strength at 28 days was 292.7 kg/cm^2 (28.7 MPa).

$$fc(t) = \begin{cases} 15.85 + 4.03\ \ln(t) & MPa \quad t < 10 \text{ years} \\ 48.9 & MPa \quad t \geq 10 \text{ years} \end{cases} \tag{4.1}$$

where $fc(t)$ = the concrete compressive strength with time in MPa, and t = the time in days.

The effect of concrete maturity is based on several studies of the long-term strength gain of concrete. From these studies, lower-bound relationships between age and strength of various concrete classes were obtained. These studies show a linear relationship between strength and the logarithm of age. For lower strength concrete at 25 years, the concrete strength was around 240% of the 28-day strengths as per Washa et al. (1975). With high-strength concrete, the strength of older concrete reached 125% to 150% of the concrete strength at 28 days.

MacGregor (1983) assumed that there was an equal probability that a structure would be loaded to failure in each month of 50 years of age, starting from the structure at 28 days old. A family of distributions presented the ratio of the concrete strength at failure f_{cf} to the 28-day strength f_{C28}. Their study showed that, for each combination of curing and concrete type, the resulting distribution of f_{Cf}/f_{C28} had a negative skewness.

A Monte-Carlo simulation was used for the reliability of a mature structure, multiplying the on-site concrete strength by a randomly selected value of f_{Cf}/f_{C28} with a mean value of 1.25 and coefficient of variation (CoV) equal to 0.07, and represented by a normal distribution.

As in the study by Washa et al. (1989), they tested concrete specimens for different ages at 1, 5, 10, 25, and 50 years. The analysis of this data is presented in Table 4.2 and the statistical parameters for the ratio between concrete strength at ages 1, 5, 10, 25, and 50 years and that at 28 days are given.

In 1976 Baykof et al. illustrated the gain in concrete strength of concrete specimens stored in wet and dry conditions with time. Table 4.1 illustrates the factor of increase in concrete strength with age for wet and dry conditions.

4.3.1.1 Concrete strength with age in different codes

Every code and standard recommends different factors to predict the concrete strength with time. The strength depends on many factors, but the

Table 4.2 Analysis of the results of Washa et al. (1989) tests for gain of concrete strength with time.

Test No.	28 days fc_{28} psi	1 yr fc_1 psi	fc_{28}/fc_1	5 yr fc_5 psi	fc_{28}/fc_5	10 yrs fc_{10} psi	fc_{28}/fc_{10}	25 yrs fc_{25} psi	fc_{28}/fc_{25}	50 yrs fc_{50} psi	fc_{28}/fc_{50}
1	3990	5195	1.30	7510	1.88	7865	1.97	7585	1.90	7650	1.92
2	5030	6530	1.30	8780	1.75	8700	1.73	8200	1.63	9220	1.83
3	4285	5805	1.35	7400	1.73	7855	1.83	7840	1.83	7610	1.78
4	4525	6015	1.33	7540	1.67	8025	1.77	7885	1.74	7920	1.75
5	6185	8195	1.32	9015	1.46	10,465	1.69	9850	1.59	10,400	1.68
6	4990	6240	1.25	7820	1.57	7885	1.58	8050	1.61	8030	1.61
7	5150	7395	1.44	7940	1.54	8525	1.66	8070	1.57	8820	1.71
8	6265	8310	1.33	10,430	1.66	10,145	1.62	9915	1.58	10,300	1.64
9	5130	7325	1.43	8860	1.73	9015	1.76	8250	1.61	7950	1.55
10	5145	6715	1.31	7640	1.48	7910	1.54	6835	1.33	7140	1.39
11	7050	8695	1.23	9995	1.42	9515	1.35	8985	1.27	9250	1.31
12	5480	7360	1.34	8405	1.53	8180	1.49	7500	1.37	7200	1.31
Mean			1.33		1.62		1.67		1.59		1.62
St. Dev.			0.06		0.14		0.17		0.19		0.20
C.O·V			0.05		0.09		0.10		0.12		0.12

Table 4.3 Ratio of F_{C28} to that at the age (days) shown, from Egyptian Code 1989.

Age (days)		3	7	28	90	360
Type of Portland cement	Normal	2.5	1.5	1.00	0.85	0.75
	Rapid hardening	1.80	1.30	1.00	0.90	0.85

main factor is the environmental condition. The following paragraphs discuss the methods given for calculating concrete compressive strength with age in different codes and standards.

A. Egyptian code

The concrete compressive strength changes with age for normal concrete in moderate temperatures. The ratio of the compressive strength at 28 days to that at a given concrete age can be calculated for normal and rapid hardening cement, as per ECP, from the values given in Table 4.3.

B. British standard

In the past, as stated by Nevil (1983), the gain in strength beyond the age of 28 days was regarded merely as contributing to an increase in the factor of safety of the structure. Since 1957 the codes of practice for reinforced and pre-stressed concrete allowed for the gain in strength to be considered in the design of structures that will not be subjected to load until having large age, except when no-fines concrete is used; with some lightweight aggregates, verifying tests are advisable.

The factors for concrete strength gain obtained from the British code of practice CP110:1972, based on the 28-day compressive strength, are shown in Table 4.4. These factors are not valid when accelerator additives are used.

Table 4.4 BS CP110:1972 factor of increased concrete strength with time.

	Time factor for concrete with a 28-day strength (Mpa)		
Months	**20–30**	**40–50**	**60**
1	1.00	1.00	1.00
2	1.10	1.09	1.07
3	1.16	1.12	1.09
6	1.20	1.17	1.13
12	1.24	1.23	1.17

It is known that concrete continuously gains strength with time, which is illustrated in Table 4.4. The additional strength is about 20% to 25% of the corresponding 28-day strength. Most of the additional strength is gained within the first year.

C. Indian standard

The Indian standard (IS) presents the relation between the concrete strength with age and the 28-day strength as given in Table 4.5.

D. European code

Based on the European code EC2, it may be required to specify the concrete compressive strength, $f_{ck}(t)$, at time t for a number of stages (e.g., demolding, transfer of pre-stress), where:

$f_{ck}(t) = f_{cm}(t) - 8$ (MPa) for $3 < t < 28$ days.

$f_{ck}(t) = f_{ck}$ for $t \geq 28$ days

More precise values should be based on tests, especially for $t \leq 3$ days.

The compressive strength of concrete at an age t depends on the type of cement, temperature, and curing conditions. For a mean temperature of 20°C and curing in accordance with EN 12390, the compressive strength of concrete at various ages $f_{cm}(t)$ may be estimated from Eqs. (4.2), (4.3).

$$f_{cm}(t) = \beta_{cc}(t) f_{cm} \tag{4.2}$$

with

$$\beta_{cc}(t) = \exp\left\{ s \left[1 - \left(\frac{28}{t} \right)^{0.5} \right] \right\} \tag{4.3}$$

where $f_{cm}(t)$ is the mean concrete compressive strength at an age of (t) days, f_{cm} is the mean compressive strength at 28 days according to Table 4.6, $\beta_{cc}(t)$ is a coefficient that depends on the age of the concrete t, t is the age of the concrete in days, and s is a coefficient that depends on the type of cement and based on EC2 is equal to 0.20 for cement of strength Classes CEM 42,5 R, CEM 52,5 N and CEM 52,5 R (Class R), 0.25 for cement of strength

Table 4.5 IS for increasing concrete strength with time.

Time per month	Factor
1	1.00
3	1.10
6	1.15
12	1.20

Table 4.6 Strength classes for concrete.

Properties	Concrete strength (MPa)													
f_{ck}	12	16	20	25	30	35	40	45	50	55	60	70	80	90
$f_{ck,cube}$	15	20	25	30	37	45	50	55	60	67	75	85	95	105
f_{cm}	20	24	28	33	38	43	48	53	58	63	68	78	88	98
f_{ctm}	1.6	1.9	2.2	2.6	2.9	3.2	3.5	3.8	4.1	4.2	4.4	4.6	4.8	5.0
$F_{ctk0.05}$	1.1	1.3	1.5	1.8	2.0	2.2	2.5	2.7	2.9	3.0	3.1	3.2	3.4	3.5
$F_{ctk0.95}$	2.0	2.5	2.9	3.3	3.8	4.2	4.6	4.9	5.3	5.5	5.7	6.0	6.3	6.6
E_{cm} (GPa)	27	29	30	31	33	34	35	36	37	38	39	41	42	44
ε_1, %	1.8	1.9	2.0	2.1	2.2	2.25	2.3	2.4	2.45	2.5	2.6	2.7	2.8	2.8

Classes CEM 32,5 R, CEM 42,5 N (Class N), and 0.38 for cement of strength Classes CEM 32,5 N (Class S).

When the concrete does not conform with the specification for compressive strength at 28 days, the use of Eqs. (4.2), (4.3) is not appropriate.

This clause should not be used retrospectively to justify a non-conforming reference strength by a later increase of the strength.

The development of tensile strength with time is strongly influenced by curing and drying conditions as well as by the dimensions of the structural members. As a first approximation, it may be assumed that the tensile strength $f_{ctm}(t)$ is equal to:

$$f\text{ctm}(t) = (\beta cc(t))\alpha \cdot f\text{ctm} \tag{4.4}$$

where βcc(t) follows from Eq. (4.3) and

$\alpha = 1$ for $t < 28$

$\alpha = 2/3$ for $t \geq 28$. The values for $f$ctm are given in Table 4.6.

4.3.1.2 Variation in concrete strength

As discussed in Chapter 3, the concrete strength in a structure has a value less than the value of the concrete compressive strength in a control specimen for several reasons, as already described. These variations are also due to different stress concentrations from the control specimen rather than the structure member and different placing procedures and compaction: for example, the effect of vertical water moving during the pouring of concrete in deep concrete members and also the higher degree of compaction for concrete near the bottom of such members due to the weight of the concrete. In general, concrete with higher strength is more affected by this rather than low-strength concrete.

The reduction in strength is presented by ACI as follows: for average quality control on the construction site, the mean strength of the control cylinders will range from 4.8 to 6.2 MPa (700 to 900 psi) greater than the specified strength. In addition, Allen (1970), and (Bloem (1968)) stated that the mean 28-day compressive strength of concrete in a structure cured with minimum acceptable curing can be taken as:

$$\bar{f}_c(structure) = \left(0.675f_c' + 7.585\right)MPa \tag{4.5}$$

but not more than 1.15 fc

Changes in material strength due to maturing of the concrete or deterioration could also be included here if desired.

4.3.2 Variability of the concrete strength in a member

Based on different research studies and as stated in ACI code, the concrete strength varies based on its location in the member, and this is obviously seen in the following figures.

Fig. 4.3 shows the variation of the concrete compressive strength as the vertical axes show the varying strength from the bottom, one-fourth of the height, one-half of the height, three-fourths of the height, and the top of the member. It can be shown that all concrete members have the maximum strength on the bottom; with slab and column, the variation is minor until between the third quarter of the height and the top of the member, as the concrete compressive strength decreases to be about 90% to 75% at the top.

In the case of a concrete beam, the concrete compressive strength is reduced gradually from the bottom to the top until reaching a reduction of around 60% at the top of the beam. The concrete wall is the worst case in reduction of strength with height: at the top it will reach about 50% reduction in concrete compressive strength.

ACI presents a contour line showing the distribution of the concrete strength inside the concrete beam, as in Fig. 4.4A. From this figure, one can find that the concrete strength will reduce to approximately 60% at the top. Fig. 4.4B presents the contour line for concrete strength inside a

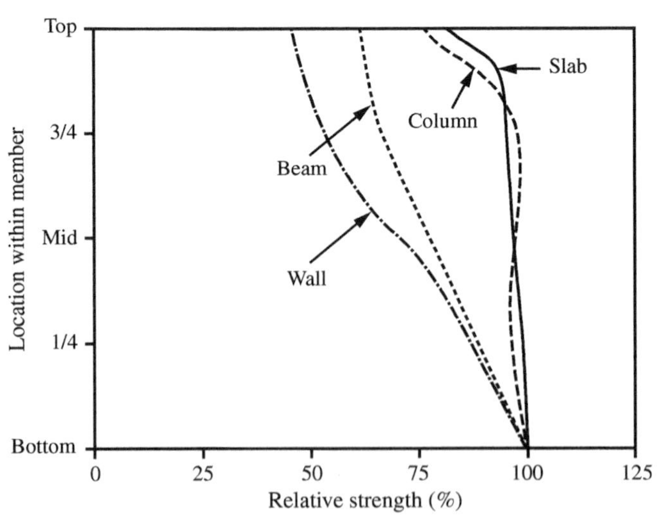

Fig. 4.3 Variation of strength along different structural member.

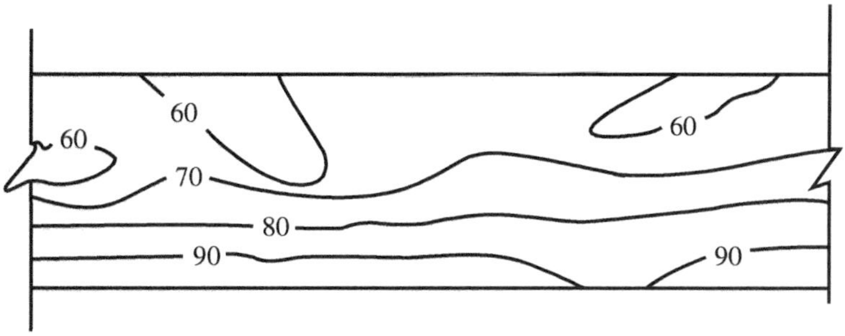

(A) Typical relative percentage strength contours for a beam

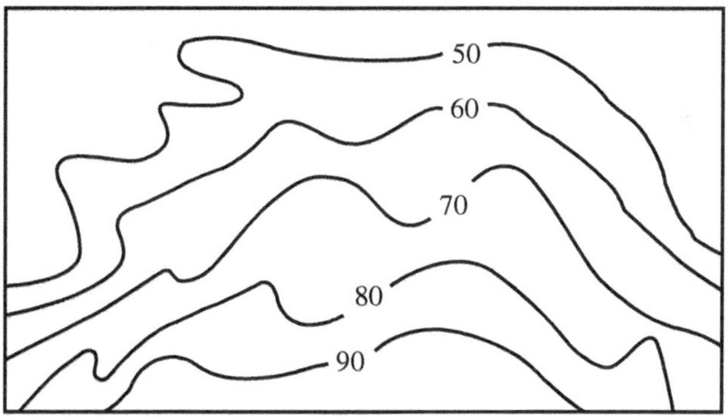

(B) Typical relative percentage strength contours for a wall

Fig. 4.4 Typical contour for relative Concrete strength.

Table 4.7 Comparison of in situ and standard cube strengths.

Structure member	Percentage of typical 28-day in situ equivalent cube strength to the standard concrete cube compressive strength at 28 days	
	Mean (%)	**Likely range (%)**
Column	65	55–75
Wall	65	45–95
Beam	75	60–100
Slab	50	40–60

reinforced concrete wall and it is obvious that the lowest concrete strength is at the upper corner of the wall.

Table 4.7 illustrates the percentage reduction in mean and likely range in standard cube compressive strength after 28 days to present the actual strength in different structure members.

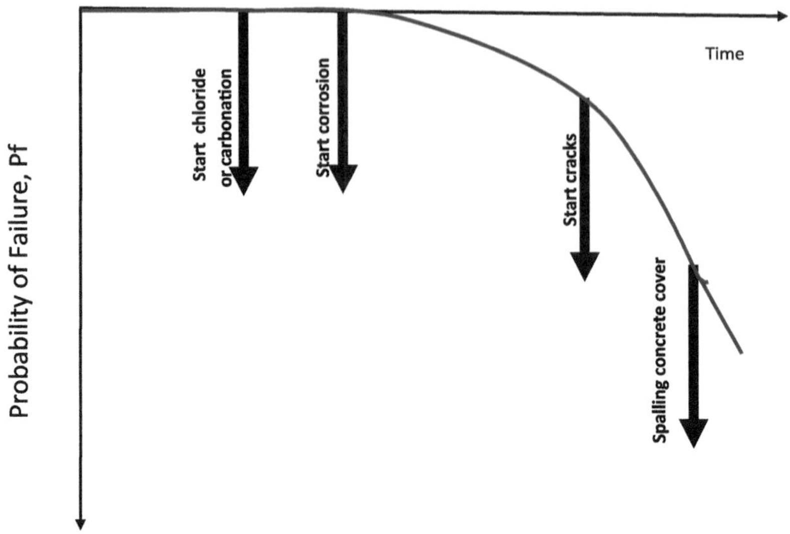

Fig. 4.5 Relation between deterioration process and probability of failure with time.

4.3.3 Corrosion of steel and concrete deterioration

Corrosion is a three-phase process. The first phase is the time period from construction until the corrosion initiation. This phase is due to the diffusion of CO_2 or Cl to cause depassivation. In the second phase, reduction of the steel section occurs; the third phase occurs due to the second phase, as deterioration of the concrete, by seeing the start of cracking and then spalling of the concrete cover. All of these phases are presented in Fig. 4.5.

The three phases are discussed in more detail in the following sections.

Fig. 4.6 presents the corrosion on steel reinforcement for different foundations in oil and gas plants under static equipment. Fig. 4.7 presents the corrosion of reinforcement steel bars for a bridge main girder.

4.3.4 Statistical data for corrosion rates

In reliability analysis of concrete structures, many researchers have used different probabilistic models to describe the corrosion initiation and rate for steel bars in concrete.

The corrosion initiation can be represented by the Poisson process, as recommended by Mori and Ellingwood (1994a). The mean Poisson parameter $\nu(t)$ can be expressed as follows:

$$\nu(w) = \begin{cases} 0 & for \quad w < t^* \\ \nu & for \quad w \geq t^* \end{cases} \tag{4.6}$$

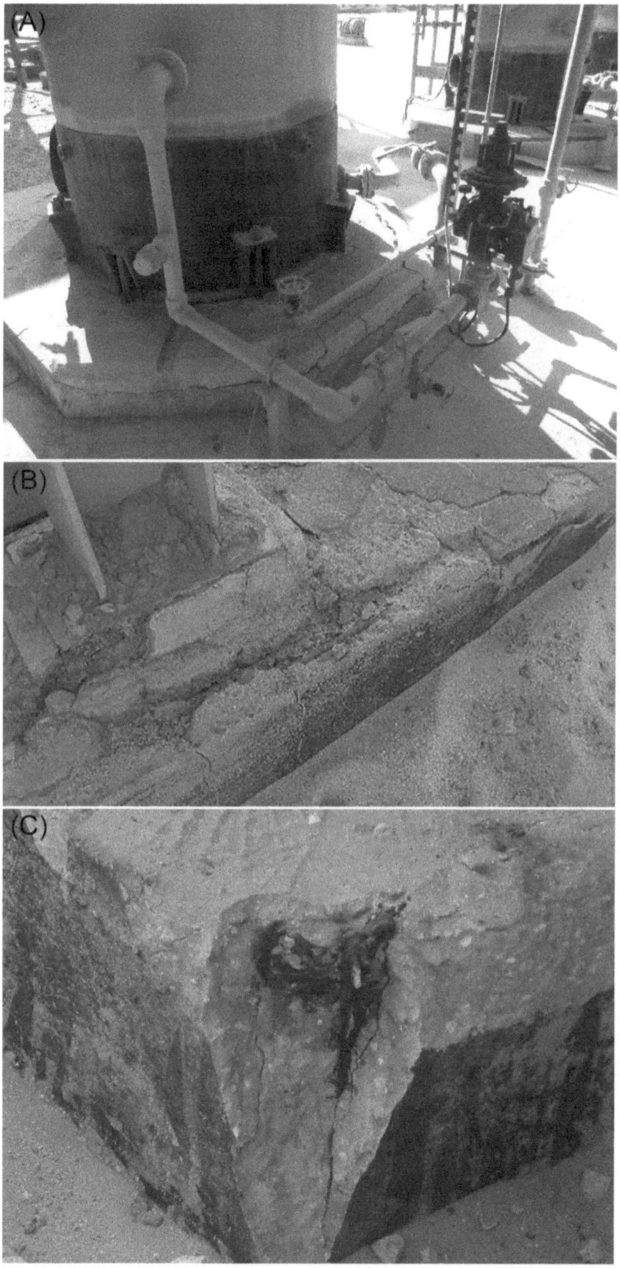

Fig. 4.6 (A) Corrosion of steel in tower foundation. (B) Corrosion of steel bars and concrete deterioration for foundation under pipe rack. (C) Corrosion of the foundation and spalling concrete covers with cracks.

Fig. 4.7 Corrosion of steel bars of a bridge main girder.

where t^* is a deterministic time, considered to be 10 years, and ν is the mean initiation rate of corrosion, equal to 0.2 mm/year.

Typical corrosion rates of steel in various environments have been reported in recent years. According to Ting (1989), the average corrosion rate C_r for passive steel in concrete attacked by chlorides is about 100 μm/year.

According to Mori and Ellingwood (1994b), the typical corrosion rate, C_r, is a time-invariant random variable described by a lognormal distribution with mean C_r of 50 μm/yr, and coefficient of variation V_{cr} of 50%.

Because the corrosion rate changes with environment, no accurate data is available to predict the real corrosion rate. Ting (1989), Mori and Ellingwood (1994b), and Frangopol et al. (1997) presented three corrosion rates of 64, 89, and 114 μm/year that cover most cases of corrosion related to different environmental conditions. The mean corrosion rate is 89 μm/year with a standard deviation around 25 μm/year.

4.3.5 Capacity loss in reinforced concrete member

The strength of any reinforced concrete member is dependent on the cross-sectional dimensions of the concrete sections, cross-sectional area of the steel area bars, the concrete strength, and steel yield strength.

To run the reliability analysis, a uniform corrosion is usually assumed. Since the reduction in thickness is the same for all its length and circumferential dimensions, the total longitudinal reinforcement area can be expressed as a function of time t as follows:

$$As(t) = \begin{cases} n\pi D_b^2/4 & for \quad t \leq T_i \\ n\pi[D_b - 2C_r(t - T_i)]^2/4 & for \quad t > T_i \end{cases} \tag{4.7}$$

where D_b is the diameter of the bar, n is number of bars, T_i is time of corrosion initiation, and C_r=rate of corrosion. Eq. (4.7) takes into account the uniform corrosion propagation process from all sides.

Regarding the corrosion effect on the reinforced concrete reliability, many parameters affect it, in addition to the environmental conditions affecting the corrosion rate.

A study by El-Reedy et al. (2002) considered the effect of the percentage of longitudinal steel bars on the reliability of a reinforced concrete column under corrosion attack. In this study the researchers considered two percentages of steel, $\rho = 1\%$ and $\rho = 4\%$, for an interior column; this reliability study used a Monte-Carlo simulation technique.

In this study different environmental conditions as per Frangopol et al. (1997) were examined; therefore different corrosion rates were considered: 0.064, 0.089, and 0.114 mm/year. Before use, the steel bars may already have rust; on the other hand, the steel bars may be taken directly from manufacture to pouring of the concrete, with no adequate time to have rust. Therefore, in these two cases, the initial time of corrosion is different, so different initial times (0, 10, 20 years) of corrosion were taken to study their effect on the reliability of columns.

4.3.6 Effect of age and steel ratio

The strength of concrete increases with age depending on the environmental conditions, as discussed earlier. Fig. 4.8 presents the reliability index for both dry and wet conditions for two steel ratios, 1% and 4%.

In general, the reliability index for a steel ratio equal to 4% is higher than for a steel ratio equal to 1% for the first 10 years, for both wet and dry conditions. However, the reliability index after 10 years for a steel ratio equal to 1% in wet conditions is less than that for a steel ratio equal to 4% in dry conditions.

As shown in the figure, it was found that with dry conditions the reliability index increased within 1 year and after that it remained constant.

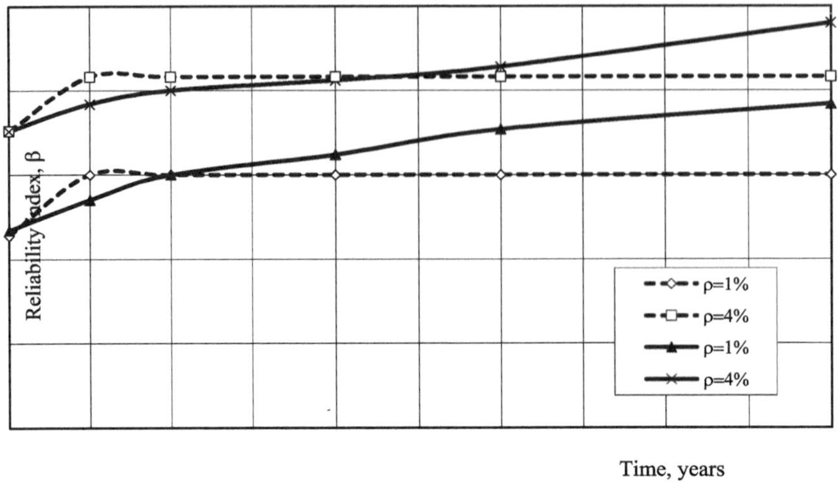

Fig. 4.8 Effect of environmental conditions and steel reinforcement ratios on the reliability of reinforced concrete columns ignoring corrosion effect.

With the steel ratio equal to 1%, the reliability index in wet conditions increased after 2 years; on the other hand, for a steel ratio equal to 4% the increase in the value of the reliability index for wet rather than that for dry conditions was after 4 years.

Based on this, the weather conditions have more influence when the steel ratio is equal to 1% than when the steel ratio is 4%, but after 10 years the percentage of increase of the reliability index for wet to that for dry conditions is the same for the two steel bar percentages.

The effect of the two percentages of steel ratio reliability index at different corrosion rates equal to 0.064, 0.089, 0.114 mm/year along the lifetime is shown in Figs. 4.9, 4.10, and 4.11, respectively. From these figures, one can conclude that the reliability index of the reinforced concrete column decreased gradually after 10 years, which is the time that corrosion began.

From this study, it is found that the reliability index for the steel ratio equal to 4% is higher than that for a steel ratio equal to 1% up to 40 years, but after that the reliability index for a concrete column with steel ratio equal to 1% is slightly higher than that with the steel ratio equal to 4%.

The rate of decrease in the reliability index of a reinforced concrete column with a steel ratio equal to 4% is greater than that having a steel ratio equal to 1%.

Considering the value of $\beta = 4.0$, as suggested by MacGregor (1983), one can find from Fig. 4.9 that repair must be done for the columns after 17 years for a steel ratio equal to 1%, and after 30 years for a steel ratio of 4%.

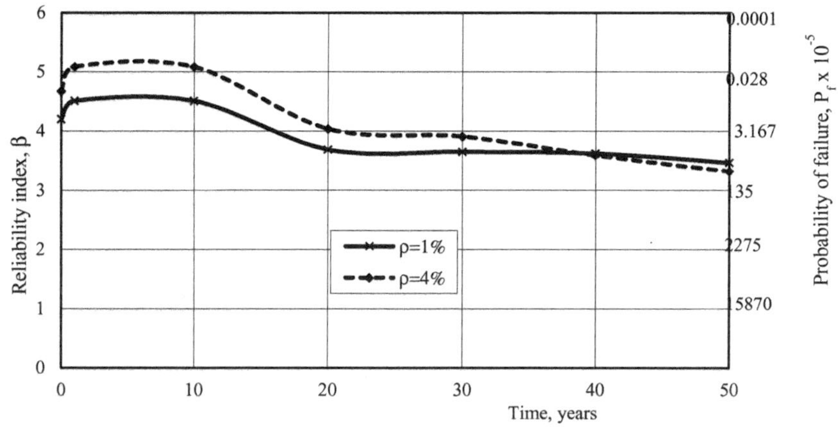

Fig. 4.9 Relation between reinforced steel ratio and the reliability index of RC column at $C_r = 0.064$ mm/year with time.

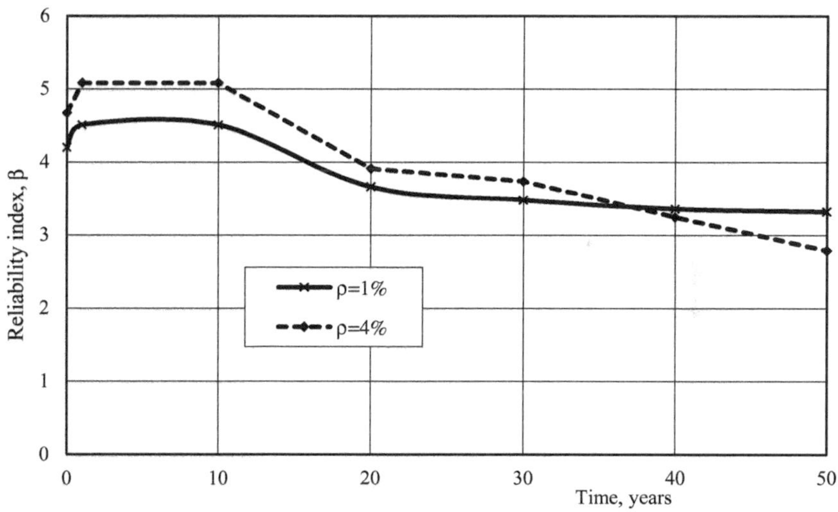

Fig. 4.10 Relation between steel ratio and the reliability index of RC column at $C_r = 0.089$ mm/year with time.

From Fig. 6.10, one can find that for a steel ratio equal to 4% the reliability index is higher than that for a steel ratio equal to 1% until 37 years, but after that the reliability index for a steel ratio equal to 1% is higher than that for a steel ratio equal to 4%.

From this figure, and by considering $\beta = 4.0$, one can find that the repair must be done after slightly less than 17 years for a steel ratio equal to 1% and at 20 years for a steel ratio equal to 4%.

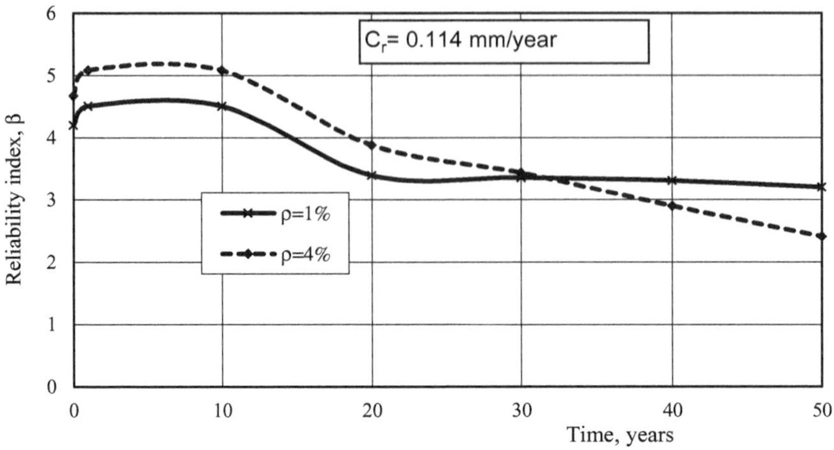

Fig. 4.11 Relation between steel ratio and the reliability index for RC column at $C_r = 0.114$ mm/year with time.

For a higher corrosion rate, with a value of 0.114 mm/year as illustrated in Fig. 4.11, the reliability index for a steel reinforced ratio in the column equal to 4% is higher than that at a steel ratio equal to 1%, until after about 30 years.

Considering $\beta = 4.0$ in Fig. 4.11, one can find that the inspection and repair of the columns must be done after 15 years with a steel ratio equal to 1% and at 20 years with a steel ratio equal to 4%.

Moreover, one can find that, at the beginning, the probability of failure of the column with steel ratio equal to 4% is lower than that with the steel ratio equal to 1% until 30, 38, and 42 years with corrosion rates of 0.064, 0.089, and 0.114 mm/year, respectively; then the probability of failure will be higher for a steel ratio equal to 4%.

Therefore, the lower steel ratio is recommended for environmental conditions that cause a high corrosion rate, as the most column strength capacity is carried by the concrete.

It is obvious that, for a steel ratio equal to 4%, the reliability index is higher than that for a steel ratio equal to 1% until 20 years. For a steel ratio equal to 4%, the rate of decreasing reliability is high.

It is found that, as a general rule, whatever the steel reinforcement ratio in the column, the reliability index increases as the corrosion rate decreases.

The three corrosion rates that were considered covered most environmental conditions. It is obvious that the reliability index of columns with later ages is highly affected by the corrosion rate, and with a steel ratio equal to 4% it is higher than with a steel ratio equal to 1%.

As a general recommendation, in wet conditions when the corrosion rate is high, it is better to consider a minimum percentage of steel in the design.

4.3.7 Start time of corrosion

The start time of corrosion is defined as the time from constructing the column until the carbonation or chloride propagates through the concrete cover until it reaches the steel bars. Then the pH value of concrete drops to 8 around the longitudinal steel bars and the corrosion will propagate inside the steel bar.

In this section, three values of start time of corrosion T_i are considered in the study by El-Reedy et al. (2002) at 0, 10, and 20 years. The value $T_i = 0$ indicates that the outer surfaces of the bars start during the construction. The other start times after 10 years and after 20 years apply for a thick concrete cover, which takes time to reach depassivation.

From Figs. 4.12 and 4.13, one can see that the reliability of a reinforced concrete column increases as a result of increasing the time of starting the corrosion. It was found that the reliability index values are slightly different in the three start times after 40 years for a steel ratio equal to 1%, but when the steel ratio is equal to 4% there is a high reliability range between start time of 20 years and start time of zero.

As a conclusion, the start time is very critical and should be carefully considered for a steel ratio of the column equal to 4%.

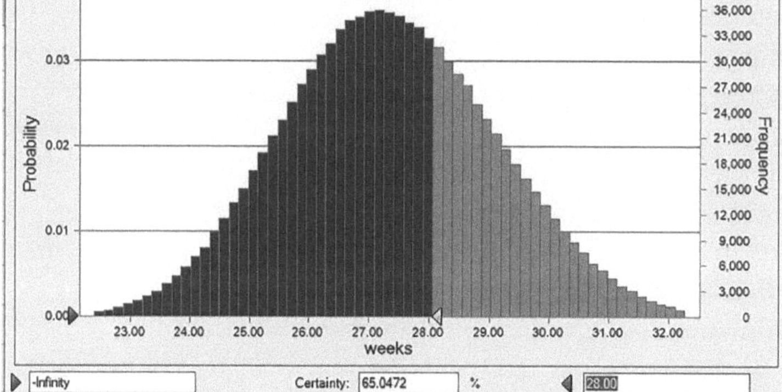

Fig. 4.12 Influence of corrosion start time on the reliability at steel ratio $\rho = 1\%$.

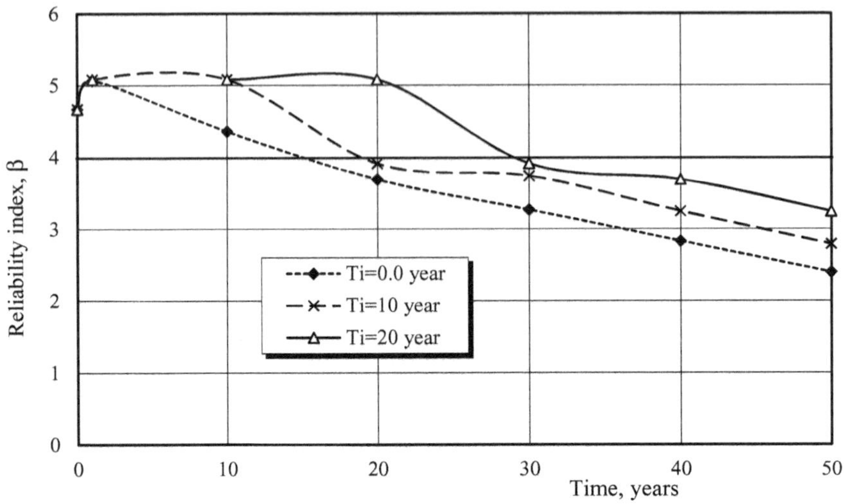

Fig. 4.13 Influence of corrosion start time on the reliability at steel ratio $\rho = 4\%$.

4.3.8 Column eccentricity effect on reliability

El-Reedy et al. (2002) studied the effect of eccentricity on corrosion attack in reinforced concrete columns. In this study, the value of eccentricity was considered by adding 5% of column width to the simulation to study the effect of different parameters on the reliability of a reinforced concrete column with an eccentric load.

The influence of eccentricity on the reliability index for corroded steel bars with time is presented in Figs. 4.14 and 4.15. In these figures, the reliability with time illustrated for a steel ratio is equal to 1% and 4%, respectively, with two values of extreme eccentricity specified by the ECP and complying with BS.

From the previous figure, one can find that the reliability index in case of no eccentricity is higher than that in the case of eccentricity that reaches 5% of column width.

The eccentricity effects on the column for different corrosion rates are illustrated in Figs. 4.16, 4.17, 4.18. From these figures, note that action should be taken to start maintenance, which is an inspection and any repair if needed, when the reliability index reaches 4.0, according to MacGregor (1983). With a corrosion rate of 0.064 mm/year, it is better to begin the maintenance at 13 years if the steel ratio of the column (ρ) is equal to 1%, and at 16 years for a steel ratio equal to 4%.

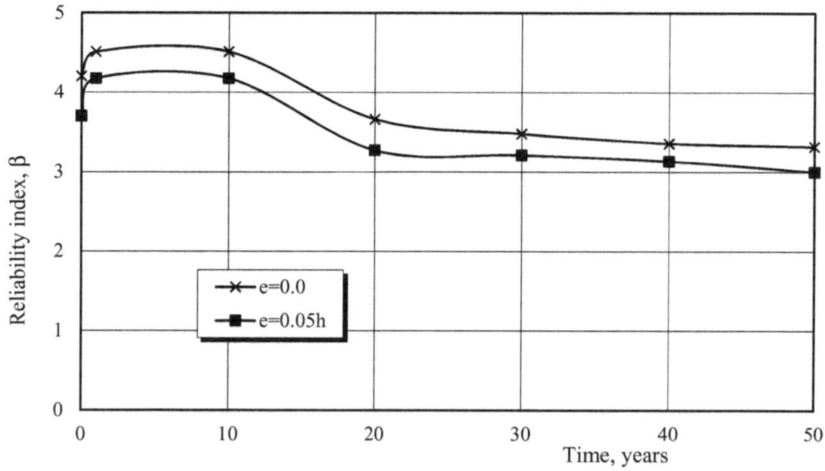

Fig. 4.14 Influence of eccentricity on the reliability index with time ($\rho = 1\%$).

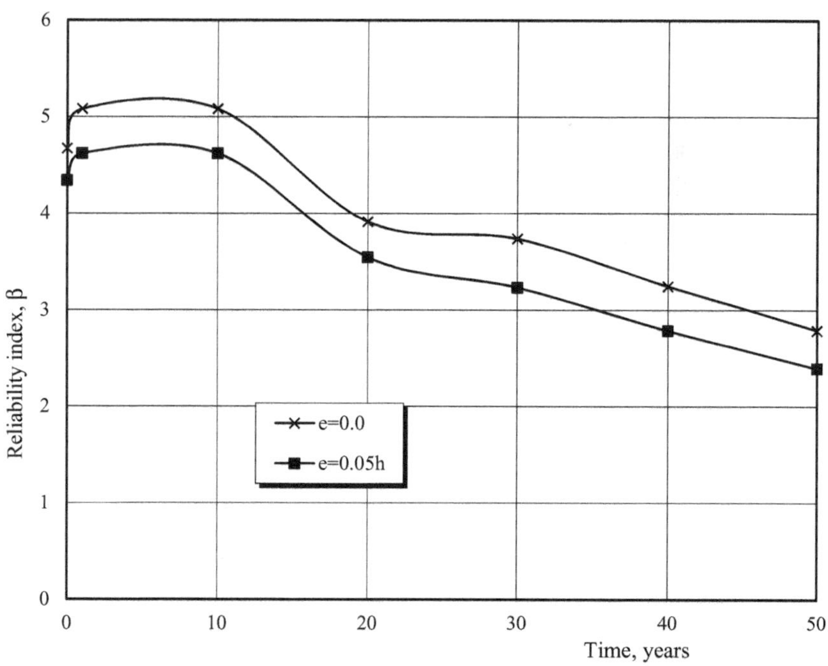

Fig. 4.15 Influence of eccentricity on the reliability index with time ($\rho = 4\%$).

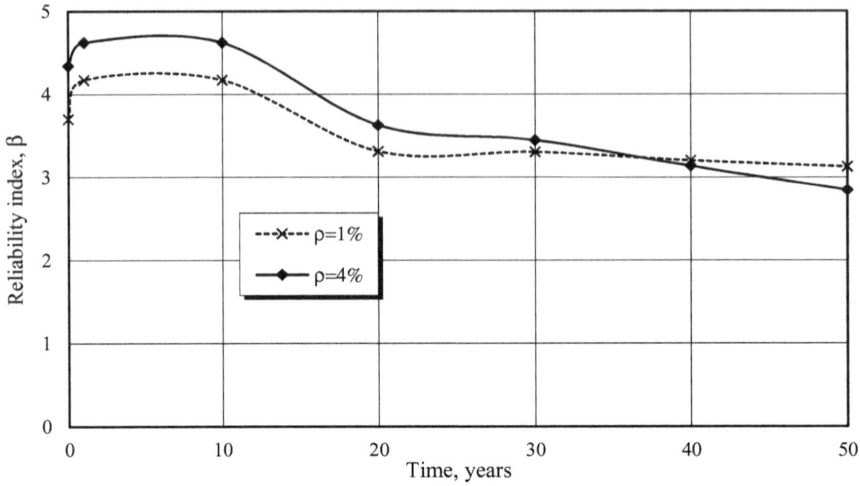

Fig. 4.16 Influence of reinforcement ratio on the reliability index for small eccentricity and low corrosion rate.

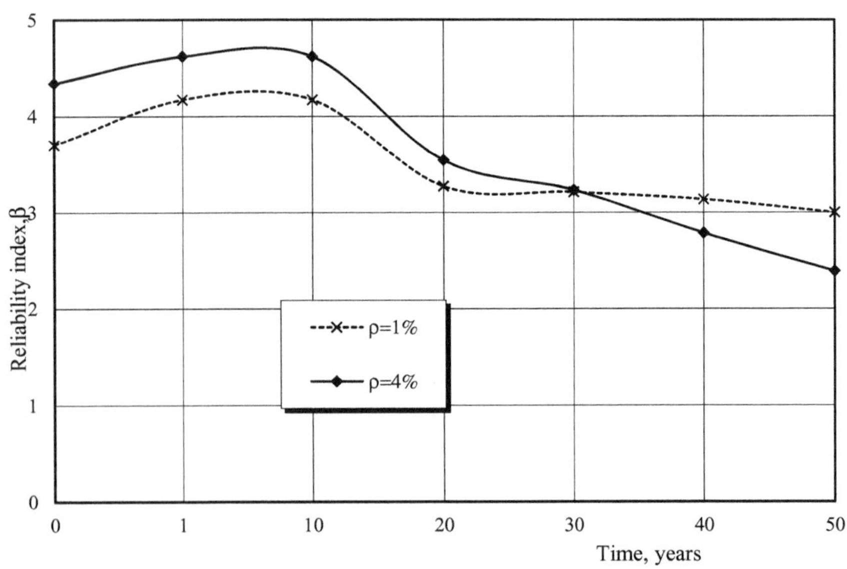

Fig. 4.17 Influence of reinforcement ratio on the reliability index for small eccentricity and medium corrosion rate.

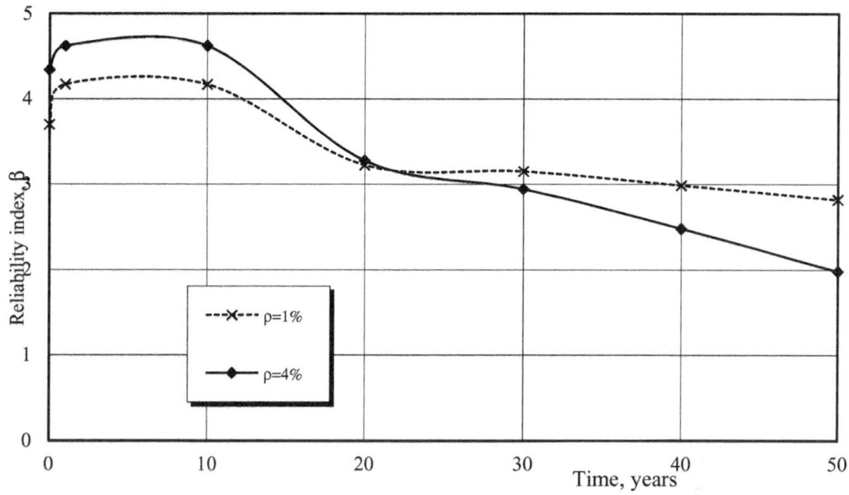

Fig. 4.18 Influence of reinforcement ratio on the reliability index for small eccentricity and high corrosion rate.

At a rate of corrosion equal to 0.089 mm/year, the maintenance start time should be slightly earlier than for the previous rates of corrosion for the two values of steel percentages.

For a corrosion rate equal to 0.114 mm/year, the maintenance should start at 12 years for a steel ratio (ρ) equal to 1%, but at 15 years for 4%; for in-between steel ratios, linear interpolation can be used.

In summary, the maintenance for a concrete column with eccentricity equal to 5% of the width should be done at least by 12 years and at most 16 years, based on the corrosion rate and steel ratio. On the other hand, the maintenance of a concentrically loaded column should be done at least at 15 years and at a maximum at 37 years, based on the corrosion rate and steel ratios.

The main information that should be available to evaluate an existing building is the corrosion rate for the steel reinforcement.

For reinforced concrete having corrosion in the steel bars that is gradually progressing, the capacity of the beam for both bending and shear will decrease. The flexural strength of a reinforced concrete beam at time t, $Mr(t)$, may be computed based on the principle of design of a reinforced concrete beam, as shown in Fig. 4.19 and Eq. (4.8).

$$M_Y(t) = A_s(t) f_Y \left(d - \frac{a}{2} \right) \tag{4.8}$$

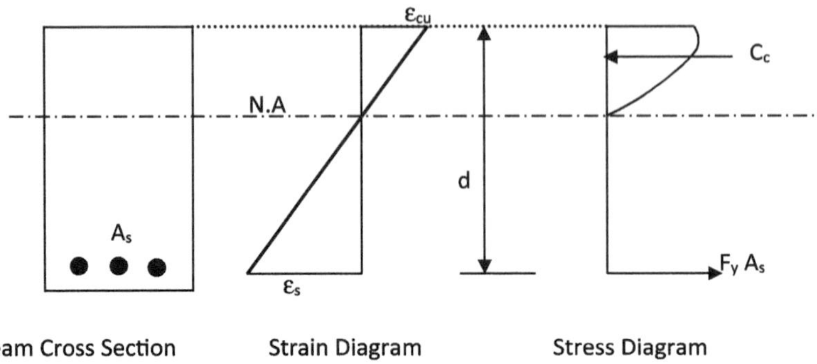

Fig. 4.19 Strain and stress diagrams of concrete beam.

where M_r (*t*), is the bending moment strength with time, f_y is yield strength of a reinforcing bar, *d* is effective depth, and *a* is depth of the equivalent rectangular stress block.

The shear strength with time can be calculated from the following equation:

$$V(t) = V_c + V_s(t) \tag{4.9}$$

In this equation it is assumed that the shear strength of concrete *Vc* is a constant value and does not change with time. Therefore shear reinforcement, like stirrups, will be considered and affected by time as given in Eq. (4.10):

$$V_s(t) = \frac{A_s(t) f_y d}{s} \tag{4.10}$$

where *As*(*t*) is the area of the shear reinforcement within the distance s, where s is the spacing between stirrups.

The loss of material due to corrosion causes a reduction in both bending and shear capacities of reinforced concrete beams. Based on Eqs. (4.8), (4.9), and AASHTO (*Standard* 1992) requirements for the design of RC girders, the moment and shear capacities can be calculated for both time-independent (intact) and time-dependent (damaged under corrosion) limit states.

4.4 Design and maintenance recommendations

The findings from the El-Reedy et al. study that was done in 2000 are as follows:

- The higher steel reinforced ratio for the concrete column has a higher reliability index than that for the lower steel ratio.

- Wet environmental conditions give a higher reliability index than that for the case of dry conditions.
- In dry conditions, the reliability index will increase about 10% after 1 year from construction.
- The maintenance for a concentrically loaded reinforced concrete column should be started at least by 15 years and at most 30 years, as per the expected corrosion rate and steel ratio.
- The reliability index increases by increasing the corrosion start time, which has more serious effects when the steel ratio equals 4%.
- The maintenance of the reinforced concrete column with eccentricity equal to 5% should be done at 12 years and at most 16 years according to the corrosion rate and steel ratio.

4.5 Pipeline assessment

After performing the inspection and defining the remaining thickness, the next step is to perform an evaluation and define the action that is required to be taken.

According to API 570 the remaining life can be defined by the following formula:

$$R_L = t_a - t_r / C_r \tag{4.11}$$

where

R_L = remaining life

t_a = actual thickness at the time of inspection, measured in mm or inches

t_r = the required thickness based on design calculation, measured in mm or inches

C_r = corrosion rate, mm/year or inches/year

From inspection, both the long-term corrosion rate (L_T) and also the short-term corrosion rate can be defined.

$$L_T = t_i - t_a / (\text{lifetime period}) \tag{4.12}$$

where t_i is the initial thickness and ta is the actual thickness

$$S_T = t_p - t_a / (\text{lifetime between previous and actual}) \tag{4.13}$$

where t_p is the thickness measurement from the previous inspection.

By knowing the history and applying the statistical data, we can define the expected corrosion rate to predict the remaining life. The pipeline is considered a series, a structural redundancy, so once the thickness reaches the limit and converts to a hole, it then needs a corrective action. Once this happens it means there is a weakness in the integrity management system.

4.6 Storage tank assessment

It is worth mentioning that in tank settlement, uniform settlement does not affect the tank but affects the ring connected to it. The main problem is differential settlement, as it affects the tank integrity and the operations in the case of a floating roof, since the tank will be converted to an ellipse rather than a circle that may hang to the floating roof and it will definitely affect the nozzles that are connected to the tank. Also a check will be needed to evaluate the tank.

In general, based on our experience in soil mechanics in sandy soil, the settlement will occur after construction, whereas in clay soil the total settlement will occur after 8–10 years. After construction of the tank, a hydrostatic test should be carried out BY FILLING THE TANK WITH WATER, WHICH IS THE MAXIMUM LOAD, so the maximum settlement will occur, and after time the minor settlement value will also occur.

To evaluate the tank for settlement, the following equations present the maximum allowable settlement as designated by API653. The following equation is from Marr et al. (1982).

The permissible out of plane settlement can be calculated as follows:

$$S_m = \left(5.5\,L^2 \cdot F_y\right)/(E \cdot H) \tag{4.14}$$

where

L is the arc length between measurement points, ft
F_y is the shell materials yield strength, ib/in^2
E is Young's modulus, ib/in^2
H is tank height, ft

This equation is presented by Andreani and Carr (2007):

$$S_m = K \cdot S_a \cdot (D/H) \cdot \left(F_y/E\right) \geq 4 \tag{4.15}$$

where

S_a is the effective settlement arc, ft
F_y is the shell materials yield strength, ib/in^2
E is Young's modulus, ib/in^2
H is tank height, ft
D is tank diameter, ft
K is a factor in Table 4.8.

Measure the bulge or depression. The permissible bulge or depression is given by the following equation:

$$B_B = 0.37R \tag{4.16}$$

Table 4.8 Relation between tank diameter and *K* value.

Tank diameter (ft)	*K* value	
	Open top tanks	**Fixed roof tank**
$D \leq 50$	28.7	10.5
$50 < D \leq 80$	7.8	5.8
$80 < D \leq 120$	6.5	3.9
$120 < D \leq 180$	4.0	2.3
$180 < D \leq 240$	3.6	Not applicable
$240 < D \leq 300$	2.4	Not applicable
$300 < D$	Not applicable	Not applicable

where

B_B is maximum height of bulge or depth of local depression, in inches;

R is radius of inscribed circle in bulged area or local depression, in feet.

For the edge settlement, the maximum settlement is presented in Figs. 4.20 and 4.21. Considering that tanks with large edge settlements are to be repaired or require a detailed analysis for the bottom, and the bottom-to-shell junction. With a settlement more than or equal to 75% of the maximum allowable settlement or larger than 2 inches, as in the following figure, the welding of tanks will be examined with a magnetic particle test or liquid penetration test.

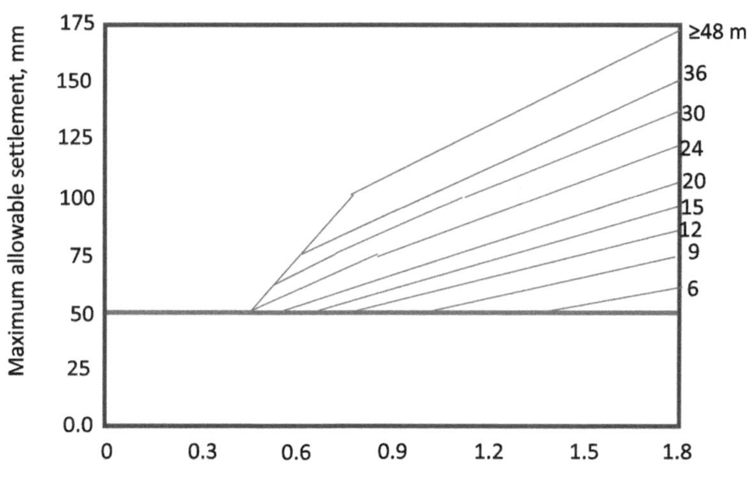

Fig. 4.20 Maximum allowable edge settlement for areas with bottom lap welds nearly parallel to the shell.

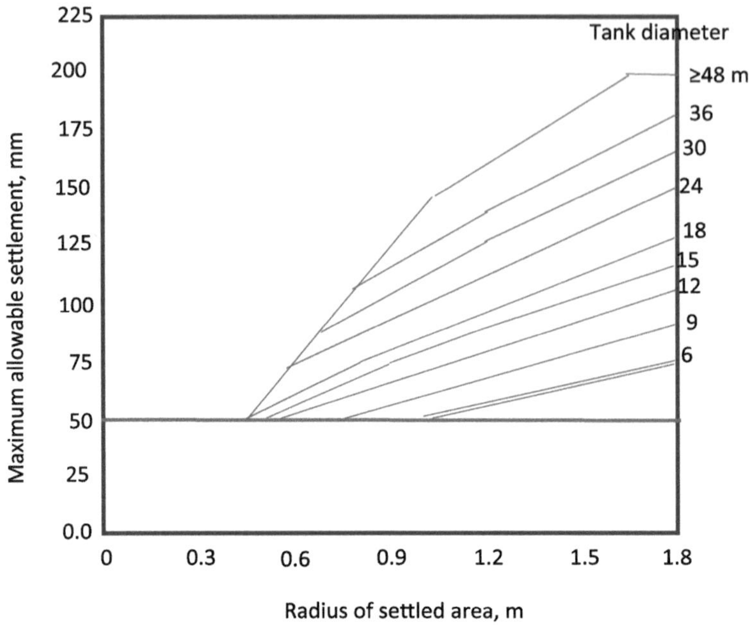

Fig. 4.21 Maximum allowable edge settlement for areas with bottom lap welds nearly perpendicular to the shell.

The API 579 fitness for service presents the method to use to evaluate the pressure vessel, piping, and tanks.

In this standard, the concept of a remaining strength factor is utilized to define the criteria to accept a component to continue on in service or not.

The remaining strength factor (*RSF*) is calculated as per the following formula:

$$RSF = L_{DP}/L_{UP} \tag{4.17}$$

where L_{DP} is the plastic collapse load of the damage component and L_{UP} the plastic collapse load for the original (undamaged) component.

With this definition of the remaining strength factor (*RSF*), the acceptance criteria will be defined through the traditional code formulas, elastic stress analysis, limit load theory, or elastic-plastic analysis.

When evaluating a thin area, the *FFS* assessment calculation steps provide a means to calculate the *RSF*. If the calculated *RSF* has a higher value than the allowable remaining strength ratio (*ARSF*), then the damaged component can be used again and placed back into service.

If the calculated *RSF* is less than the allowable value, then the component will be repaired, rerated, or another mitigation action will be taken to minimize the severity of the operating environment.

With this in mind, the rerated maximum allowable working pressure (MAWP) can be calculated from the remaining strength factor *RSF* as follows, for RSF less than the allowable remaining strength factor (ARSF):

$$\text{MAWP (rerated)} = \text{MAWP (RSF/ARSF) for RSF} < \text{ARSF} \tag{4.18}$$

$$\text{MAWP (rerated)} = \text{MAWP for RSF} \geq \text{ARSF} \tag{4.19}$$

For tankage, the maximum fluid height (MFH) can be rerated by this formula:

$$\text{MFH (rerated)} = \text{H}_\text{f} + (\text{MFH} - \text{H}_\text{f})(\text{RSF/ARSF}) \text{ for RSF} < \text{ARSF} \tag{4.20}$$

$$\text{MFH (rerated)} = \text{MFH or RSF} \geq \text{ARSF} \tag{4.21}$$

where H_f is distance between the bottom of the flaw and the tank bottom.

The recommended value for the allowable remaining strength factor (A*RSF*) is 0.9 for designs using the following codes:

- ASME Section 1
- ASME Section VIII Division 1
- ASME Section VIII Division 2
- AS 1210
- BS PD 5500
- ASME B31.1
- ASME B 31.3
- API 620
- API650

The fitness for service API 579 classified the mode of failure into six cases, as follows:

- Brittle fracture
- Corrosion and erosion
- Cracks like flaw
- Creep damage
- Fire damage
- Mechanical damage

In general API 579 fitness for service covers the evaluation of any component based on its deficiencies and it covers assessment for 10 conditions, as follows:

- EXISTING EQUIPMENT FOR BRITTLE FRACTURE
- GENERAL METAL LOSS
- LOCAL METAL LOSS
- PITTING CORROSION

- HYDROGEN BLISTERS AND HYDROGEN DAMAGE ASSOCIATED WITH HIC AND SOHIC
- WELD MISALIGNMENT AND SHELL DISTORTIONS
- CRACK-LIKE FLAWS
- COMPONENTS OPERATING IN THE CREEP RANGE
- FIRE DAMAGE
- DENTS, GOUGES, AND DENT-GOUGE COMBINATIONS
- LAMINATIONS

For fitness for service API 579, the assessment procedure consists of three levels. Level 1 will be done based on the available inspection results and by simple calculation, to be on the conservative side. For Level 2 many inspections are done to obtain accurate data; this will be less conservative and fatigue should be considered as a factor if applicable. For level 3, in addition to a high level of inspections, analysis by the finite element method can also be done to ensure an accurate assessment.

Assessment in case of dents and gauge

A. Dent Assessment

The Level 1 assessment procedure for determining the acceptability of a dent is shown here

1. Determine the parameters *D*, future corrosion allowance (*FCA*), either *rd t* or *nom t*, and *LOSS*, which is the amount of metal loss at the time of assessment.

$$dp = 0.7\,\text{do}$$

where *dp* is the dent depth at pressurized conditions and do is at unpressurized conditions. The minimum specified ultimate strength is equal to 414 MPa, which may be used if the materials are unknown.

2. Determine the wall thickness to be used in the assessment using Eq. (4.22) or Eq. (4.23), as applicable.

$$t_f = tn - \text{LOSS} - \text{FCA} \tag{4.22}$$

$$t_f = t_a - \text{FCA} \tag{4.23}$$

where t_f is the future wall thickness, t_n is the nominal thickness of the component, t_a is actual uniform thickness away from the damage established by thickness measurements at the time of assessment, FCA is future corrosion allowance, and LOSS is the amount of uniform metal loss at the time of assessment.

3. If Eqs. (4.24), (4.25) are satisfied, proceed to STEP 4. Otherwise, the Level 1 assessment is not satisfied

$$Lm \geq 1.8\left(D t_f\right)^{0.5} \tag{4.24}$$

$$L_w = 2\, t_f \geq 25\,\text{mm} \tag{4.25}$$

where L_w is the distance to the nearest weld joint and L_m is the distance to the nearest major structure discontinuity.

4. If the component is not in cyclic service and Eq. (4.26) is satisfied, proceed to STEP 5. Otherwise, the Level 1 assessment is not satisfied.

$$d_d \leq 0.07\,\text{D} \tag{4.26}$$

where D is cylinder outside diameter; d_d is the depth of the dent when the component is not pressurized.

5. Determine the *MAWP* for the component using the thickness from STEP 2. If the *MAWP* is greater than or equal to the current design condition, then the component is acceptable for continued operation. Also, if t_f is higher than t_{min} you can operate. Otherwise, the Level 1 assessment is not satisfied.

- Metal loss for tanks

In the case of metal loss of tanks, the following equation can be applied to define the minimum thickness and the maximum allowable liquid height.

The minimum thickness and *MFH* in metric units are shown here. In these equations, the tank nominal diameter *D* is in meters, the design fill height *H* is in meters, the allowable stress *S* is in MPa, the wall thicknesses min *t* and *t* are in millimeters, and the metal loss *LOSS* and future corrosion allowance *FCA* are in millimeters.

$$t_{\text{min}} = 4.9\text{D}(\text{H} - 0.3)\text{G}/\text{S} + \text{FCA} \tag{4.27}$$

In the case of metal loss of piping, the following equations are written in terms of outside diameter of the pipe, *D*; therefore the equations do not need to be adjusted for metal loss and future corrosion allowance, which occur on the inside surface. If metal loss has occurred on the outside diameter (e.g., corrosion under insulation) *D* would need to be modified to account for this metal loss.

It is required to obtain the thickness corresponsing to maximum allowable working pressure (MAWP).

For straight pipes subject to internal pressure, the minimum thickness and *MAWP* equations for straight sections of pipe subject to internal pressure are as follows:

(a) Circumferential stress (longitudinal joints):

$$t_{\text{min}} = 0.5\,\text{PD}/(\text{SE} + \text{PY}) + \text{MA} \tag{4.28}$$

Table 4.9 Y values.

	Temperatures (°C)					
Material	**≤482**	**510**	**538**	**566**	**593**	**≥621**
Ferritic steels	0.4	0.5	0.7	0.7	0.7	0.7
Austenitic steels	0.4	0.4	0.4	0.4	0.5	0.7
Other ductile metal	0.4	0.4	0.4	0.4	0.4	0.4
Cast iron	0.4	–	–	–	–	–

$$\text{MAWP} = 2\text{SE}\left(t_f - \text{MA}\right)/\left(\text{D} - 2\text{Y}\left(t_f - \text{MA}\right)\right) \tag{4.29}$$

$$\sigma_m = (\text{P}/\text{E})\left[0.5\,\text{D}/\left(t_f - \text{MA}\right) - \text{Y}\right] \tag{4.30}$$

where MA is mechanical allowances (thread or groove depth); for threaded components, the nominal thread depth (dimension h of ASME B.1.20.1) will apply, P is the internal pressure, S is the allowable stress, E_w is weld joint efficiency or quality factor from the original construction code; if unknown use 0.7.

Y is a coefficient from ASME B31 piping codes used for determining the pipe wall thickness. This coefficient can be obtained from the following table that is valid for min 6 o $t < D$. As shown in Table 4.9, the value of Y can be interpolated for intermediate temperatures.

(b) Longitudinal stress (circumferential joints):

$$t_{\min} = 0.25\text{PD}/(\text{SE} + \text{PY}) + t_s + \text{MA} \tag{4.31}$$

$$\text{MAWP} = 4\text{SE}\ \left(t_f - t_s - \text{MA}\right)/\left(\text{D} - 4\text{Y}\left(t_f - t_s - \text{MA}\right)\right. \tag{4.32}$$

$$\sigma_{ml} = (\text{P}/\text{E})\left(0.25\text{D}/\left(t_f - t_s - \text{MA}\right) - Y\right) \tag{4.33}$$

where t_s is supplemental thickness for mechanical loads other than pressure that result in longitudinal stress; this thickness is usually obtained from the results of a weight case in a stress analysis of the piping system.

The final values of $t_{\min}$ will be the minimum from circumferential and longitudinal and MAWP is the minimum for both and the maximum stress σ_m is the maximum from both.

4.7 Assessment for onshore steel structure

The assessment and inspection of the steel structure in onshore facilities, which typically is the pipe rack, follow the same procedures and techniques used for the topside of the offshore structure. However, for the onshore

Fig. 4.22 Corrosion in the bolted connection.

structure there are many bolted connections that are crucial to be addressed. As shown in Fig. 4.22, the corrosion starts on the bolt.

As shown in the figure, the bolted connection is starting to corrode. However, the part of the bolt inside the steel cannot be seen in order to determine the status of its corrosion, which is considered a high risk.

If huge corrosion is seen, a temporary support must be made and these bolts replaced. As a mitigation action in case of a location near the sea or other harsh environmental condition, it is recommended after the erection to put grease in these bolts to prevent propagation of oxygen inside the bolt, in the location of the bolt inside the steel.

If anchor bolts are used in a harsh environment, they require more precautions and attention. The anchor bolt is for tanks with a height greater than the diameter, to resist the wind effect on the tank. As shown in Fig. 4.23, during construction if there is inaccurate alignment between the bolt and the hole, it will be affected by bending stresses during installation. As a rule of thumb, for steel under tension, corrosion will start very fast. As shown in the figure, the corrosion has started, and it needs mitigation by coating, activating a cathodic protection system, or being readdressed in another way. This is a case for any integrity management system to be proactive.

Fig. 4.24 presents the start of corrosion for steel towers in a petrochemical plant in a harsh environment. These need attention and represent another target for the integrity management system. If these become totally corroded, it will be a crisis and can lead to plant shutdown if there is no standby for this tower, which is the usual case.

Fig. 4.23 Anchor bolt for steel tank.

Fig. 4.24 Start of corrosion of anchor bolt for a steel tower.

4.8 Assessment of offshore structure

The criteria for assessment of platforms was still under development at the time of preparation of this book. Only the draft of API SIM was available, which focuses on the assessment of existing structures. The acceptance criteria determine whether a structural assessment has confirmed the platform fit for continued operation in its present condition.

The ultimate structural capacity of an offshore platform may be characterized either in terms of the platform reserve strength ratio (RSR) or in terms of the probability of an increase in environmental load due to wind, wave, or current causing platform collapse. The RSR is defined as the ratio of the environmental load set causing collapse to the environmental load set with a 100-year probability of exceedance. Therefore a platform with an RSR of 1.0 will have a deterministic probability of failure in the region of 1 in 100 (10^{-2}).

Note that designs of offshore platforms under 100-year actions from wave and wind based on APIRP2A will have an ultimate capacity, or design capacity, about 1.8 to 2.5 times the design load. This design capacity represents the explicit factors of safety in the design code as well as the implicit margins due to different factors such as mean steel strength, conservatism in foundation capacity assumptions, and presence of additional steel as a result of temporary load conditions in the case of construction, transportation, and installation, and the increased capacity associated with structure analysis in a three-dimensional space frame structure.

Fig. 4.25 shows a comparison of two geographical regions, region (1) and region (2). The plot origin represents the design point for both regions (1) and (2). In both cases the design capacity of the structures is assumed to be in the range of 1.8 to 2.5 times the 100-year (storm condition) loads indicated by the shaded region of the graph. This reserve, mentioned earlier for 100-year design, is referred to as the RSR. The two curves shown on the

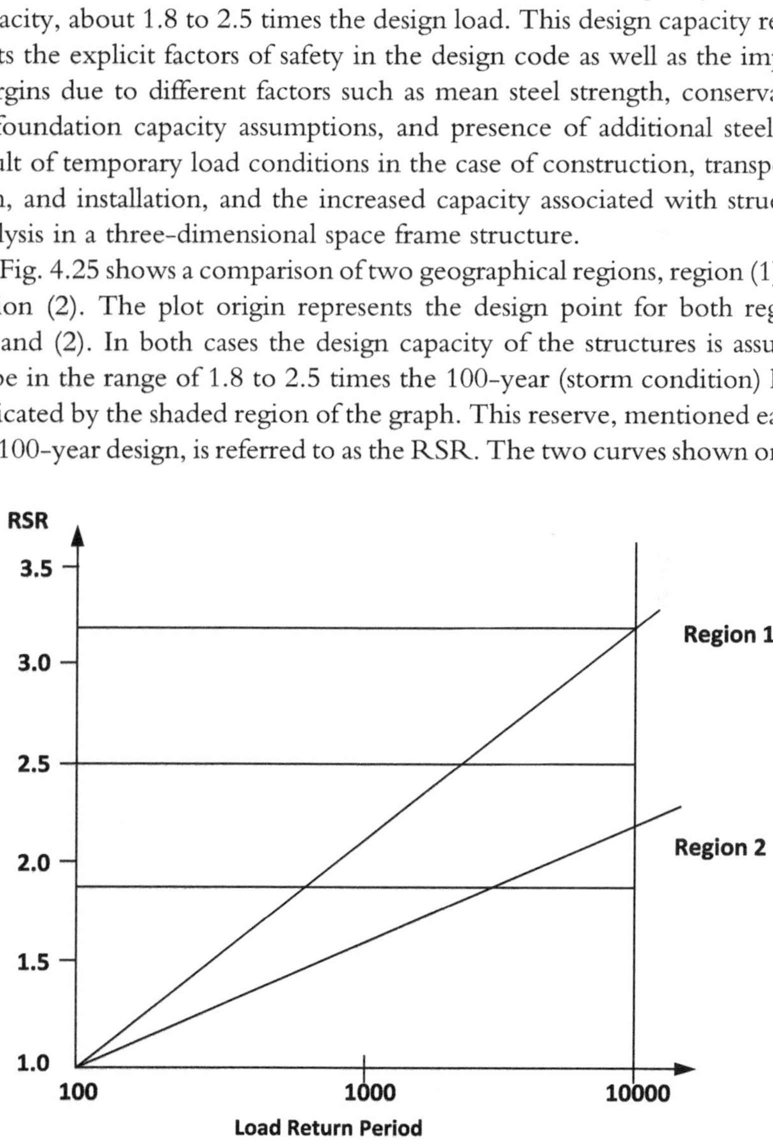

Fig. 4.25 Comparison of regional hazard curves.

figure represent the relationship between the load associated with the 100-year environmental event and the load from n-year events.

These curves are referred to in API SIM as region hazard curves. This figure can represent that a typical platform structure in region (1) with a shallower hazard curve would be expected to have sufficient reserve strength to survive the 10,000-year regional environmental event, an RSR of approximately 2.3 being required. On the other hand, a typical platform structure in region (2) with a steeper hazard curve would not be expected to survive the 10,000-year environmental event, since it would require an RSR of approximately 3.4 to do so, which is greater than that explicit in API RP2A.

This example shows that, in regions that have different hazard curves, the same RSR will be associated with significantly different structural reliability or probabilities of failure. Based on this, it is clear that every region in the world has its own environmental hazard curve, so the acceptance criteria will be different from one region to another to obtain the optimum reliability against its environmental load.

The probability of structural failure of the range of offshore platforms that exist in the region in question should be understood very well. This can be accomplished by performing a structural reliability analysis of the platforms selected as the fleet representatives. Sensitivity studies may be required to establish whether certain characteristics are significant or not.

To account for uncertainties in the derivation of both the loading and response, the probability of failure should be estimated using reliability analysis.

4.8.1 Non-linear structure analysis in ultimate strength design

A normal structural analysis procedure will be based on an idealized linear-elastic model to define internal forces at each structure member. After that, the member section size is selected based on design codes that achieve the required safety factor.

In the ultimate strength analysis approach, non-linearities associated with the plasticity and large deformations of components under extreme load are considered in the element modeling. The non-linear analysis software will track the interaction between components as member end restraints are modified and forces are redistributed as a result of local stiffness changes due to the member reaching plasticity. The sequence of non-linear events for many elements will lead at the end to a global collapse mechanism, and then the associated system capacity shall be defined.

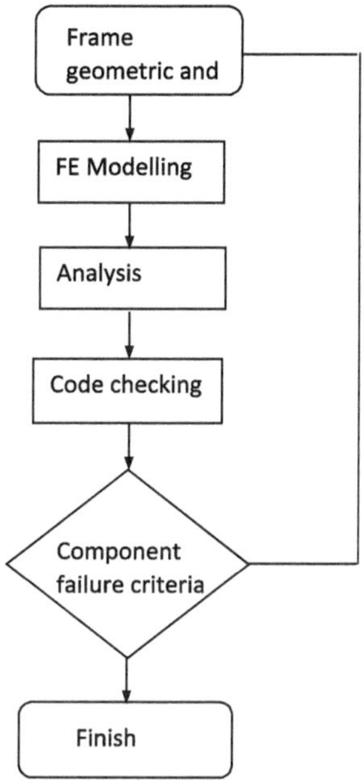

Fig. 4.26 Procedure for conventional analysis design.

Thus, while the typical linear design process checks for the adequacy of each individual component, the non-linear ultimate strength analysis models the performance of the system as a whole.

The comparison between the elastic structure analysis and non-linear structure analysis is shown in Figs. 4.26 and 4.27.

In general, non-linear analysis techniques have four basic types:

- General-purpose non-linear beam column models
- Plastic hinge beam column models
- Phenomenological models
- Shell finite element models

4.8.1.1 General purpose non-linear beam column models

Beam elements generally do not incorporate geometric non-linearity at the element level and multiple beam elements along a member are required to accurately model buckling responses. The equation of equilibrium is usually evaluated in the deformed condition, which is the largest displacement.

The stiffness is integrated numerically from the stress distribution at points across the section due to the combined action of axial forces and moments.

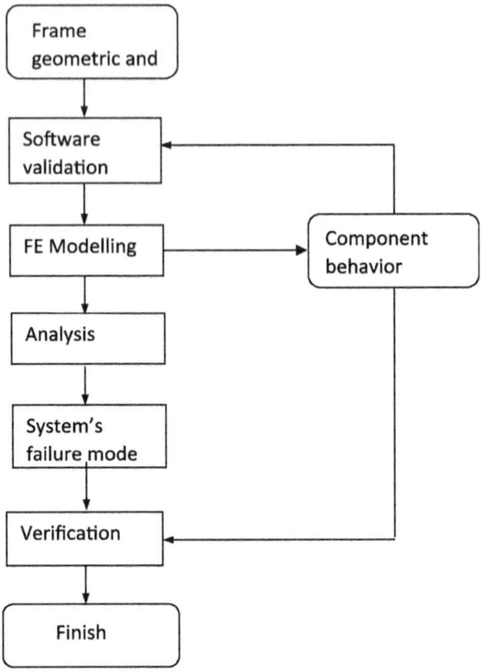

Fig. 4.27 Non-linear ultimate strength analysis.

If detailed stress-strain material modeling formulations, including strain hardening, are available, these may require definition in true rather than engineering format. Components may also be modeled by shell elements or with a combination of shell elements, as for joins damage, it will be modeled by using beam element to simulate the member to run the FE analaysis.

4.8.1.2 Plastic hinge beam column models

Certain techniques have been developed specifically for the ultimate strength analysis of frame structures. The elements have been derived to model beam column behavior and each member requires only a single element to model the buckling response.

The plasticity may be presented by modeling the propagation of yield through the section and along the element sections or by the formation of annualized hinges.

The plastic hinges are defined using the basic relationships of plasticity and yield surface.

Depending on the element formulation, hinges may form at the calculated locations of first fiber yield for a representative end and mid-point location.

Elastic plastic behavior analysis or gradual strain hardening may be accommodated related to the energy dissipated in the element. The specification can be adjusted to account for the imperfection or to calculate responses to characteristics rather than mean component strength given by test data.

4.8.1.3 Phenomenological models

In a phenomenological model, component responses are prescribed with force and deformation relationships that can be empirically related to element geometry or determined through analysis. A single element represents the member behavior.

In some software, the type of failure must be anticipated for each component and loading mode prior to analysis and the element type and its nonlinear characteristics defined accordingly, e.g.,

- Elastic members
- Buckling members
- Beam members
- Frame members
- Joints

Material non-linearity and initial imperfections are embodied in the phenomenological representation of test data. Section yielding may be determined from the full-section forces and a specified interaction surface.

4.8.1.4 Shell FE models

Any member or joint components can be modeled by using shell elements or with a combination of shell element as for joints, damage representation of members, and beam elements, even in the case of a greater modeling and analysis demand.

Steel usually yields before it fractures, because the uniaxial yield stress is significantly less than the stress required for a cleavage fracture. However, in three-dimensional stress fields the individual stress components can be much greater than the uniaxial yield stress. Without yield occurring in the vicinity of a crack tip, a simple elastic calculation of the stress distribution combined with the von Mises equation predicts that the stresses may reach 2.5 times the uniaxial yield stress before yield occurs.

Fracture is much more dependent on the largest stress and is less affected by the triaxial stress pattern. Therefore in the triaxial stress field the yield stress may exceed the fracture stress so that fracture becomes the failure mechanism. The effect is particularly important in the center of thick plates, owing to the through-thickness stresses generated in the plane strain conditions by the Poisson's ratio effect.

Brittle fracture is more likely to occur in cold conditions and at high rates of loading for examples under impact. The toughness grade of material specified for any application should be carefully selected to account for these factors as well as the likely stress levels, crack growth through fatigue, inspection procedures, and consequences of failure.

4.8.1.5 Modeling the element

The sliding action of piles within legs should be modeled with the approximate constraint conditions, which allow unrestrained differential axial displacement and rotations but couple the lateral displacements of piles and legs.

Grouted piles can be modeled as composite leg-pile members. On the other hand, leg and pile members can be modeled as separate elements with full coupling between end degrees of freedom.

For ungrouted piles, the lateral displacements for pile and leg are usually coupled at each horizontal elevation, but some differential movement can occur between elevations.

Conductors provide limited contributions to the strength of the steel frame system. In most cases, they can be modeled as pure load-attracting members; as long as their load contributions are correctly captured, the conductors can be omitted from the strength calculations. However, for structures with limited foundation resistance, conductors can contribute significantly to the foundation stiffness and collapse strength of the structure.

In that case, the conductor should also be modeled and analyzed as a structural element and included in the integrated structure soil model. The conductor guide framing at mudline may then be highly loaded and may need a more detailed inclusion in the structural model as primary framework.

4.8.1.6 Conductor connectivity

Conductor guides should be modeled in sufficient detail to transfer the required loads into the primary framework. Local over-stresses in conductor guides should not cause concern if they occur in areas where the model has been purposely simplified and the surrounding primary members have sufficient capacity.

If conductors are included as structural members, the sliding action of conductors within the guide frames should be modeled. Differential rotations should also be unrestrained.

The contract action between curved conductors and their guide frames due to the imposed conductor curvature and friction may require specific consideration.

4.8.2 Structural modeling

The structures must be modeled in sufficient detail to ensure that the non-linear analysis program adequately captures the relevant global and local failure modes and load redistribution.

The models for component strength, such as member compressive strength and joint strength, are semi-empirical. They have a theoretical basis, but are formulated to conform to experimental data. In general, all theoretical formulations need some calibration in order to represent the behavior of "real" structural components with sufficient accuracy.

Moreover, it should be possible for an engineer to select specific failure criteria and have the analysis tools calculate the structure's strength based on those criteria. In such a case, the requirement for the analysis tool is not to present theoretically correct solutions for the structure, but rather to present a consistent strength estimate based on the engineer's specifications.

This implies that the analysis tools should be able to represent different failure criteria, from the theoretical, "idea," solution to a characteristic lower-bound solution as specified by different codes of practice.

Some non-linear analysis programs have built-in features to calibrate component failure modes to specific criteria. For other programs, the engineer must give special consideration during the modeling of the structure in order to make the program represent the required failure modes or limiting criteria. Which consideration to take depends on the component as member, joint, or foundation and on the mathematical formulation. If there is any doubt about the FE formulation, a simple model should be subjected to a well-defined load and deformation path. This will allow the results to be judged and calibrated against engineering practice.

Instead of modeling the structure out of purely geometric considerations, the modeling must consider the analysis tool that will be used for the non-linear analysis, and the mathematical formulation that is embedded within the program.

This discussion gives a set of modeling recommendations to help make the non-linear analysis tool produce reliable results that conform to recognized failure criteria and design formulations.

Load modeling is treated in Chapter 6, while Chapter 7 discusses the actual analysis execution.

Frame modeling of the space frame model should describe the three-dimensional geometry of the platform.

The model for ultimate strength assessment can usually be significantly simpler than models required for design and fatigue analysis. The primary framework essential in maintaining overall integrity of the structure for the in-place condition must be included in the structure model. Secondary structures and members generating dead loads and/or environmental loading need only be represented in sufficient detail to introduce the relevant loads on the primary structure.

The analytical models should consist primarily of beam elements. The structural members of the framework may be modeled using one or more beam elements for each span between the nodes of the model of the framework.

The primary framework of the structure comprises those members that provide the stiffness and strength of the structures. These are usually the legs, the piles, the vertical diagonal members, and the main plan frame bracing members.

4.8.2.1 Secondary framework

The secondary framework consists of members that don't contribute to global stiffness and strength of the framework; in general their structural contribution may be neglected and they do not need to be included in the model as structural members. Boat-landing/fenders, spider-deck, walkways, and others are examples of secondary members. The secondary framework should be modeled in sufficient detail to transfer the required loads into the primary framework.

Some local over-stresses in the secondary framework should be accepted if they occur in areas where the model has been purposely simplified and where the adjacent primary members show sufficient capacity. This should, however, be subject to separate justification in each case.

The following secondary framework should be included in the model:

(i) Members or joints that are essential for transfer of reaction loads from conductors and appurtenances, etc., to the main structural elements.

(ii) Members or joints that are highly loaded by local wave action; a separate assessment may be done on the local behavior. The global assessment may be performed with a simplified model if it is demonstrated that the load can be carried and transferred by the secondary framework.

Secondary members associated with launch framing, mud-mats, conductor support during transportation, and others should be included in the model if they share in the system capacity with primary members.

When neglecting the structural contribution of secondary members, their load-attracting properties, that is, loading due to self-weight or hydrodynamic loading, should still be accounted for and included in the appropriate loading condition.

Conductors and other structure members such as launch runners, mud-mats, J-tubes, risers, skirt pile guides, and other should be included in the model if they contribute significantly to the overall strength of the structure or foundation. Otherwise, they may be disregarded as structural elements.

4.8.2.2 Dented beam and cracked joint

For a dented member, which usually occurs due to a dropped object on the bracing, the strength of the member will be reduced by the following formula in the case of normal strength and bending strength:

$$\left(\frac{N_U}{N_P}\right) = \exp\left(-0.08\frac{D_d}{t}\right) \tag{4.34}$$

$$\left(\frac{M_U}{M_P}\right) = \exp\left(-0.06\frac{D_d}{t}\right) \tag{4.35}$$

where D_d is the denting depth and t is the wall thickness.

The damaged member should be modeled in sufficient detail to assess the impact of the damage on the global behavior of the structure. A lower bound on the remaining structure's strength can usually be obtained by removing the affected member(s) from the model. A less conservative strength estimate is obtained by modeling the damage in the non-linear analysis. Some non-linear analysis tools include special formulations to model dented or distorted members. Alternatively, the damage can be modeled explicitly by shell elements, or a reduced cross-sectional area can be specified in the damaged zone.

The same philosophy can be applied for a cracked joint by eliminating the member from the model. A less conservative approach can be taken by reducing the strength of the affected joint by some fraction. Note that the presence of a crack will also limit the ductility and the cyclic capacity of the cracked joint. These properties should be evaluated by refined analysis that indicates that a cracked joint is loaded up to its static capacity.

Due to the crack, there will be a reduction factor, F_{JR}:

$$F_{JR} = \left(1 - \frac{A_c}{A}\right)\left(\frac{1}{Q_\beta}\right)^{mq} \tag{4.36}$$

where A_C is the cracked area and A is the weld length multiplied by the thickness, T. Q_β allows for the increased strength observed at b values 0.6 where β=brace diameter/chord diameter. Q_β is known as the geometrical modifier, usually used in design codes to account for the increasing capacity of uncracked tubular joints at high $_\beta$:

$$Q_\beta = 1 \text{ for } \beta \leq 0.6 \tag{4.37}$$

$$Q_\beta = 0.3/\beta(1 - 0.833\beta) \text{ for } \beta > 0.6 \tag{4.38}$$

Mq is the power allocated to Q_β and depends on the approach used to estimate the capacity of the uncracked joint:

For tubular joints containing part-thickness flaws, m_q=0.

For tubular joints containing through-thickness flaws, validated correction factors giving lower-bound estimates of the collapse load are at present limited to joints with b ratios less than 0.8 and the following configurations:

- K-joints with a through thickness crack at the crown subjected to balanced axial loading.
- axially loaded T and DT joints with a through thickness crack at the saddle.

4.8.3 Determine probability of structural failure

It is necessary to understand the probability of structural failure in the range of offshore platforms that exist in the region in question. This can be done by selecting a number of platforms that represent the fleet and then applying the structural reliability analysis on these platforms. As discussed previously, sensitivity analysis may be required to identify whether certain parameters are significant or not and their weight.

The platform probability of failure shall be calculated by using reliability analysis to consider the uncertainties and variabilities for the load and strength. The analysis procedure is outlined in the flowchart in Fig. 4.28.

4.8.4 Establish acceptance criteria

In order to establish the assessment acceptance criteria, it is necessary to establish the relationship between the probability of failure of the representative structures and their reserve strength, defined in terms of RSR.

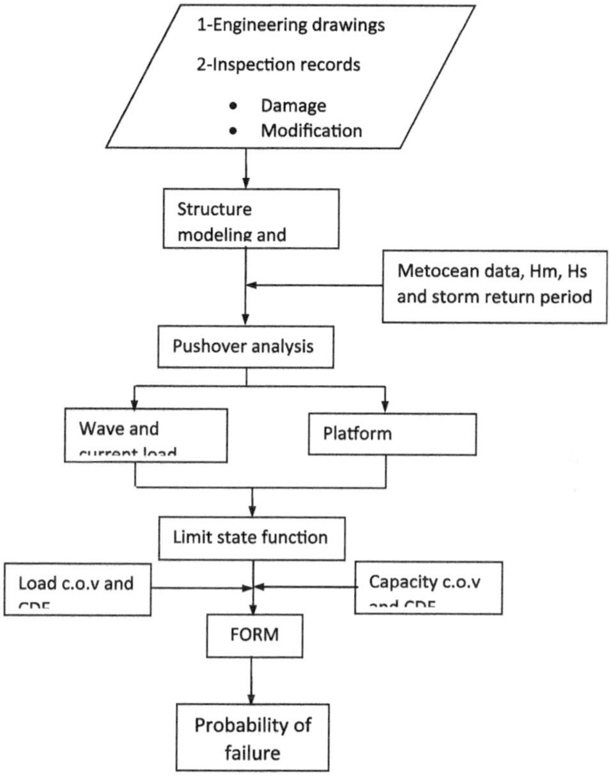

Fig. 4.28 Probability of failure calculation procedure.

In the reliability analysis procedure described earlier, the structural capacity of the platform, R, is assumed to be a random variable with a log-normal distribution. The mean value of R is estimated from an ultimate strength (or pushover) analysis. By arbitrarily moving the mean structural capacity to represent a range around the calculated capacity, it is possible to estimate the probability of failure of a range of designs of the same structural type. This allows the relationship between RSR and POF to be approximated and plotted. It should be noted that the approach is an approximation and assumes the load on the structure remains constant. In fact, it may be argued that increasing RSR would perhaps also increase load; assuming larger members and therefore increased fluid drag loading—and conversely reducing the RSR—may reduce the load. If this effect was felt to be significant for the structures in question the effect on the load could also be estimated and included by adjustment of the estimated mean load.

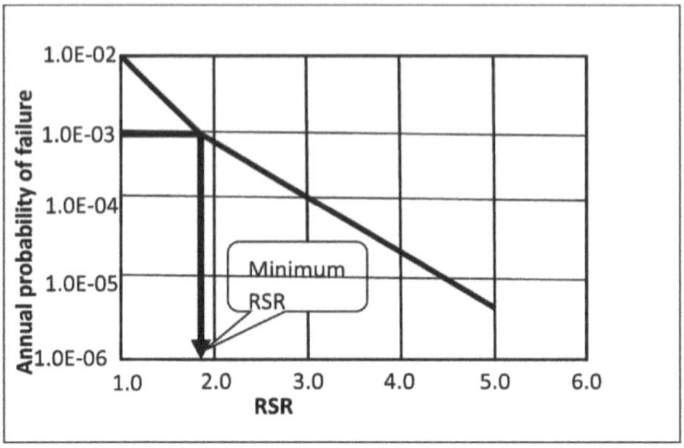

Fig. 4.29 Relation between RSR and the probability of failure.

Fig. 4.29 illustrates, for example, a minimum RSR corresponding to a probability of failure of 1×10^{-3}, representative of potential assessment criteria for this (fictitious) region for medium failure consequence.

4.8.4.1 Topside assessment

A study was done by MSL (2000) to address the topside and jacket inspection results. The database includes data from the MMS, CAIRS, and multiple Gulf of Mexico platform operators. Details of over 3000 underwater inspections have been catalogued and almost 5000 anomalies recorded. An assessment of the underwater inspection has been carried out and separately reported. If there is no integrity management system, the situation shown in Fig. 4.30A and B can be reached.

The data relevant to topsides inspection was extracted and was carefully screened, manipulated, and reorganized into a more useful form for assessing the reported incidents. The original data were filtered and broken down into both anomaly type and structural component. During this process it was found necessary to divide existing categories further, such as corrosion into coating and corrosion. This distinction is useful since the coating of a component is the first to deteriorate and should be reported as a separate incident from corrosion (metal loss), which takes place after the coating is depleted. After reviewing the revised data, an apparent variance was established from the original data.

You can find from the table that 25% of the anomalies findings are due to handrails and the major reasons are corrosion or missing parts.

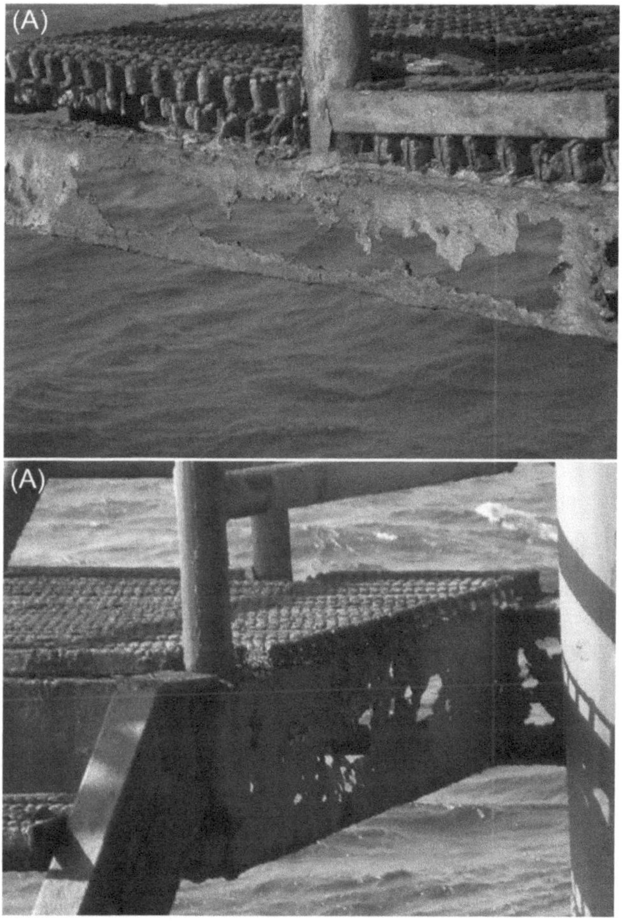

Fig. 4.30 (A) Totally corroded beam. (B) Corrosion of stair beam.

In general, for all the equipment on the topside, this study done by Health and Safety Executive (2001) is very useful and presents the results seen in Table 4.10.

HSE reviewed the main causes of leaks and it was found that the main reasons for failures were due to piping failures, valve loss of containment, flange or joint leaks, and leaks of instrument tapings.

One can find that pumps, compressors, vessels, tanks, and heat exchangers were much less likely to have a leak and the reason was mainly corrosion, erosion, pinhole, or related incidents. The incidents causing a leak are around 72% in piping compared with 4% in vessels and tanks. In general, piping failure is the major issue, as shown in Table 4.11.

Table 4.10 Analysis for anomalies in GoM.

Component	Corrosion	Coating	Dents/ bows	Impact	Separation or missing	Miscellaneous	Leak	Total	Total %
Boat landings	73	47	12	29	43	10		214	12.9
Conductor	1				1			2	0.12
Cranes	3				3	1		7	0.42
Grating	90	19	21	4	39	7		180	10.85
Handrails	183		53	2	130	54		422	25.44
Helidecks	56	2	27	5	24	5		119	7.17
Other	7				1	4		12	0.72
Piping	6	3					3	12	0.72
Platform CP	18					8		26	1.57
Risers cp	16	1			1	24		42	2.53%
Stairs	60	8	8	1	20	3		100	6.03
Structures	77	36	36	1	57	10		217	13.08
Swing ropes	32				68	9		109	6.57
Total	656	119	167	42	390	146	3	1659	100
Total %	39.54	7.17	10.07	2.53	23.51	8.8	0.18	100	

Table 4.11 Different facilities failure and its percentage of occurrence.

Mode of failure	Percentage of occurrence	Main reason
Pipes failure	23%	Corrosion or erosion
Valve leak	20%	Leaking gasket or O-ring
Flange or joint leak or failure	19%	Leaking gasket or O-ring
Instrument taping and fitting	18%	No specific reason
Pumps, compressor of fans	11%	Seal failure
Vessels and tanks	3%	Leaking gasket or O-ring
Heat exchangers	4%	Leaking gasket or O-ring
Fired heaters	2%	Corrosion or erosion

4.9 Assessment for FPSO

If it is required to increase the load or add any facilities on the vessel deck, it is normal to do an inspection and a structural analysis of the deck and to provide strengthening to withstand the extra load, in addition to coordinating with the vessel class to get their approval as per the normal procedure. The main challenge is that it is required to do an extension of the lifetime of the vessel; this approach was studied by Moan and Ayala-Uraga (2003).

The main methodology is called reliability-based service life extension (RBSLE). The target of this methodology is to do an evaluation and assessment of the effects of many parameters, such as the design criteria, inspection procedure, and the repairs, and include the uncertainties and variabilities associated with different variables that affect the FPSO reliability.

The reliability assessment should consider corrosion and fatigue effects and this can be done by collecting the data of previous and current environmental conditions and all the data on inspections, maintenance, and repairs from the history of the FPSO.

The next step is to do screening by applying a simple approach, and perform a detailed RBSLE and analysis, and then prepare a plan, based on the collected data and analysis, by presenting the necessary actions required to guarantee the vessel safety during the extended service life.

Studies carried out by Yamamoto and Yao (2001) and Guedes Soares and Garbatov (1998, 1999) focused on the reduction in the wall thickness due to corrosion in the ultimate strength ship hull. In these studies it was found that the section of stiffeners decreased due to crack growth and thickness reduction, which is a function of the rate of corrosion for every element.

A study was carried out by the Tanker Structure Cooperative Forum (TSCF) in 1997 that collected the data from 32 tankers to estimate corrosion rates for different structural components and areas of the tankers. These data were organized into five tank categories with four groups of structural components for every tank category.

Based on the TSCF study, Table 4.12 presents the corrosion rate for different FPSO longitudinal structure elements for water ballast tanks.

Typically, corrosion rate data is lacking. In the study the corrosion rate was considered a random variable because of its uncertainty.

Research performed by Guedes Soares and Garbatov (1996) assumed a corrosion rate represented by a probability normally distributed with a coefficient of variation (CoV) of 0.1 for all the members. On the other hand, a study by Akpan et al. (2002) used the mean values recommended in the TSCF and assumed a Weibull probability distribution with CoV varying from 0.1 to 0.5.

For corrosion rate values, Paik et al. (1998) performed a study for determining corrosion rate wastages in bulk carriers with information from 44 existing marine vessels. It was found that the corrosion rates data fitted a Weibull distribution for 16 different types of structural members, with mean values varying from 0.01 to 0.12 mm/yr and a very large scatter, with CoV varying from 0.8 to 2.0.

Melchers (2000), based on testing results of corrosion wastage under (immersed and splash zone) marine conditions, suggested that the coefficient of variation of the corrosion rate increases with time exposure, as presented in Fig. 4.31.

In 2003 Wang et al. studied an extensive database for thickness measurements of 140 tankers; as a result of this study it was found that the corrosion rates have mean values in the range from 0.02 to 0.06 mm/yr, but with a very

Table 4.12 Corrosion rates for segregated ballast tank for some longitudinal structure element, from TSCF.

Structural component	Side	General corrosion (mm/yr)	Pitting/Groove
Deck plating	1	0. 1–0.5	0.05–2.0
Deck longitudinal	2	0.25–1.0	
Side shell longitudinal	2	0.20–1.0	
Side shell plating	1	0.06–0.10	
Bottom shell plating	1	0.05–0.25	0.5–1.5
Longitudinal bulkhead plating	1	0.05–0.30	1.0–3.0

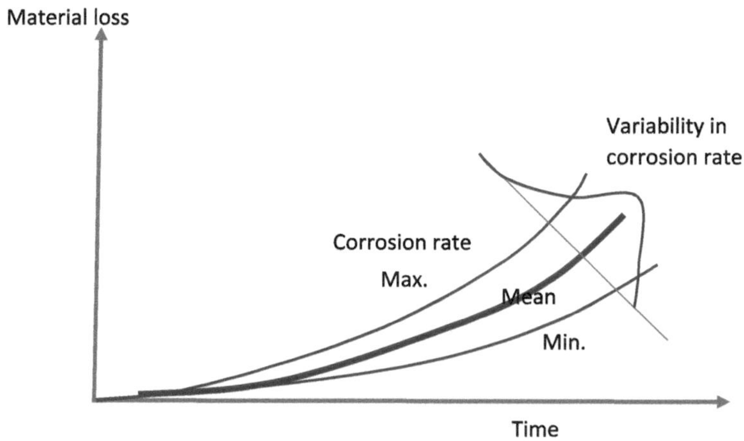

Fig. 4.31 Variable corrosion rate with time as per Melchers (2000).

large variation. It was also observed that the maximum corrosion rates was 0.55 mm/yr for bottom plating and deck longitudinal.

From these studies, it was concluded that the corrosion rates are affected by the geographical location of the marine vessel as there is a difference in the corrosion rate for vessels located in the North Sea, as compared to vessels located in West Africa or the Gulf of Mexico.

Based on the Moan and Ayala-Uraga (2003) study, in principle, a component that is initially designed for a fatigue life of 20 years with an allowable damage will exhaust its service life at the end of 20 years in the case of a harsh environment. By applying reliability methods, there is a chance to extend its use beyond its initial service life if the vessel is moved after 10 years to a more benign environment.

In this case the remaining fatigue life under these new conditions can extend its service life to about 30 years, if an equivalent design value is assumed for the latter conditions. On the other hand, if the structure initially spends 10 years in a benign climate and is moved to harsh conditions, the remaining fatigue life would then be 17 years in the harsh environment.

Guedes Soares and Garbatov (1996) performed a fatigue reliability assessment for converted FPSO. This study revealed that the fatigue damage accumulated during the life spent as a tanker dominated in the assessment of the residual fatigue life as an FPSO.

In their study, it was found that for an FPSO operating in benign environments the main governing criteria concerning fatigue strength is the damage that the structure had before conversion.

References

Akpan, U.O., Koko, T.S., Ayyub, B., Dumbar, T.E., 2002. Risk assessment of aging ship hull structures in the presence of corrosion and fatigue. Mar. Struct. 15, 211–231.
Allen, D.E., 1970. Probabilistic study of reinforced concrete in bending 67, 989–993.
Andreani, J., Carr, N., 2007. Final Report on the Study of out-of-Plane Tank Settlement. Report to API SCAST, May,.
Baykof, F., Sigalof, Y., 1984. Reinforced Concrete Structure. Mier, Mosco, Russia.
Bloem, D.E., 1968. Concrete strength in structures. ACI J. 65, 176–187.
Cady, P.D., Weyers, R.E., 1984. Deterioration rates of concrete bridge decks. J. Trans. Eng. ASCE 110 (1), 34–44.
El-Reedy, M.A., Ahmed, M.A., Khalil, B.A., 2002. Reliability analysis of reinforced concrete column under corrosion attack. In: Proc. of ACI 5th International Conference, Cancun, December, pp. 77–91.
Frangopol, D.M., Lin, K.Y., Estes, A.C., 1997. Reliability of reinforced concrete girders under corrosion attack. J. Struct. Div. ASCE 123 (3), 286–297.
Guedes Soares, C., Garbatov, Y., 1996. Reliability of maintained ship hulls subjected to corrosion. J. Sh. Res. 40, 235–243.
Guedes Soares, C., Garbatov, Y., 1996. Fatigue reliability of the ship hull girder accounting for inspection and repair. Reliab. Eng. Syst. Saf. 51, 341–351.
Guedes Soares, C., Garbatov, Y., 1998. Reliability of maintained ship hull girders subjected to corrosion and fatigue. Struct. Saf. 20, 201–219.
Guedes Soares, C., Garbatov, Y., 1999. Reliability of maintained ship hulls subjected to corrosion and fatigue under combined loading. J. Construct. Steel Res. 52, 93–115.
Health & Safety Executive, 2001. Corrosion Risk Assessment and Safety Management for Offshore Processing Facilities. Offshore Technology Report OTO 1999/064,.
Kilareski, W.P., 1980. Corrosion induced deterioration of reinforced concrete – an overview. Mater. Perf. Nat. Assn. Corrosion Eng. 19 (3), 48–50.
Lin, K.-Y., 1995. Reliability-Based Minimum Life Cycle Cost Design Of Reinforced Concrete Girder Bridges (Ph.D. thesis). Dept. In: of Civ., Envir., and Arch. Engrg. Univ. of Colorado, Boulder, CO.
MacGregor, J.G., 1983. Load and resistance factors for concrete design. ACI J. 80 (4), 279–287.
Marr, W.A., Ramos, J.A., Lambe, T.W., 1982. Criteria for settlement of tanks. J. Geotechn. Eng. Div., Proc. Am. Soc. Civil Eng., ASCE 108 (GT8), 1017–1039.
Melchers, R.E., 2000. Probabilistic models of corrosion for reliability assessment and maintenance planning. In: Proceedings of 20th International Conference on Offshore Mechanics and Arctic Engineering, Rio de Janeiro, ASME. paper S&R-2108.
Moan, T., Ayala-Uraga, E., 2003. Reliability-based assessment of FPSOs for service life extension. Technical Report OTD 2003-01. Houston.
Mori, Y., Ellingwood, B.R., 1994a. Maintaining reliability of concrete structures I: role of inspection/repair. J. Struct. Eng. ASCE 120 (3), 824–845.
Mori, Y., Ellingwood, B.R., 1994b. Maintaining reliability of concrete structures II: role of inspection/repair. J. Struct. Eng. ASCE 120 (3), 846–862.
MSL Services Corporation, 2000. Joint Industry Project, "rationalization and optimization of underwater inspection planning consistent with API RP2A section 14". Final Report, Document Reference CH104R006 Rev 0, November,.
Nevil, M.A., 1983. Properties of Concrete. Pitman, London.
Paik, J.K., Kim, S.K., Lee, S.K., 1998. Probabilistic corrosion rate estimation model for longitudinal strength members of bulk carriers. Ocean Eng. 25 (10), 837–860.
Takewaka, K., Matsumoto, S., 1980. Quality and cover thickness of concrete based on the estimation of chloride penetration in marine environments. In: Malhotra, V.M. (Ed.),

Concrete in Marine Environment. American Concrete Institute (ACI), Detroit, MI, pp. 381–400. SP-109.
Thoft-Christensen, P., 1995. Advanced bridge management systems. Struct. Eng. Rev. 7 (3), 149–266.
Ting, S.C., 1989. The Effects of Corrosion on The Reliability of Concrete Bridge Girder. Ph.D. thesis, Dept. of Civ. Engrg., Univ. of Michigan, Ann Arbor, MI.
Washa, G.W., Wendt, K.F., 1975. Fifty-year properties of concrete. ACI J. Proc. 72 (1), 20–28.
West, R.E., Hime, W.G., 1985. Chloride profiles in salty concrete. Mater. Perf. Nat. Assn. Corrosion Eng. Houston Tex 24 (7), 29–36.
Yamamoto, N., Yao, T., 2001. Hull girder strength of a tanker under longitudinal bending considering strength diminution due to corrosion. In: Corotis, et al. (Eds.), Structural Safety and Reliability. Swets & Zeitlinger.

Further reading

Allen Marr, W., Ramos, J.A., Lambe, T.W., 1982. Criteria for Settlement of Tanks. Proc. Am. Soc. Civ. Eng. 108.
American Association of State Highway and Transportation Officials, 1992. Standard Specifications for Highway Bridges, fifteenth Ed. American Association of State Highway and Transportation Officials (AASHTO), Washington, DC.
American Association of State Highway and Transportation Officials (AASHTO), 2008. Bridging the Gap. http://www.infrastructurereportcard.org/fact-sheet/bridges.
American Bureau of Shipping, 2003. 2 American Bureau of Shipping, Houston TX, USA.
Broomfield, J.P., 1997. Corrosion of Steel in Concrete. E&FN Spon, London.
El-Sayed, H.A., Kamal, M.M., El-Ebiary, S.N., Shahin, H., 1987. Effects of reinforcement corrosion and protective coatings on the strength of the concrete/steel bond. Corros. Prevent. Control 34 (1), 18–23.
George, W.W., Jesse, C.S., Steven, M.C., 1989. Fifty-year properties of concrete made in 1937. ACI Mater. J. 86 (4), 367–371.
Kilareski, W.P., 1980b. Corrosion induced deterioration of reinforced concrete – an overview. Mater. Perf. Nat. Assn. Corrosion Eng. Houston Tex 19 (3), 48–50.
Morinaga, S., 1988. Prediction of service lives of reinforced concrete buildings based on rate of corrosion of reinforcing steel, special report of institute of technology. Shimizu Corporation, No. 23, Tokyo, Japan.
Schiessl, P., 1997. "Corrosion of Steel in Concrete", Report of the Technical Committee 60-C SC. Chapman and Hall, London.
TSCF, 1997. Guidance Manual for Tankers Structures. Tanker Structure co-Operative Forum & IACS. Whitherby Publishers, London, UK.
Wang, G., Spencer, J., Sun, H., 2003. Assessment of corrosion risks to aging ships using an experience database. In: Proc of Offshore Mechanics and Arctic Eng; Cancun, Mex. OMAE. paper 37299.

CHAPTER 5

Repair methods for offshore and onshore structures

5.1 Offshore structures repair

After performing the assessment of an offshore structure, the next step is to do the repair or strengthening of the structure.

5.1.1 Jacket repair

There are many methods for strengthening and repairing an offshore structure platform. The main element of repair is the clamp, as shown in Fig. 5.1, which depicts the repair process. The shape of the traditional clamp that connects to the leg of the platform can be seen in the figure.

New K-bracing between the two levels.

In the case of severe corrosion in the platform leg, the repair procedure is as presented in Fig. 5.1. To reconnect the leg in step 2, the bracing must be removed; also the half leg is fixed by installing the clamp that is connected to the bracing.

In case of damage to the bracing, or if there are deformations on the bracing due to buckling or other causes, the typical repair will be as shown in Fig. 5.2. As shown, clamps are used to connect the deformed bracing in the middle of the bracing to the leg.

Repair of a corroded horizontal member is as shown in Fig. 5.3. The main principle of the strengthening of this member is to add a new horizontal member, which is connected to the corroded member by clamps.

The clamps can also be used to strengthen the jacket face, as shown in Fig. 5.4.

These clamps must be properly designed for their task. To be clear, these clamps are not an "easy fix." Fig. 5.5 is a close-up photo of a clamp after fabrication.

5.1.2 Dry welding

Welding is often regarded as the best strengthening, modification, and repair ~~(SMR)~~ technique and no doubt would be used even more often if it did not

Asset Integrity Management for Offshore and Onshore Structures
https://doi.org/10.1016/B978-0-12-824540-8.00005-3

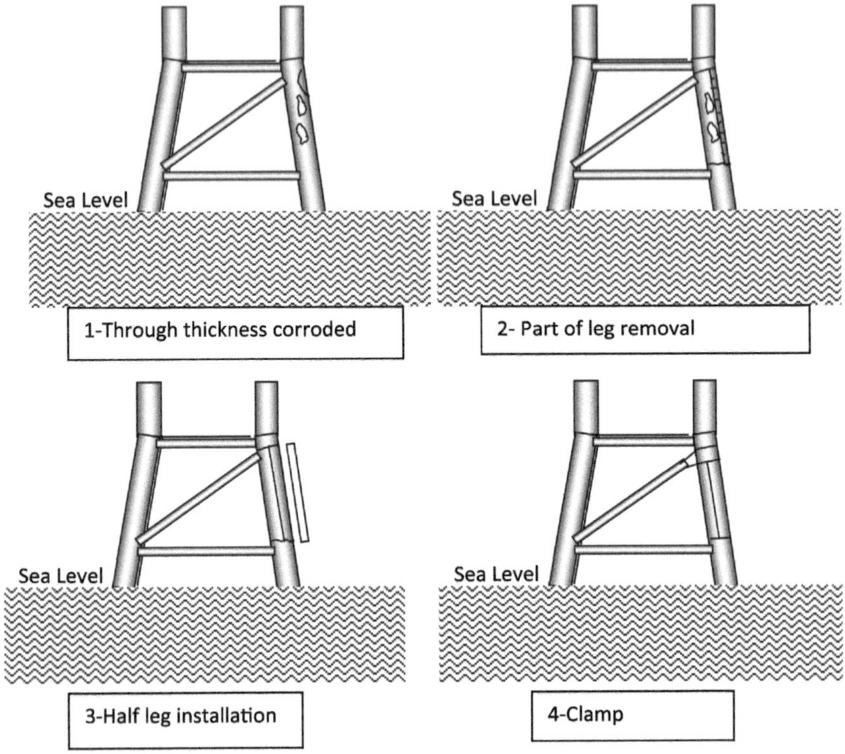

Fig. 5.1 Repair procedure.

present such operational difficulties in its execution. There are several welding techniques and a number of welding processes that can be considered:

- Dry welding topsides;
- Dry welding at or below the sea surface at 1 atm using cofferdam or pressure-resisting chambers. All normal welding processes can be used, but gas tungsten arc welding (GTAW), shielded metal arc welding (SMAW), and, to a lesser extent, flux cored arc welding (FCAW) are the main methods used in practice;
- Hyperbaric welding using habitats. The main processes used are GTAW and SMAW, although FCAW and GMAW are sometimes employed.

Repairs using both cofferdam and hyperbaric habitat welding techniques have proven track records. Since 1970, particularly in the North Sea, hyperbaric welding has been used as an underwater SMR technique.

Most of the time required for an underwater welding repair is taken up in the preparatory work rather than the welding process, and this must be planned with considerable care. Especially around the more complex node

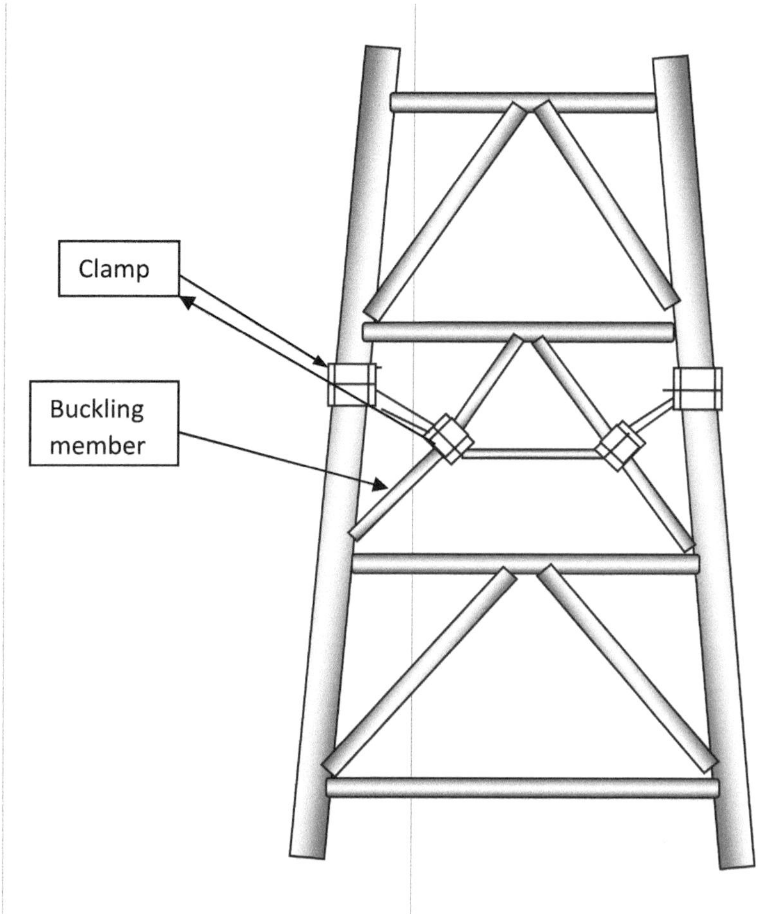

Fig. 5.2 Bracing repair.

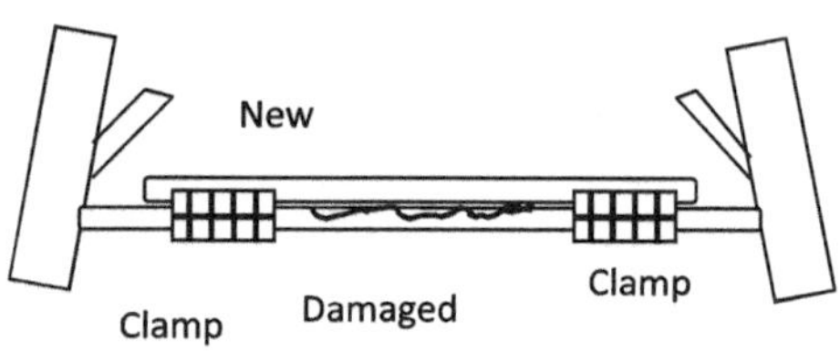

Fig. 5.3 Corroded member repair.

geometries, assembling and sealing the welding chamber can take as long as the welding operation. Some optimizations are needed such as separating the welding chamber into a few pieces to minimize the subsea assembly work, but large chamber components increase the divers' risk, specifically if the work is in tidal or splash zone conditions. In most cases, the as-is condition of the

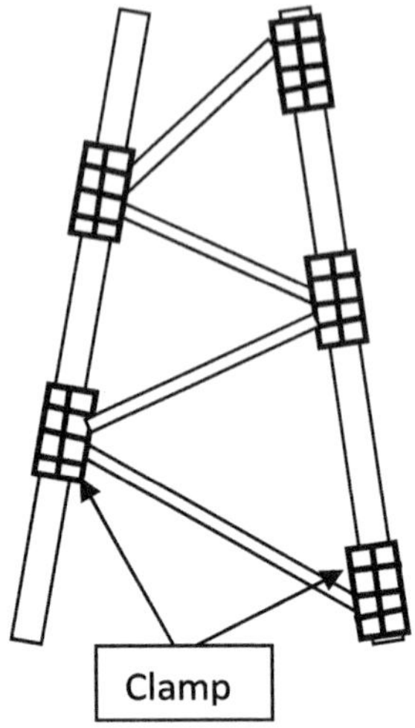

Fig. 5.4 Repair by clamps.

Fig. 5.5 Offshore clamp.

structure geometry is not exactly the same as the design, so it is essential to do a survey of the location before building the welding chamber.

One of the first and most important tasks is to clean portions of the structure to guarantee that an effective seal can be achieved, allowing the chamber to be dewatered. The welding chamber must be made sufficiently large to enable the welders to have effective access to the weld site.

5.1.3 Dry welding on topsides

Dry welding is a routine method used for the repair of the topsides structure above water level. The area where dry welding is undertaken is normally designated as a temporary hazardous area and all routine safety must be followed. Hot works are a consideration when welding topsides. Platform shut-in may be required, depending on the location of the welding relative to the well bay and equipment/piping.

The most widely used form is SMR. The only limitation is the requirement for hot works.

5.1.4 Dry welding at or below sea surface

Since a large body of welding technology exists relating to normal atmospheric pressure, a logical approach to underwater welding repair is to duplicate surface welding conditions by providing a 1-atm environment at the repair site. This method is limited to shallow water depths. Two methods are available to achieve this:

- Cofferdam: This essentially is a watertight structure surrounding the repair location and is open to the atmosphere. The structure can be open topped, or have a closed top with an access shaft to the surface.
- Pressure resistant chamber: The worksite is surrounded by a chamber constructed as a pressure vessel, capable of withstanding the water pressure at the depth of the repair location. Once the chamber is in place and sealed to the structure, it is dewatered and the pressure can then be reduced to 1 atm. The repair crew can transfer to the welding chamber in a 1-atm environment, within a diving bell, to perform the repair.

Dry welding is also possible at pressure below the sea surface using a hyperbaric habitat or chamber. The chamber is filled with gas equal to the hydrostatic head at the weld depth. The following factors govern the selection of habitat:

- The extent of welding required
- The repair site geometry complexity
- Depth of repair
- Welding process and ancillary equipment
- Environmental conditions

Given that conditions within the cofferdam or welding chamber duplicate those on the surface, any normal welding process could be used. In practice, GTAW and SMAW predominate, with minor usage of FCAW.

The primary limitation for atmospheric welding below sea surface is depth. Differential pressure at depth precludes the use of cofferdams or pressure-resistant chambers due to size and related cost. Thus the main advantage is that a good quality weld is guaranteed, but there are water depth limitations and it is potentially costly.

5.1.5 Hyperbaric welding

Hyperbaric welding is the most widely used dry weld repair technique for primary structures and pipelines. The repair site is again enclosed within a working habitat, which is dewatered by filling the habitat with gas. Since the gas and water will be at equal pressure at a point close to the bottom of the chamber, the maximum differential pressure will be at the top of the chamber, and obviously depends on the height of the chamber. This differential pressure (normally a few tenths of a bar (tens of kilopascals)) is easily resisted by lightweight habitats and simple flexible seals, making deployment and sealing of the work chamber operationally feasible. Hyperbaric welding is only limited in depth by access.

A typical hyperbaric welding operation will require the following items of equipment:

- Purpose-built cofferdam or pressure-resistant chamber
- Diving support
- Environmental control equipment
- Pre- and postweld heating equipment
- Welding equipment (often GTAW and one other)
- Weld inspection equipment
- Equipment to remove marine growth and grit blast
- Temporary holding clamps to take weight of additional members and maintain root gaps, as needed
- Crane capacity

The problem with hyperbaric techniques is that the environmental pressure at which the weld is carried out is essentially that of the worksite. These elevated pressures affect the gas/slag/metal reactions for all welding processes, and the high-density gas enhances the rate of heat loss from the weld. Hyperbaric welding research is mainly concerned with ensuring that for any specific environmental pressure and composition, welding parameters can be specified that will ensure the production of welded joints with properties acceptable to the certification authorities responsible for the structure on which the weld is being made. Given that the welding process has to be specially optimized for hyperbaric conditions, the number of techniques used has been limited. The great majority of the welding is carried out using gas tungsten arc welding (GTAW) and shielded metal arc welding (SMAW) techniques, with small amounts of flux cored arc welding (FCAW) and gas metal arc welding (GMAW).

A variety of habitats have been used, dependent on such factors as the extent of welding required, the complexity of the repair site geometry, depth of repair, welding process and ancillary equipment, and environmental conditions. Generally, designs of dry hyperbaric habitat fall into one of the following four groups:

- Lightweight steel habitats: These are of stiffened plate construction and are fabricated in two or more sections to allow their placement around jacket members. They may have an open grate floor with an access hole, or be fitted with a closed floor and access shaft. The latter is used in shallow depths where the shaft acts as a surge tube, thereby reducing the volume and pressure changes in the habitat, which otherwise could affect diver physiology.
- Inflatable flexible habitats: Where the differential pressures are expected to be low, flexible habitats of sufficient strength are practicable. The skin of the habitat takes up a shape dictated by the skin membrane stresses and the depth-dependent differential pressure, and is the same shape that would be obtained onshore by turning the habitat upside down and filling it with water.
- Mini habitats: These habitats are of small construction with just enough room for the arms and sometimes the head of the welder/diver. These, in essence, only protect the welding head and a small area around the weld. The clear plastic box, fitted with sponge or flexible rubber seals, moves with the head. These devices have not undergone as much development as either large habitat welding or wet welding.

The hyperbaric conditions require that the welding has to be specially optimized for the elevated pressure, which affects the gas, slag, and metal reactions for all welding processes, and the high-density gas enhances the rate of heat loss from the weld.

The following equipment and tools are required due to this special repair process for the hyperbaric welding operation:

- Purpose-built habitat
- Saturation diving support
- Environmental control equipment
- Pre- and postweld heating equipment
- Welding equipment
- Weld inspection equipment
- Equipment to remove marine growth and grit blast

- Temporary holding clamps to take weight of additional members and maintain root gaps, as needed
- Crane capacity

5.1.6 Platform underwater repair

This platform is a four-leg jacket platform operating in 21 m of water depth. The jacket structure is horizontally braced at four levels with K diagonal braces.

The platform was inspected immediately after Hurricane Andrew. The inspection discovered that two midpoint joints, located at EL (−) 12 m level between the four legs, were damaged, as shown in Fig. 5.6.

Divers performed the wet welding repair work of replacing the two damaged joints.

The repair procedure is as follows:

- The new joint should be designed and fabricated as illustrated in Fig. 5.6.
- It is mandatory to approve the welding procedure and it can be done by using specimens from the jacket structure to measure the amount of carbon.
- Wet welding specimens are tested.
- Afterwards, remove the two damaged joints.
- Grit blast the remaining structural members in preparation for welding.
- Final dimensions are measured to ensure the new joints fit.

Fig. 5.6 Remove the corroded part and install another.

- Install new joints.
- Inspect all welds by magnetic particle.

There are different repair options. Mechanical clamps, dry hyperbaric welding and wet welding repairs were considered as repair options. It was estimated that wet welding would save 40% and 60% when compared to dry hyperbaric welding and mechanical clamps, respectively.

5.1.7 Platform "shear pups" repair

This platform consists of two 8-leg jackets in 8 ft. of water that are connected above water by a common deck structure and by horizontal braces. The platform sustained several buckled members as well as small cracks at several members above and below water joints following a large storm in 1997.

A small crack in the bracing member was identified by MPI. The damaged members were repaired using traditional slip-sleeve replacement members. The cracked joints were repaired using "shear pups," as shown in Fig. 5.7, to strengthen the joints. The shear pups consisted of one-third circumference pipe pieces about 24″ long that "piggy-back" on the brace at the joint in order to provide an additional path for loads from the brace into the joint. The shear pups are wet welded into place. The shear pups were considered a better repair than mechanical clamps due to the complexity of the joint configuration, which made it difficult for a clamp. In addition, the cracks were small at these locations, identifiable only by MPI, or were on the order of only a few inches or less.

The repair procedures were as follows:

- As with any wet weld, take a material sample of the jacket to be sure that wet welding can be performed.
- Clean joint for fitting the shear pup and wet welding.

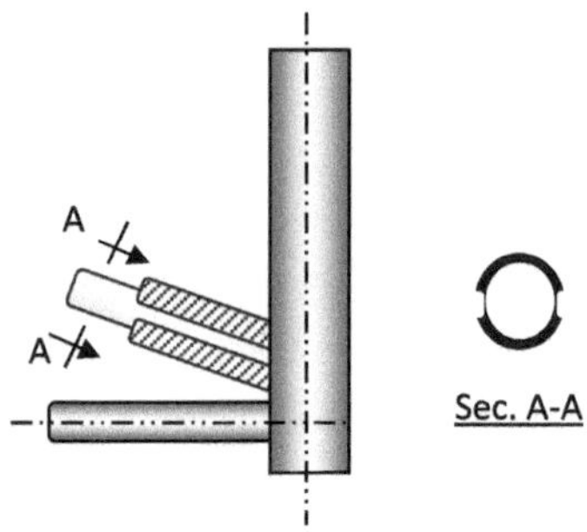

Fig. 5.7 Shear pup.

- Multiple shear pups with around 600 mm (24 in.) are cut and prepared on-deck considering that it will be an approximate dimension at the ends of the joint connection.
- Ends of the shear pup opposite the joint are also copped via transition to a narrow section in order to eliminate the potential for fatigue cracking.
- Diver fits generic shear-pup into place and notes additional copping needed, if any, as shown in Fig. 5.7.
- Final copping on shear pup above water.
- Diver wet welds shear pup into place and inspect welds.

5.1.8 Underwater repair for platform structure

The platform is a four-leg, fixed steel jacket platform operating in 43 m of water. The jacket is a vertical diagonal braced structure.

The underwater inspection revealed a buckled vertical diagonal. The buckled area was located approximately in the center of the member. The replacement of the member and wet welding work for the buckled vertical diagonal member were performed.

Repair Procedure

- Remove vertical brace leaving a 2-ft stub at top and a 6-in. stub at bottom.
- Install single telescoping vertical diagonal member over stubs.
- Wet weld the sliding sleeve as shown in Fig. 5.8.

5.1.9 Case study 2: Platform underwater repair

The platform was a fixed-steel jacket platform operating in 7.8 m of water depth. The jacket structure was braced with vertical diagonal bracing members.

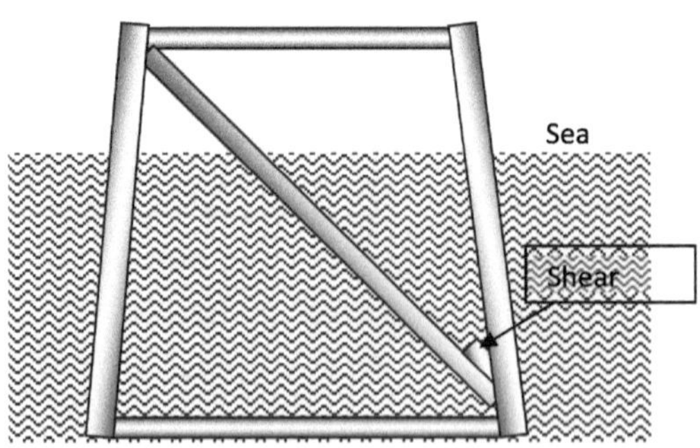

Fig. 5.8 Sketch of proposed wet weld repairs.

An inspection revealed a 10¾-inch diameter vertical member, with a 30-in. crack located at midpoint around the circumference, as shown in Fig. 5.9. An additional 6-in. diameter hole was located on the horizontal bracing member. Contractors performed the wet welding replacement work of the vertical diagonal member and the dry welding work of the corrosion hole patch.

5.1.10 Clamps

Repair clamps are normally constructed from low carbon steel and consist of a reinforcing sleeve that may include brace attachments. For ease of installation, the clamps can be split or hinged. Whether used for repair member or node, a clamp will typically take on the same general appearance as the part of the structure it is to reinforce. In general, typical clamps weigh between 0.5 and 10 tons, although clamps weighing some 50 tons are known, and they may require grouting, bolting, or some other structural bonding technique.

Clamps may be used to repair a member by installing an additional brace, and it is normally used for a connection to a riser or caisson if there has been loss of support to such an appurtenance.

However, the principal use of steel repair clamps is in repairing primary structural joints. There are four clamp types typically used in offshore platform repair. The classification between them is based on the installation and fixing method rather than on terms of usage.

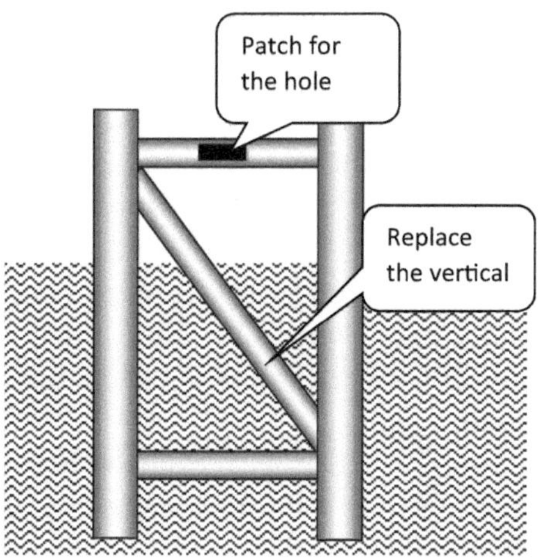

Fig. 5.9 Repairs of the members.

- The stressed mechanical (friction) clamp uses long stud bolts to produce a friction grip on the repaired structural elements.
- The unstressed grouted repair clamp relies upon grout shear strength to affect load transference, working in a similar way to a grouted pile sleeve connection.
- The stressed grouted clamp is a hybrid of the previous two clamp types with load being transferred partly by grout bond but mainly by friction. It is popular for its ability to tolerate dimensional variation while achieving load transfer in a reasonable sleeve length.
- The stressed elastomer-lined clamp is similar to a mechanical clamp except for an elastomer that is usually a neoprene liner that lies between the member and the clamp. Such clamps are often selected for repairs to caissons and other secondary structural elements.

The basis of any clamp design must be the establishment of the forces in the structural elements under operating and storm loading conditions. It is normal to evaluate such forces for the undamaged condition and then design the clamp so that load transfer is accomplished within the body of the repair clamp.

The technology surrounding clamp installation is highly developed and a number of specialist companies are now able to provide the necessary equipment and materials on a contract basis.

5.1.11 Stressed mechanical clamps

A stressed mechanical clamp, also called simply a friction clamp, is a steel-to-steel friction clamp that is connected by long tension stud bolts. It comprises two or more segments of closely fitting stiffened saddle plates, stressed directly onto a tubular section by means of long stud bolts. The strength of a mechanical connection is obtained from steel-to-steel friction generated by compressive forces normal to the tubular/clamp saddle interface, applied by the external bolt loads.

These clamps are generally used for the strengthening and repair of damaged members or connecting new members. Since very close tolerances are required for close contact between the clamp and a tubular member, these clamps are unsuitable for repair of tubular joints.

The major advantage of a mechanical clamp is that large forces can be transferred through friction over a short clamp length, limited only by the hoop resistance of the member.

The time of installation for stressed mechanical clamps varies depending on the complexity of the clamp (for instance, number of clamp segments), space limitations, and water depth of the repair site. Typically, installation times vary between 2 and 6 days, depending on size and complexity.

Over the past 20 years there have been numerous applications of stressed mechanical clamps worldwide. Their use has principally been to repair or strengthen jacket components damaged due to boat impact, fabrication flaws, fatigue cracking, and corrosion. In this type of clamp, full contact is required between the clamp and member surface interface, and the clamp offers minimal translation or angular tolerances. Thus extremely accurate offshore surveys are needed for its proper construction. Noting this, the clamps require very tight tolerances in fabrication. The connection between clamp and tubular member is susceptible to crevice corrosion. Therefore it needs periodical inspections to confirm bolt tension.

This is one of the quickest clamps to deploy and has good transfer capacity, but it requires detailed survey and good fabrication tolerances. It is ideal for clamping on intact tubular members to strengthen, replace, or add members.

For diver installation of a stressed mechanical clamp, the following items are required:

- Diving spread and divers
- Crane for lifting and placing in the required position
- Rigging for installation
- Tools for underwater cutting and grinding and to remove any obstruction
- Tools for stud bolt tensioning
- Video camera to monitor and record installation.
- Tools required to remove marine growth and grit blast

5.1.12 Unstressed grouted clamp connections

Short bolts are used for this grouted sleeve type clamp. An unstressed grouted clamp or sleeve connection comprises sleeves placed around a tubular member or joint with the annular space filled with grout. The sleeves may be split, as in the case of a clamp, or continuous as in a sleeve connection. For split sleeves, short bolts are provided and these are tightened prior to injection of grout into the annulus.

The bond between the grout and steel will transfer the load between the tubular member and the clamp. The length of the clamp can be increased if it is required to increase the clamp capacity. The function of shear keys, which are a form of weld beads, is to increase the clamp capacity, but it should be considered that the underwater welding option is very expensive.

Unstressed grouted clamps and connections offer a versatile means for strengthening or repair of tubular joints and members, since they require less accurate offshore surveying. Both angular and translation tolerances can be readily accommodated by the annulus. The loads that affect the member or the joint and the accessibility, with enough space allowed, are the main factors that affect the decision to use an unstressed grouted connection or clamp.

The time taken for installation of this type of clamp varies depending on the complexity of the clamp and the number of pieces in which the clamp is installed. The number of bolts that have to be tensioned and the amount of grout will influence the time scales as well. As an example, two X joints strengthened using unstressed grouted split sleeve connections can take about 50 diving hours, as follows: 20 h for survey and cleaning, about 20 h for clamp installation and bolting, 5 h for grouting, and 5 h for inspection.

It is important to note that curing time should be considered in the time schedule to ensure that the unstressed grouted clamp or sleeve connection is not subjected to loading before the grout has gained sufficient strength. Sometimes temporary clamps are inserted to handle this issue.

Several applications of unstressed grouted clamps and sleeve connections are evident. Pile or sleeve connections for numerous jackets make use of this technique. For repair or strengthening, unstressed grouted clamps or sleeve connections have often been used to overcome fatigue cracks and damaged members due to boat impact.

The advantage of this technique is that it has reasonable transfer capacity, with good tolerance for fit-up, and is ideal for clamping on joints and members. It is a very useful solution in the case of strengthening dented members. On the other hand, without the use of shear keys, the required connection length may be an unacceptably long time. Grout seal, if improperly fitted, often results in leakage of grout, resulting in loss of friction.

The following equipment is required for diver installation of an unstressed grouted connection or clamp:

- Diving spread and divers
- Crane for lifting and placing in position
- Rigging for installation
- Underwater cutting and grinding tools and tools to remove any obstructions

- Bolt torque or tension tools for split-sleeve clamp
- Grouting spread
- Video camera to monitor the installation
- Tools to remove marine growth and grit blast

5.1.13 Stressed grouted clamps

A stressed grouted clamp is formed when two or more segments of strengthened saddle plates are stressed by means of long stud bolts onto a tubular member after grout has been injected and allowed to cure in the annular space between the clamp and the tubular member. This type of clamp is a hybrid between a stressed mechanical clamp and an unstressed grouted clamp. The strength of the clamp is obtained from a combination of "plane pipe" bond and grout/steel friction developed because of the compressive force applied normal to the grout/tubular surface interface by the stud bolt tension.

Stressed grouted clamps offer the benefits of stressed mechanical clamps of high strength-to-length ratio, and the benefits of unstressed grouted clamps regarding the ability to absorb significant tolerances.

As an example, eight stressed grouted clamps were fully installed on a platform in the GoM in a total of 18 days. About one-third of this time was used in removal of obstructions while cleaning; clamp installation, grouting, and stud bolt tensioning took about equal time. It is important to recognize, however, that installation times are very dependent on the complexity of the clamp, access at the repair site, water depth at the repair site, and environmental conditions that may severely limit dive time and greatly influence weather downtime.

Stressed grouted clamps can be viewed as representing the strength advantages of stressed mechanical clamps with the tolerance advantages of unstressed grouted clamps; it is therefore not surprising to note that stressed grouted clamps are the most popular form of clamp concept used widely these days.

Grout seals, if improperly fitted, often result in leakage of grout, resulting in loss of friction.

This type of clamp has good transfer capacity and good tolerance for fit-up and is ideal for clamping on joints and members. It is particularly good for repair of joints.

The required equipment and tools are the same as for unstressed grouted clamps.

5.1.14 Stressed elastomer-lined clamp

These types of clamps are used for stub connections to appurtenances and to seal holed caissons. The elastomer-lined clamps are used in secondary members, as the stiffness is not critical to its performance, because this liner is flexible, which will reduce the repair system efficiency. Thus it is not used in main members.

The time duration for installation varies, depending on the complexity of the clamp (for instance, the number of clamp segments, space limitations, and water depth of the repair site. In general, the installation time duration is similar to that for stressed mechanical clamps.

It is worth mentioning that the stressed elastomer-lined clamps are similar to stressed mechanical clamps, the only difference being that an elastomer lining is bonded to the inside faces of the clamp saddle plates. The lining is made up of solid polychloroprene, which commercially is called neoprene sheet.

The strength of the clamp is derived from external bolt loads, which impart compressive force normal to the interface of the liner and the tubular member. This type of clamp is not recommended for transfer of forces between structural components. It is recommended for the attachment of components such as guides for appurtenances.

The elastomer lining offers a degree of angular and translation tolerance, eliminating the need for very accurate offshore surveys, as required for stressed mechanical clamps.

The relatively low stiffness of the liner gives rise to significant stud bolt load fluctuations due to elastomer relaxation. Fatigue of stud bolts therefore requires careful consideration. Large-scale tests have demonstrated this type of clamp does not transfer structural forces. Furthermore, typical coefficients of friction between liner and steel are considerably lower than previously assumed values. These clamps require periodic inspection to confirm bolt tension.

This is one of the quickest clamps to deploy. It requires a reasonably detailed survey and has poor axial and bending load transfer capacity. It is deal for clamping on intact tubular members or for adding members and appurtenance supports.

The required equipment for installation is the same equipment as for the stressed grouted clamp.

5.1.15 Drilling platform stabilization post-hurricane Lili

The drilling platform, located in 71 m water depth, was severely damaged when Hurricane Lili passed through the GoM.

In 2003 in hurricane season, the platform was stabilized during doing the plug to the well to be abandoned by doing strengthening to the platform by adding template and installed piles.

The leeward pile was severed at 6.3 m below the base of the jacket leg. The pile severance was located at mudline due to the lean of the platform. In addition, the remaining three piles were assumed to be damaged at or below the mudline due to the platform rotation.

The diving activity performed member cleaning, clamp installation, and grouting operations. Once the grout had sufficiently cured within the clamp annulus, the stud bolts were tensioned to the required tensile loads. Heavy lifting and pile installation was carried out.

Note that several repair methods were considered, such as guy wires to piles, props and attachments to the undamaged production platform. Analyses indicated that templates were required in addition to severed pile reinstatement to provide sufficient torsional resistance against potential hurricane wave loads. The strengthening scheme consisted of two templates, a stressed grouted clamp and five piles.

The repair procedures were as follows for this platform:

- Clean exposed pile.
- Cut holes in base of severed pile and place support clamp.
- Install sleeve clamp around exposed pile using support clamp.
- Close clamp and hand tighten several bolts.
- Check seals for leakage and place grout.
- Once grout has reached required cube strength, stress the 32 2-in. stud bolts to required tensile load.
- Lower piles through sleeves to self-penetration depth.
- Hammer piles to the required 200-ft penetration.
- Grout pile sleeves.

5.1.16 Grouting

Using grouting to fill structural members is a cheap and effective solution to several repair and strengthening problems. It is most beneficial in these conditions:

- For compressively loaded dent-damaged elements where the grout prevents any further deformation of the tubular section. In such cases the grout filling need only be extended to the immediate region of the damage.
- To improve the strength and fatigue performance of a tubular joint by grouting the chord in the location of the entire tubular joint.

No benefit accrues under the following situations:

- Tensile loads in the element (unless there is a problem with hydrostatic/tension interaction collapse).
- Unchanging compressive or bending loads in element (because a repair can only carry a load that is applied after the grout has set).
- Partially filled compressive elements, because there is no guaranteed load transfer mechanism between the grout and the steel.

Grout work should be carried out through a competent company that delivers a grouting design mix to reduce the heat of hydration effects, grout shrinkage, and grout port locations and ensure complete element filling if deemed essential.

In general, grout-filled elements and joints will be stiffer than their pure steel counterparts and as such they may attract more load in a statically indeterminate frame. In seismic zones, the additional mass associated with grout filling may need to be considered in the structural analysis. On any structure the additional weight imposed by grout may constitute a significant load, particularly if the grouted joint or element lies in a horizontal plane, as is the case with conductor bracing.

Based on typical offshore time scales, a grouting operation should be achievable with 2–3 days offshore work.

5.1.16.1 Joint grouting

In general, joint grouting is performed by filling of a chord with grout in the region of a tubular joint. Grouted joints have the chord member fully filled with a cementitious grout material. Double-skin joints are those in which the chord member contains a pile and a grouted annulus.

For a member rather than the jacket leg, it is easy to fill a chord member over its full length, as this method avoids cutting windows in the member to put seals to localize the grout plug. The grout can be placed through small diameter inlets and outlets, which can be drilled and tapped into the tubular wall. Jacket legs need only be filled up to the level required in view of the quantity of grout needed.

Filling tubular chord members with a cementitious material will increase its strength and improve ductility, and the radial stiffness of the chord member will increase due to grout. The grout restricts local chord wall deformations, leading to reduction of deformation-induced bending stresses and associated SCFs, so it will improve fatigue life.

Therefore, grout filling of tubular chord elements is used to improve the static strength of the joint and to extend the fatigue life of the connections made at the joint. The repair method has the advantage of introducing no additional wave and current loads on the platform, but the local dead loads can be increased.

Only grout filled joints have been considered. It is therefore assumed that the material is cementitious and not reinforced. If the joint is simply pumped full of grout, then a simple grout-filled joint is created. In some instances, there will be a concentric pile within the joint, as is the case with many leg sections, and the resulting construction is then termed double skinned. The grout is assumed to completely fill the available annulus in the joint.

Ring stiffening is sometimes used at the joint to increase the resistance of the chord wall to applied member forces. Both enlarged can diameter and stiffening will cause problems in ensuring complete grout filling.

In most cases, grouting the joint takes about 3 to 4 days, as this time should be enough for installing grout bag seals, allowing seals to set and cure, and then grouting the plug.

Void formation is a potential problem at ring-stiffened joints and at joints with an expanded can diameter. Increased strength and fatigue performance of tubular joints is exhibited.

There is no increase in hydrodynamic load but this technique does add mass to the structure.

If a large volume of grout is used, excessive heat while setting will need to be addressed. Consideration of mix design and heat loss is required.

The following equipment is required for a typical grout-filling operation:

- Diving spread
- Drilling and tapping tools for grout ports
- Underwater cutting and grinding equipment if temporary seals are to be placed
- Grouting spread
- Monitoring equipment such as video or camera
- Marine growth-removing equipment

5.1.16.2 Grout filling of members

Grout filling of a member increases both its cross-sectional strength and its overall stability. This process provides a relatively easy way of strengthening tubular members, particularly compression members with or without bending. The grout filling the compression member increases the axial load capacity and prevents local buckling.

As a direct effect of grout filling, the additional mass of the member, the additional earthquake loads due to increased inertia and the extra stiffness of the member and tubular joints along its length need to be considered in the design. This technique is most beneficial to tubes with low L/D ratio and high D/t ratios.

It is important that the grout completely fills up the tubular, as small voids close to the tubular inner wall may reduce the load-carrying capacity of the strengthened member significantly. Void formation is also a potential problem for all ring-stiffened tubulars.

If there is no damage to the member but an increase in axial load-carrying capacity is sought, then a few preliminary calculations will indicate if there is likely to be sufficient increase in member capacity. If the member is short and has a low (kl/r) ratio, then the axial load capacity in compression will be increased if the element is completely filled and load transference can be made from the node capping plate to the grout body. If the element has a high D/T ratio, then there may be an improvement in load-carrying capacity due to the prevention of local instability.

The design mix should be done professionally to prevent the generating of excess heat during setting, as a large volume of grout will be used.

On the other hand, grout filling has little benefit for tension members. In addition, full grout filling must be achieved and the increase in mass due to the filling of a member with grout may result in overstress of the member under seismic and/or in-place conditions.

This technique increases strength of a member but will not increase hydrodynamic load, but does add mass to the structure.

The member grouting operation requires the same equipment as specified earlier for joint grouting.

In cases where the annulus between pile and leg is filled by grouting, the load is transferred to the pile from the structure across the grout. Based on experimental work that was done on grouting the pile annulus, the load transferred is a combination of bond and confinement friction between the grout and the steel surfaces and the shear keys are used, they bear part of the transfer load.

To have a uniform annulus or space between the pile and the surrounding structure, centralizers should be considered. As per API RP2A the minimum annulus width of 38 mm (1½ inches) should be provided, if the grout is the only means of load transfer. The dimensions of shear keys should be considered, to have adequate clearance between pile and sleeve.

Note that packers should be used as necessary to confine the grout. Good quality control and good installation of the grout should be ensured, to avoid any possibility of dilution of the grout and to minimize the formation of voids in the grout.

5.1.16.3 Allowable axial force calculation

If there is no reliable detailed data that can assist with the use of other values of connection strength, then the allowable axial load transfer shall be considered to be the smaller value from the pile or leg of the force calculated by multiplying it by the contact area between the grout and steel surfaces. The allowable axial load stress f_{ba} is calculated by a reasonable value that must be greater than or equal to the calculated applied axial force.

The value of the allowable axial load transfer stress f_{ba} should be taken as 0.138 MPa for loading conditions 1 and 2 and 26.7 psi (0.184 MPa) for loading conditions 3 and 4, with the following loading conditions based on API:

$$1 - E_o + D_L + L_{max}$$

$$2 - E_o + D_L + L_{min}$$

$$3 - E_D + D_L + L_{max}$$

$$4 - E_D + D_L + L_{min}$$

where E_o is the operating environmental condition, D_L is the dead load, L_{max}, L_{min} are the maximum and minimum live loads, respectively, and in both cases the live load is appropriate to the normal operating condition of the platform. E_D is the design environmental condition in extreme environmental conditions.

Where shear keys are used at the interface between steel and grout, the value of the nominal allowable axial load transfer stress f_{ba} should be taken as:

$$f_{ba} = 0.138 + 0.5 f_{cu} \cdot \frac{h}{s}; \text{MPa} \tag{5.1}$$

for loading conditions 1 and 2, and should be taken as:

$$f_{ba} = 0.184 + 0.67 f_{cu} \cdot \frac{h}{s}; \text{MPa} \tag{5.2}$$

for loading conditions 3 and 4, where:

f_{cu} = unconfined grout compressive strength, MPa

h = shear key outstanding dimension, mm
s = shear key spacing, mm.

Shear keys designed according to Eqs. (5.1), (5.2) should be detailed in accordance with the following requirements:

1. Shear keys may be circular hoops at spacing "s" or a continuous helix with a pitch of "S_s". Shear keys should be one of the types indicated in Figs. 5.10 and 5.11.

3. For driven piles, shear keys on the pile should have sufficient length to guarantee that, after driving, the pile length in contact with the grout has the required number of shear keys as designed.

4. The cross section for each shear key n and weld shall be designed to transfer that part of the connection capacity that is attributable to the shear key for different loading conditions.

The shear key and its welding should be designed as per the allowable steel and welding stresses to be capable of transferring an average force equal to the shear key bearing area multiplied by 1.7 f_{cu}, except for a distance of 2 pile diameters from the top and the bottom end of the connections, where 2.5 f_{cu} should be used.

The following limitations should be observed when designing a connection:

- $17.25\,\text{MPa} \leq f_{cu} \leq 110\,\text{MPa}$
- Sleeve geometry $D_s/t_s \leq 80$

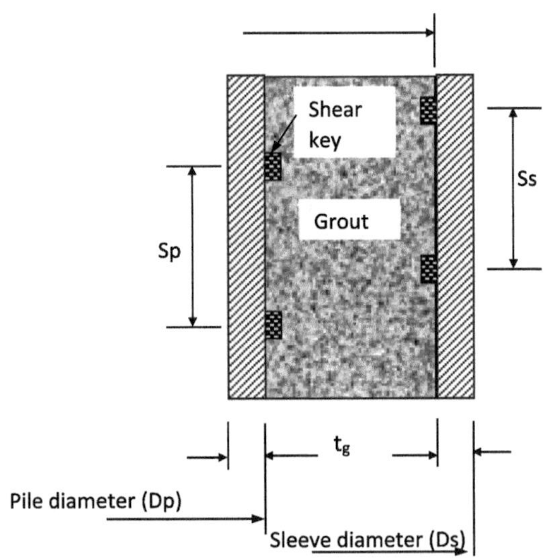

Fig. 5.10 Grouting the annulus between piles and legs.

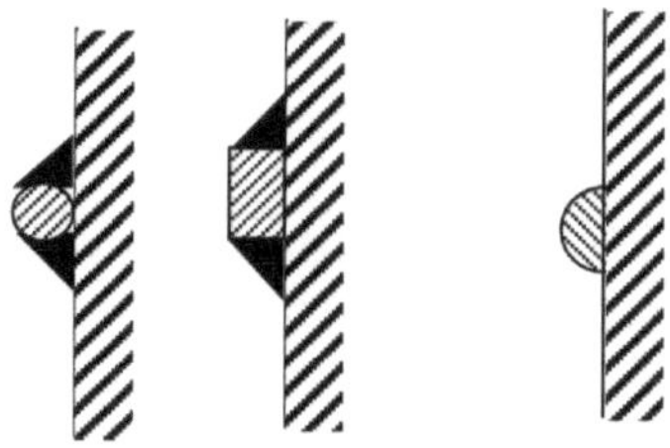

Fig. 5.11 Recommended API shear key detail.

- Pile geometry $D_p/t_p \leq 40$
- Grout annulus geometry $7 \leq D_g/t_g \leq 45$
- Shear key spacing ratio $2.5 \leq D_p/s \leq 8$
- Shear key ratio $h/s \leq 0.10$
- Shear key shape factor $1.5 \leq w/h \leq 3$
- f_{cu} (h/s) ≤ 5.5 MPa

5.1.17 Composite technology

The definition of composite technologies is a combination of two or more materials. Composite materials have been extensively used onshore; however, the use of composites for offshore structural applications has been somewhat limited. Composites have been used widely for pipeline and piping on topsides and composites are seeing greater use for tertiary steel structures such as handrails and grating.

Two main groups of composites have been used in offshore applications:

- Reinforced epoxy grout—A rebar-reinforced epoxy grout typically used for the repair of conductors.
- FRP (fiber reinforced polymer) is a structure containing an arrangement of unidirectional fibers or woven fiber fabrics embedded within a thin layer of light polymer matrix material. The fibers, composed of carbon or glass, provide higher strength. The matrix in most cases is polyester, epoxy or nylon, which binds and protects the fibers from damage and mainly transmits the stresses between fibers. The matrix also provides a bond and thus an interface for the transfer of forces between the parent structure and the composite component. FRP can be molded or pultruded to form many shapes for topsides applications.

Composite applications can consist of either one or a combination of the groups described. One variation of the grouted clamps/sleeves is covered in the clamps section of this chapter.

The use of composites to date, particularly for offshore applications, has been limited largely by lack of design guidance, designer lack of familiarity, combustibility concerns, lack of performance data, and fabrication scale. These have in the main been addressed, with specialist contractors offering off-the-shelf solutions and data providing proof of performance. A range of resins are available for different applications, including low generation of smoke and toxicity in fires.

The application of composites is different in repairing on topsides as compared to subsea applications. Composites technology to date has shown little success in subsea use when the laminates are formed in-situ. Preformed components can be used, but they are typically limited to nonstructural applications. The use of composites on topsides are as follows:

- Repair wraps to tubular members
- Wraps to deck legs for corrosion protection
- Grating
- Stair treads and ladders
- Handrails
- Mudmats
- Fire and blast walls
- Access platforms

Many of the topsides composite applications can be utilized as part of strength and repair or for new build. A number of notable new-build offshore facilities have made extensive use of composites for a range of applications. As much as 10% of the topsides weight has been represented by composite components.

5.1.17.1 Reinforced epoxy grout

Rebar reinforced epoxy grout is typically used for the repair of conductors. Reinforced epoxy grout provides a method of repair that requires little preparation onshore prior to implementation of the repair offshore. The repair technique can readily conform to the structure geometry without the requirement for a detailed survey. The main advantages of using epoxy grout are:

- Transportation to site and installation is very easy.
- Can be implemented without the requirement for hot works.
- Minimal lift equipment is required, particularly if rebar cage is assembled in situ.
- Significantly improved corrosion resistance compared to steel.

- Prevents further deterioration, particularly corrosion, in the case of conductors.
- Limited requirement for design with minimal fabrication.

Reinforced epoxy grout has been mainly limited to axially loaded members such as conductors under compression loading and is mainly used for above-water applications. In addition, it is easily transported and utilized for repair to conductors.

In general, the following equipment and support may be needed:

- Grouting spread
- Composite materials
- Other components depending on repair type, such as temporary outer jacket to provide formwork
- Installation aids

5.1.17.2 FRP composites

The linear elastic response of FRP to axial stress is one of the main attractions of its use in structural engineering. The response of FRP to axial compression is subject to the volume of fibers, the fibers' properties and resin, and the interface bond strength. If there is fiber buckling due to rapid side sway or deflection, FRP composite compression failure occurs.

The base fiber matrix is available in many forms from random mixed fibers to woven fiber blankets consisting of either glass fibers or carbon fibers. The woven matrix provides the necessary bidirectional fiber strength depending on the strength requirements and application.

FRP's response to transverse tensile stress is very much dependent on the properties of the fiber and matrix, the interaction between the fiber and matrix, and the strength of the fiber-matrix interface. Generally, however, tensile strength in the direction perpendicular to the fiber direction is very poor so the greatest tensile strength is in the direction of the fibers. The shear strength of FRP is difficult to quantify. To avoid this, in some products on the market, the direction of fiber will be in two directions and the other in the other two diagonal directions. So, it can carry the load in four directions strongly. This type of FRP will be used for repairs on-site, but the prefabricated FRP as grating is usually in one direction.

Besides FRP having a higher strength, the other major advantages are its durability and corrosion resistance. In addition to that, the high strength-to-weight ratio is of significant benefit; a member composed of FRP can support larger live loads since its dead weight does not contribute significantly to the loads that it must bear. Other features include ease of installation, versatility, electromagnetic neutrality, excellent fatigue behavior, fire resistance, and possible maintenance-free use. So, FRP increases reliability due to good corrosion resistance and structural superiority through weight savings, higher stiffness, and ability to better tailor the structure to the load.

The composite repair method takes a shorter time offshore and involves light material handling. This method often proves to be more cost effective than other repair solutions.

FRP has the following limitations, which can be easily overcome through design and application and a search for the most advanced product on the market:

- Brittle behavior
- Susceptibility to deformation under long-term loads
- UV degradation and photodegradation (from exposure to light)
- Temperature and moisture effects

Limitations also lie in the level of previous experience of the contractor. However, the number and experience of contractors have grown due to ongoing research to enhance the properties of FRP.

Increased reliability due to good corrosion resistance and structural superiority through weight savings, higher stiffness and ability to better tailor the structure to the load.

5.1.18 Example of using FRP

Davy and Bessemer are two shallow water monopod platform structures located in the Southern North Sea that commenced production in 1995. These two platforms used FRP. The fiberglass components consisted of 10% by weight of composites, including the following:

- Office and equipment modules
- Diesel and water storage tanks
- Pultruded glass/phenolic gratings for floors
- Ladders
- Walkways
- Handrails
- Enclosures and heat protection walls
- Glass reinforced epoxy was also widely used for the pipework and tubular

The main benefit was the reduction in weight, which was important for these lightweight structures, as it will reduce the cost, reduce the topside weight, and eliminate the cost of regular maintenance in the future.

5.1.19 Case study for conductor composite repair

An eight-leg jacket drilling platform was operating in 55 m of water depth in the GoM. The platform was installed in 1964.

The original conductors were heavily corroded (Fig. 5.12). A 5-year conductor maintenance plan identified the required conductor restoration work.

The cost analysis revealed a composite repair system to be the most cost-effective repair method, consisting of the following steps:

- Underwater and above-water repair site inspection.
- Plan staging area for equipment and material.
- Surface preparation including removal of excess scale; removal of grout if it is not structurally sound; and grit blasting to near white metal.
- Install shear lug, rebar cage, and translucent FRP jacket outside rebar cage.
- Pump epoxy grout into FRP jacket from bottom up.
- The original conductors were heavily corroded as in Fig. 5.12.

5.1.20 Fiberglass access decks

The use of fiberglass is a very important option to reduce the weight of the mature platform; FRP can be applied to many auxiliaries. For 20 years, fiberglass use for grating and handrails has been growing in offshore structures. It is used on new platforms or in replacing existing ones.

Fiberglass access decks and stair towers are lighter than the steel equivalents and easier to install. This lighter weight will solve many integrity problem for the platform.

Fig. 5.12 Severely corroded conductor.

Fiberglass structural decks can be designed to meet any load requirements. The decks can be custom built and installed offshore with no welding and/or heavy lift equipment required.

Fiberglass access decks consist of pultruded structural sections that are bolted together. Fiberglass grating and handrails complete the full deck.

The benefits of fiberglass access decks can be installed without hot work and there is no long-term maintenance or painting. They can be readily transported to the site and handled using lighter lifting equipment compared to steel decks. The main advantage of the fiberglass access deck is that it does not corrode, eliminating maintenance and costly repairs or replacement of steel decks.

In some cases the small decks are made from fiberglass, as shown in Fig. 5.13.

Fiberglass access decks, primarily well access decks, have been used throughout the world on projects ranging from minimal structures to deep water TLPs and Spars.

Fiberglass grating products and resins are available to cover a range of requirements including:

- Strength
- Chemical resistance
- Impact resistance
- Fire resistance

A number of attachment options are available, including bolting, welding, clips, and friction welding. The choice of attachment depends on the location (wave zone or nonwave zone) and whether or not hot works are possible.

Fig. 5.13 Deck from FRP.

Fiberglass grating systems have been proven to perform even after being subjected to hurricane wave forces. Grating is typically designed to last for the life of the structure, without replacement.

Fiberglass grating does not corrode, which is the primary cause for the replacement of steel grating and plating (Fig. 5.14).

Fiberglass access stairs and treads are lighter than the steel equivalent and easier to install. Fiberglass stairs can be designed to meet any access requirements. Fiberglass stairs consist of pultruded structural sections with fiberglass stair treads. Fiberglass handrails complete the full stair. The fiberglass stairs can be replacement items for steel stairs. They can also be used for new design or new access requirements. Fiberglass stairs are installed using lighter lifting equipment compared to the steel equivalent.

It is important to highlight that fiberglass treads were proven to perform even after being subjected to hurricane wave forces.

Fiberglass stairs and handrails are being used in many offshore structures worldwide as they do not corrode, which is the main factor affecting their integrity and this reduces the cost of frequent replacement (Figs. 5.15 and 5.16).

Fiberglass stairs and treads have been used throughout the world on offshore platforms.

Fiberglass handrails can be used for the jacket walkways, boat landings and deck perimeters.

Handrail sockets are typically fabricated from stainless steel and are welded to the structure. If hot work is not permitted, then bolted attachments are available. The attachment of the handrails can be modified for installation where hot work is not permitted.

Fig. 5.14 FRP grating.

Fig. 5.15 Nonwave zone fiberglass stairs.

Fig. 5.16 Stair tread covers.

Fiberglass handrail systems have been proven to perform even after being subjected to hurricane wave forces. Handrails are typically designed to last for the life of the structure, without replacement.

Fiberglass handrails do not corrode, which is the primary cause for the replacement of steel handrails.

5.1.21 Fiberglass mudmats

Fiberglass mudmats provide an economical alternative to traditional steel systems. Due to the high potential flexural strength, as high as $427\,N/mm^2$, coupled with the high flexural stiffness (EI), as high as $64{,}121\,N/mm^2$, the mudmats can span two to three times further than unstiffened steel plate mudmat skins. As many as one-half to two-thirds of the beams required to support a steel mudmat skin can be eliminated when using fiberglass mudmats.

In addition, fiberglass mudmats, shown in Fig. 5.17, reduce steel fabrication tonnage and eliminate the need for mudmat cathodic protection that steel mudmats require. The service life of any mudmat system is very short, and after the piles have been driven and welded to the jacket structure, mudmats provide no further service; however, a steel mudmat system continues to draw from the platform's cathodic protection system.

A fiberglass mudmat system is significantly lighter than steel, ranging from 44 to 58 kg/m^2 in air and 49 to 58 kg/m^2 submerged.

The mudmat constructed of fiberglass has the following advantages:

- High flexural strength
- High flexural stiffness
- Reduced number of support beams
- Reduction in fabrication costs
- Reduction in anodes
- Saves weight
- Easily installed—no bolting.

Fiberglass mudmats have been utilized on approximately 20 jackets in the Gulf of Mexico since the late 1990s by a number of companies.

Case study A: For repair of a flare jacket

This flare was constructed in 1970. After performing the survey, cracks on the bracing due to corrosion were found, as shown in Fig. 5.18. The flare jacket leg brace clamps were designed in 1984 but not fitted until 1994. They were ungrouted, loose, and heavily corroded. The vertical diagonal braces in all three elevations in the vertical interval between the upper bay were totally corroded with extensive holes through most of the perimeter. The horizontal braces in all three elevations were totally corroded

Fig. 5.17 FRP mudmat system.

Fig. 5.18 Corrosion on the bracing member.

with extensive holes through most of the perimeter. After visual inspection it was shown that the roller bearings at deck level supporting the flare bridge were highly corroded and were not fit for purpose, as shown in Fig. 5.19A and B. In addition, the dead load deflection of the last span of the flare bridge had caused a rotation of the roller shoe beam, with the result that the flare bridge load was carried only on the inner one or two rollers and the thermal expansion/contraction was causing excessive loading of the jacket structure.

Prior to the start of diving operations, the current condition of the submerged structure should be assessed by ROV to provide a general inspection of the structure to avoid any potential hazards to divers. There should be a particular focus on the first bay up and vertical diagonal braces and horizontal plan bracing members.

Face brace members on elevations should also be generally inspected. If cyclic lateral movements of any braces or leg clamps are detected in a running sea, remedial measures should be taken, up to and including removal of members that could present a hazard and that would not further reduce the strength and stability of the structure.

Main access will be required to the underside roller bearing deck to attach beam clamps to support the weight of the new brace members during installation. Scaffolding to enable safe access to these locations from the flare bridge and to provide a working platform shall be constructed prior to the start of repair operations.

Repairs shall be carried out in the following sequence:

1. Pile Wall Thickness Survey;
2. Refurbishment of Roller Bearings;
3. Grouting of Leg/Pile Annuli;
4. Cutting of Legs or Crown Weld Shims;
5. New Horizontal Braces for the Two Levels.

Case Study B: Case study for repair bearing support

The bearing support of the bridge is shown in Fig. 5.19A; it can be seen that it is severely corroded, as also shown in Fig. 5.19B.

The severe corrosion of the bearing is shown in Fig. 5.19A and B. The procedure for refurbishing the roller bearings is critical, as it is based on removal of the old corroded rollers and cleaning up the existing steel work and bronze bearing bushings.

1. Machine sleeves to final outside diameter to fit, on the basis of the surveyed gaps above the existing rollers.
2. Install temporary lateral bracing to support flare-bridge lower chord during refurbishment of rollers.
3. Install hydraulic jacks to lift and support the flare-bridge (4 × 50 ton) foreseen as required and jack up flare-bridge by 30 mm to unload rollers.
4. Place heat-shield under flare-bridge lower beam underside.
5. One at a time, flame cut the rollers 50 mm from inside face on each side.
6. Cut or burn off keeper plates from outside ends.
7. Remove all corrosion on outer parts by needle-gun and grit-blast to reduce the diameters of the protruding parts so that they can pass through the bushings without causing damage.
8. Hammer drift the remaining roller portions through the bronze bushings from outside and remove remaining roller portions.
9. Clean up and remove corrosion products and rust scale from surfaces and apply paint system to bearing support steelwork.
10. Inspect, clean, and grease the bronze bushings. The bearings were believed to be intact. If there is wear or corrosion then they should be cleaned up as far as possible and the surfaces of the bushings polished without increasing the diameter by more than 2 mm.
11. Insert new rollers from outside by pushing the jacks and using the hammer across through opposite bushings and fit new stainless steel sleeves (in three sections each) onto new rollers as they are inserted.

Fig. 5.19 (A) Corrosion on the bridge bearing stud. (B) Corrosion on the bridge bearing.

12. Achieve symmetrical projecting ends.
13. Install new keeper plates; weld keeper plates to ends of rollers.
Finally, grease rollers and apply paint system to roller ends.

5.2 FPSO repair

FPSOs have a special characteristic if the deterioration is on the deck so the repair by the usual procedure is to replace the thickness of the steel that was corroded or increase the steel thickness to achieve the required strength.

The other parts in case of the corrosion on the tanks itself for oil storing or in case of ballast water storing tanks. As discussed before, the UT thickness measurement is the main tool to define if the reduction of the thickness is within the limits or needs to be replaced. After replacing it, it requires coating and also adding artificial anodes, as shown in Fig. 5.20.

Fig. 5.20 After repair of FPSO tanks.

5.3 Pipeline repairs

Pipelines are main elements for any oil and gas facilities offshore or onshore, as they transfer the hydrocarbon products to and from the facilities. Any defects in a pipeline will directly affect production and also can harm the environment. So, there are many methods to repair offshore and onshore pipelines.

Abdullah et al. (2019) summarized most of the methods that are used in repairing pipelines:

1. External repair method
   a. conventional clamps
   b. composite clamps
   c. grouted clamps
   d. lightweight clamps
   e. welded sleeve
   f. FRP for pipelines
2. Internal repair methods
   a. expandable steel patch
   b. plastic new pipe inside the old pipe
   c. using a robot for repair
   d. removing and replacing the defective parts.

The applicable methods are presented here, considering that grouted clamps are used for the repair of the jacket, as discussed in detail in the next sections.

For FRP, the same process and concept used to repair concrete structures are used. Fig. 5.21A and B shows the use of FRP wrap to repair the nozzle of a steel storage tank.

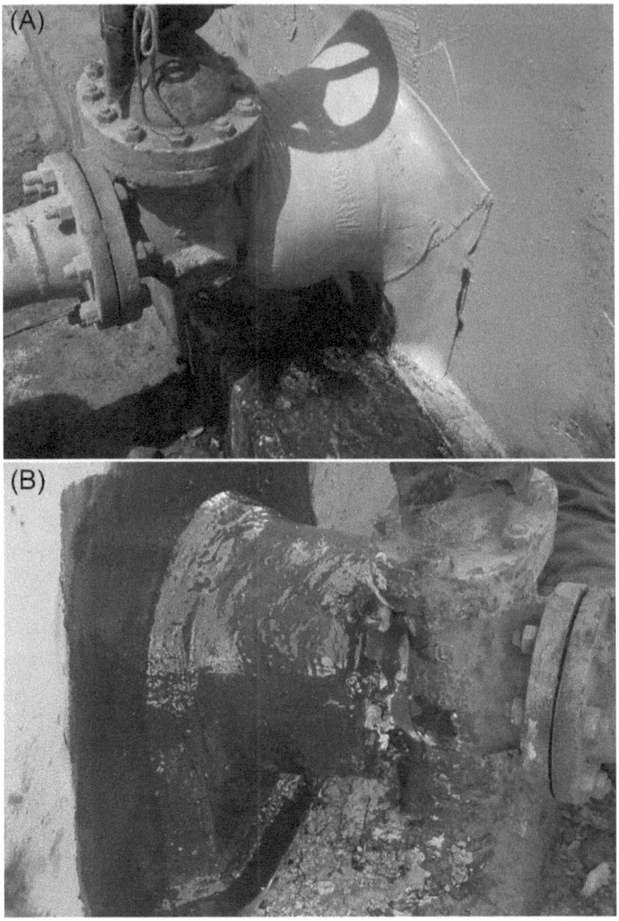

Fig. 5.21 (A) Wrapping to repair tank nozzle. (B) Wrapping of tank nozzle.

1. Conventional clamp (mechanical clamp)

A mechanical clamp is a traditional and quick repair method. It consists of two half shells, as shown in Fig. 5.22, with two end plates that are tightened around the pipe outer diameter by a bolting connection as per API 6H.

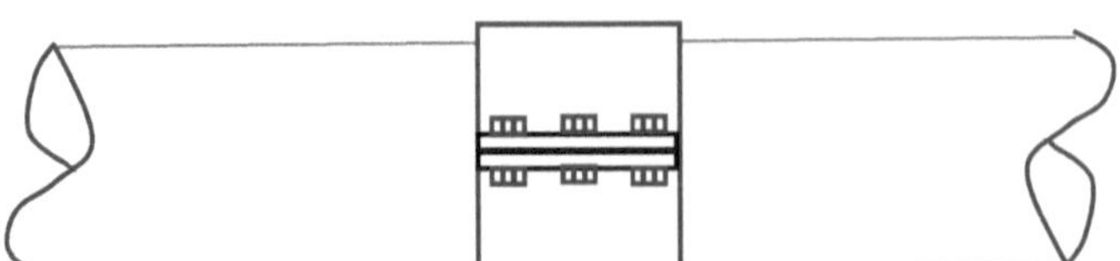

Fig. 5.22 Traditional clamp repair.

Antaki (2003) stated that these mechanical clamps are commonly used in power and process piping systems. Certain clamps can be welded to the pipe, in which case the clamp will be pressure tested and the welds will be examined as usual by radiographic and UT, as per the approved inspection procedure. This repair method is defined in API 570 as an acceptable temporary method.

2. Fillet welded patch

For local flaws on the pipeline, the best repair method is to use a fillet welded patch, as shown in Fig. 5.23. This repair method will be implemented by fillet welding a circular or square steel patch to attach to the pipe at its outer diameter.

This traditional repair method is recommended by API 510. Taking into consideration that API 570 for pipelines patches are not used if the hoop stress is more than 20% of the yield strength. The patch is chosen to be a similar or higher-grade material than the pipe materials and also has a thickness comparable to the pipe, so that the membrane stresses in the patch and in the welds do not exceed the design code allowable. It is important to have a rounded corner to limit local stresses.

The API 570 recommendation is to use this method of repair if the pipe base material has a yield strength equal to or less than 278 MPa (40 ksi).

Both ASME B31.4 and B31.8 prohibit the use of welded patches on high-strength pipe. ASME B31.4 limits the material to API 5LGrade X42 and B31.8 to material with a maximum specified yield of 278 MPa (40 ksi).

In the case of hazardous liquid inside the pipelines, ASME B31.4 recommended the patch length to be not less than 150 mm (6 in.) and can be used for pipe diameter sizes less than or equal to 12 in. Fig. 5.23 presents this technique of repair.

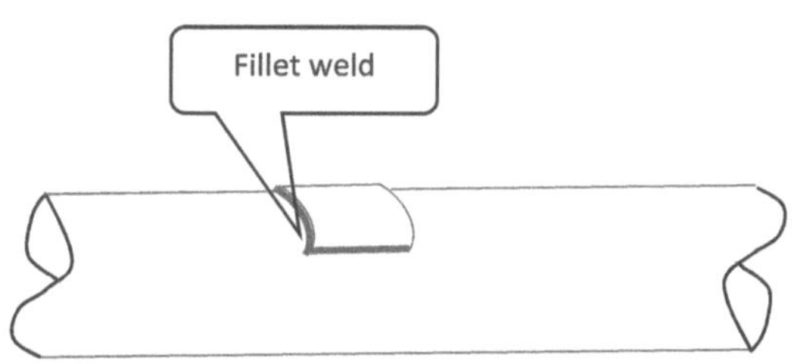

Fig. 5.23 Fillet welded patch.

3. New plastic pipe inside the old pipe

Another repair technique mainly used in onshore pipelines is inserting a plastic pipe inside the original steel pipe. This method is usually used in gas distribution to repair aging buried lines without interrupting gas flow. In this repair method, a new plastic pipe is inserted into the existing larger diameter metal pipe.

The inserted plastic pipe will be pressure tested in place while flow continues without interruption through the new pipeline. The annulus between the old steel pipe and new plastic pipe inside it will be sealed off by using injection of high-density polyurethane.

4. Sleeve

A degraded pipe can be repaired by using a split sleeve as illustrated in Fig. 5.24, with or without the two circumferential fillet welds to be sealed at the ends of the sleeve.

In the pipeline industry, textbooks state that a sleeve without the two end closure welds is referred to as type A, and a sleeve with the two end closure welds is referred to as type B.

A type A sleeve, with no end closure welds, serves to reinforce a locally corroded area and compress the pipe wall to avoid the formation of an outward bulge that precedes failure by overpressure burst or pressure cycling fatigue in ductile material.

The type B sleeve, with the two end closure welds, serves the same role as the nonwelded sleeve but also provides leak-tight containment in the case of failure of the pipe.

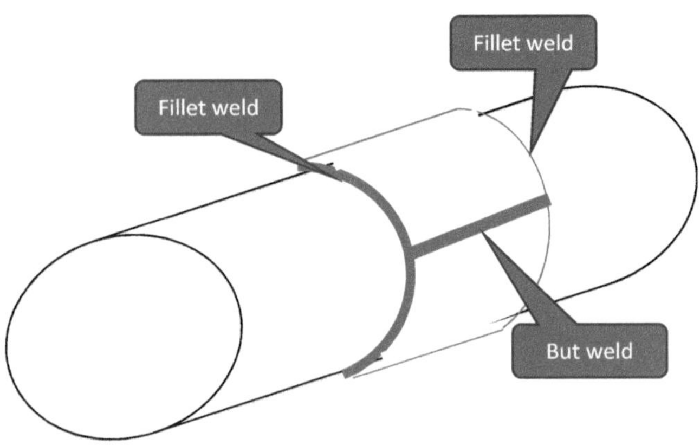

Fig. 5.24 Sleeve repair.

The circumferential weld could also be sized and designed to carry the thrust force, if a full separation of the degraded pipe occurs. The standards 49 CFR 192, 49 CFR 195, API 570, API 2200, ASME B31.4, and ASME B31.8 accept repairs of pipelines using welded split sleeves.

5. Offshore pipeline repair

Sometimes, in the case of rupture of the pipeline, it is required to remove part of the pipeline and insert a new one. The steps are as follows:

1. Cutting the pipe from the two ends
2. Removing it
3. Insert part of flange connectors on both side
4. Measure the distance to fabricate the pipe spool on the barge; this is called the metallurgy step
5. Insert the spool pipe with flanges in both ends
6. Adjust the two flanges and then tighten the bolt in both ends.

This is a common method now for repair. The advantage of the smart flange is that it can swing if the line is not at the same level, so it will make the installation easier, saving time and effort.

5.4 Onshore structures

In most cases the repair and strengthening of concrete structures is more complicated than executing a new concrete structure, as in this case an engineer with many years of experience is not required, as the senior engineer can have reasonable experience of 5 to 10 years. On the other side, repairs of concrete structures are more challenging and need competent engineers and consultant engineers with many years of experience, as some associations require at least 25 years' experience to register as a consultant engineer for repairs.

When the building already exists, new solutions are needed for deterioration problems that are compatible with the building characteristics, owner requirements, and economics.

The repair process for reinforced concrete structures is very important and dangerous as well, and requires extensive care in choosing the suitable repair methods and tools.

The code and specifications precautions to protect the structure have not been considered for some structures in the design phase or execution phase, or the necessary protection methods were not considered in the design phase, or possibly the designer chose the best protection method but the execution was not handled very well. The next chapter presents the different types of protection methods and best ways of execution.

In all the previous cases, the result is the same, which is structure deterioration, and this will be obvious from seeing the cracks and falling of the concrete cover of the different reinforced concrete elements in the structure.

When cracks and falling of the concrete cover appear, the situation has become serious. When shortages occur in a cross-sectional area of a concrete member, it requires the work of repair to be started as soon as possible to avoid the structure condition worsening and becoming more likely to collapse over time.

The first and most important step in any repair is to evaluate the structure and determine what part of the structure needs to be repaired. This depends mainly on the deep assessment of the structure, and finding the answers to these questions in the assessment report:

- What is the reason for corrosion, chloride, carbonation, bacteria, etc.?
- Will the cracks and the deterioration of the structure be increased or stop?
- What is the expectation of the extent of concrete deterioration and the corrosion propagation in the steel bars?
- What is the impact of the current condition and expected future deterioration on the structure safety?

Finding the answers to these questions will require the use of assessment methods for the structure and different inspection methods that have been illustrated in the previous chapters. These results will decide the cause of corrosion, the current deterioration of the structure, and the expected condition in future.

The process of repair and restoration of concrete structures, from the assessment phase to the execution phase of the structure, requires a high level of experience. Since the simplest stage is the evaluation stage, when it is prepared by inexperienced engineers it may lead to the wrong choice of repair method and cause a major problem.

In 1994 RILEM submitted a special report about repair strategy that matched with most cases of repair, summarized as follows:

1. Reestablishment of the deteriorated member
2. Performance of the required repair to the concrete member to restore its capability to carry the design full load.
3. Repair of a particular portion of the concrete member followed up by continuous periods of time
4. Strengthening of structure by an alternative system to carry part of the loads

In most cases, the strategy of repair is either complete repair of the concrete member by adding steel and increasing its section; sometimes the slab is destroyed and reconstructed, which mainly happens in renovation of the toilet (W/C) slab, as it is a simple slab in four directions in most cases, or in doing just a partial repair of a concrete member, for example just a redo of the concrete cover.

These strategies are common in the rehabilitation of concrete members and it depends mainly on the structure system, external environmental factors, and the existing condition, which is the degree of structure degradation.

As a rule of thumb, if the rehabilitation of the building costs around 75% of the cost of building a new structure, the rehabilitation option is not economical.

5.4.1 Main steps to execute repair

In general, there are several regular steps in the repair of all structures exposed to corrosion. The first and most critical step is to provide temporary support to the structure by performing structural analysis and designing and defining the suitable location of the temporary supports.

The second step is to remove the cracked and delaminated concrete.

After that, it is important to clean the surface of the concrete member from dust and debris and then clean the steel bars by removing rust from it.

After removing the rust by using brush or sand blasting, paint the steel bars by epoxy coating or exchange the steel bars for new ones, and then pour new concrete.

The final step is to carry out the painting of the surface of the concrete member as an external protection.

This is briefly the repair process. These steps will be explained in detail in the following sections.

5.4.2 Structure strengthening

The first critical step is to design and establish the temporary support before starting any repair work. The following procedure is required to design and construct a suitable temporary support.

1. Evaluate the whole structure
2. Define the method of transferring the loads.
3. Determine the type and location of the member that requires repair
4. Determine the amount of materials to be removed and repaired.

Based on the criticality of this issue, the repair of a concrete or steel structure in general should be done by a competent structure engineer who is capable of evaluating the existing structure and reaching a decision from a small amount of data. Thus success mainly depends on the engineer's previous experience and having a sense of load distribution and the performance of the aging structure.

Therefore the responsible engineer should design the temporary support based on the collected data and the previous analysis and be cautious in the execution phase of the temporary supports.

The choice of methods for removing the defective parts will be based on the location of the concrete member in the building; any concrete member weakness will directly have a serious impact on the adjacent members as per the structure system, considering that the process of breaking itself will produce a high level of vibration that will affect the surrounding members.

The temporary support should be properly designed to carry the loads and transfer and distribute them to another member safely.

5.4.3 Demolish the delaminated concrete

In general, many methods can be used to demolish the concrete cover and any cracked part of the concrete member.

The method of removing the concrete cover depends mainly on the characteristics of the member, if it is inside the building or outside, and the size and volume of the concrete that is required to be removed. For example, removing the delaminated concrete cover for a bridge deck is different than a beam inside a building, and so on.

If the corrosion in steel reinforcement is due to chloride propagation into the concrete, most specifications recommend removing about 25 mm behind the steel reinforcing bars and ensuring that most concrete around the steel bars has no traces of chlorides after performing the repair process.

There are many methods commonly used for demolishing and removing the defective concrete; these methods include:

1. Manual method

This is a common method in traditional building, as it is one of the simplest methods; the hammer and chisel are usually preferred to be made of copper. This method should be done by experienced labor to remove the defective concrete part only. It is considered one of the cheapest methods as well as easiest, but is slower compared with mechanical methods.

However, it is worth considering that the mechanical methods produce loud noises and vibration and have special requirements for trained labor.

Using the manual method, it is difficult to remove concrete behind the steel bars. This method is used in the case of small spaces and the preferred use is in the event of a corrosion due to carbonation and attacking chlorides from outside, so there is no need to break concrete behind steel.

Note that any laborers can manually break the concrete, but laborers should be chosen who have previous experience with repair work, as they must be sensitive in breaking the concrete to avoid causing cracks to the adjacent concrete members. The adjacent member may be in good condition, so it is important to avoid any cracks that can occur to it during this process.

2. Pneumatic hammer methods

These hammers operate by compressed air and weigh between 10 and 45 kg; they can demolish about 0.025 to 0.25 m^3 per hour. Those used in the roof or walls weigh about 20 kg.

The worker that uses this tool needs proper training. According to Vorster et al. (1992), the compressed air hammers have a few initial costs, which were discussed by the Strategic Highway Research Program (SHRP), on a research program of highways in which the terms of the contract are governed by the contractor, which will be used in breaking.

In cases where only a small area has to be removed, the use of pneumatic hammers is more economical than for large areas, as for a large area it is preferred to use the water gun, which will be discussed in the following section.

However, with clients who do not specify the area that needs to be removed in the tender and the cost calculation is based on square meters, in this case the risk will be low because the machine's initial cost is small, so in this situation it is preferred to use a pneumatic hammer.

In summary, pneumatic hammers are useful in the case of small areas of concrete to be removed but are not preferred for use in removing a large area of concrete.

3. Concrete removal by water jet

From 1970 the water jet method has been commonly used, since it first appeared on the market. It relies on the presence of water at the work site, depending on the removal of a suitable depth of concrete in a large area. It can handle the removal of fragmented concrete and cleaning steel bars, and also it removes part of the concrete behind the steel bars.

The water jet is a manual tool used by technically experienced workers who have dealt with a hose; it pushes water under high pressure so it requires very high safety precautions for the workers who use it and the whole site around it.

Potable water shall be used and any water with higher chloride should be avoided, as it will be a dramatic problem, as discussed before.

A water gun consists of diesel engines and is related through a pressure pump and connected by a hose to bear high pressure resistance of water and water pressure at the gun nozzle to 300–700 kg/cm^2. It requires at least 400 kg/cm^2 to cut the concrete and the rate of water consumption is about 50 L/min.

The performance rate of a water jet to break is between 0.25 m^3/h and in the case of the use of a small pump, it can reach up to about 1 m^3/h in the event of the use of two pumps or one big pump.

4. Grinding machine

Grinding machines for demolishing the concrete cover are usually used on flat surfaces like the bridge deck. All precautions should be considered to avoid reaching the steel bars to avoid damaging them.

If any contact occurs between the grinding machine and the steel reinforcement, it will cut the steel bars and damage the machine.

In the United Kingdom they do not use this method in most cases, as their construction procedures usually include placing insulation film to prevent water entry; however, in the United States this tool is used sparingly in bridges.

The rate of removal of the concrete by this machine is very rapid: it removes about 1 m^3/min when using a machine with a cutting part of 2 m width.

5.4.4 Clean concrete surface and steel reinforcement

In this stage the main task is to remove any remaining broken concrete by a cleaning process and, at the same time, assess the steel and clean up and remove corrosion from the roof.

5.4.5 Bond new concrete with old concrete

The stage of preparing the surface by pouring the new concrete is one of the most important stages of the repair process, specifically regarding the bond between the new concrete and the old concrete.

Before applying the primer, the coating that provides the bond between the existing old concrete and the new concrete for repair, the concrete surface must be well prepared based on the materials used.

In all cases the concrete surface must be clean and must not contain any oils, broken concrete, soil, or lubricants. If any of these are present, the surface must be cleaned completely through sand blasting, water, or cleaning using brushes done manually. This stage is crucial, regardless of the type of material used to bond the new concrete with the old concrete, as the lack of attention to the preparation of the surface might affect the repair process as a whole. However, this stage of repair is a less expensive stage in the repair process.

One of the advantages of using a water gun for removing the concrete cover is that there is no need to prepare the surface to remove the delaminated concrete, because the surface will be wet enough and clean. A normal process is to use air pressure to clean the surface of the concrete and remove any debris or concrete fragmented to prepare the concrete surface to have a strong bond with new concrete repair.

If using cement mortar or concrete in the repair, the concrete surface will be sprayed by water until saturation can be reached. The necessary stage of saturation can be attained by spraying water on the concrete surface for 24 h or through wetting burlap. The spraying of water must be stopped or the burlap removed for about 1 h to 2 h, to ensure that the surface is dry before applying the coating, also considering the weather conditions during this time.

The use of epoxy to bond new concrete with old concrete with a brush must follow all specifications and precautions stated by the manufacturer.

The preparation of a concrete surface to achieve a better bond between the old and new concrete is stated in the American specifications ACI 503.2-7. In this specification they mention that the preparation of surfaces to receive epoxy compound applications must be given careful attention, as the bonding capability of a properly selected epoxy for a given application is primarily dependent on proper surface preparation. Concrete surfaces to which epoxies are to be applied must be newly exposed, clean concrete free of loose and unsound materials. All surfaces must be meticulously cleaned and be as dry as possible, and be at proper surface temperature at the time of epoxy application.

The epoxy materials that perform the bond are specific in the ASTM C881-78, which must be well-defined by the supplier specifications and identical with the different circumstances surrounding the project, particularly where the temperature that with the increase of temperature change might occur in a resin, as well as the nature of loads that is carrying by the required repair member.

The use of epoxies will be common for different processes during the repair; therefore the worker safety for those who use such materials should be considered, as the workers must wear their personal protective equipment (PPE) such as gloves and special glasses for safety issues, because these epoxies are very harmful to skin and cause many dangerous diseases if any one is exposed to them for a long time. Note also that the epoxies are flammable with temperature, and that must be considered during storage and operation.

5.4.6 Clean steel reinforcement bars

After removal of the concrete covers and cleaning of the surface, the next step is to measure the diameter of the steel bar to define the amount of its reduction and do an assessment based on that.

In the first assessment, as a rule of thumb, if the cross-sectional area of the steel bars is reduced equal to or more than 20%, then additional reinforcing steel bars should be added. The amount of extra steel bars will be determined based on the redesign calculation. The addition of steel bars should have a development length based on the code of design when it is connected with the existing steel bars.

It is usually preferred to connect the new steel bars by drilling new holes in the concrete and connect the additional steel on concrete by putting the steel bars in the drilled hole filled with epoxy. But in most cases, the steel bars are completely corroded and need to be replaced.

In the case of beams and slabs that need to have additional steel reinforcement bars added, the preference is to connect the steel bars with concrete by drilling new holes in the concrete and making the bond of the steel bars in the holes by using adhesive epoxies.

For repair beams, the additional steel bars are fixed in a column that supports this beam. For slabs the steel bars are fixed in the sides of the beam that is supporting the slab, as shown in Fig. 5.25. Fig. 5.26 presents adding steel for slabs and beams. For the doweling, holes are drilled with a depth of around 70 to 80 mm, and the dowel will be fixed in the holes by epoxy.

The depth of the hole may differ depending on the type of the epoxy and the supplier assumption and recommendations. The hole diameter will be

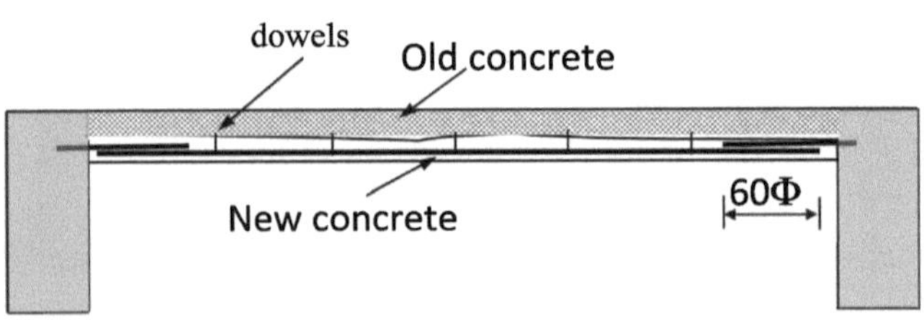

Fig. 5.25 Steel reinforcement installation for slab repair.

Fig. 5.26 Adding steel for slab repair.

20 mm greater than the bar diameter to ensure the steel bars are fixed completely.

Typically 12 or 16 mm is used, and it is preferred to increase the bar diameters and decrease their number to reduce the number of holes in the beams. The dowel is put in the drilled hole, as shown in the Fig. 5.25, with development length 50 times the bar diameter, which will be overlapped with the new steel bars and the fixing between them by the steel wire or welding. The bars are fixed in the two directions by putting a small dowel from the slab to be in the intersection between the steel bars in the two directions and these small dowels are fixed to the slab by epoxies also.

The structure system of the balcony is a cantilever, which has the lowest number of redundancy members and it exposed directly to the weather, as it is an external member. Thus it is the first member affected by corrosion in coastal areas.

In most cases, from different structure diagnoses, it is found that the corrosion usually occurs on the upper steel reinforcement, which is more critical, as the main steel bars in the cantilever are the upper steel.

Fig. 5.27 shows a common method to repair the cantilever. In this method, the floor finishing tiles, or wood, or others are removed with the concrete cover and then the reduction on the upper steel bars is inspected, as mentioned earlier, in the same steps.

If it is found after inspection that it is required to add a new upper steel reinforcement, then place the upper steel bars and make them extend into the adjacent span by a length equal to 1.5 times the cantilever length, as shown in Fig. 5.27.

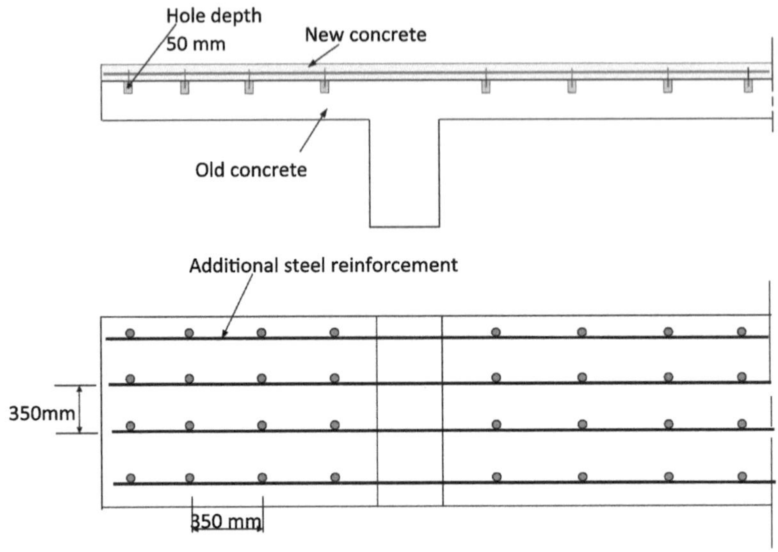

Fig. 5.27 Cantilever repair.

The dowels will be installed vertically and fixed by the concrete slab in a depth of 50 mm with epoxies. The distance between the dowels is about 350 mm and the new main steel bars are attached by it.

The redundancy of the cantilever structure system is very low, so it needs more care in doing the repair design and execution, such as the wood form and temporary supports and the part of the building that will be overloaded during the repair process, in order to define the load path during the repair process, considering that the structure is mature, so a good assessment is very important.

As for the beam repair process, note that the work of repair of the beams is also the same as the slab repair process, but with the difference that fixing the dowels is in the columns, as shown in Fig. 5.28, and the stirrups are fixed through making holes with depth around 50–70 mm as per most epoxy manufacturer recommendations; on both sides of the beam an epoxy is used to stabilize the stirrups, about 8 mm per 200 mm, as shown in Fig. 5.28.

After that, an epoxy coating is applied as a normal procedure, as an adhesion between old and new concrete, and then the new concrete is poured manually or by shotcreting as shown in Fig. 5.29.

As shown in the figure, the beam is strengthened by increasing the cross-sectional dimension and adding steel reinforcement.

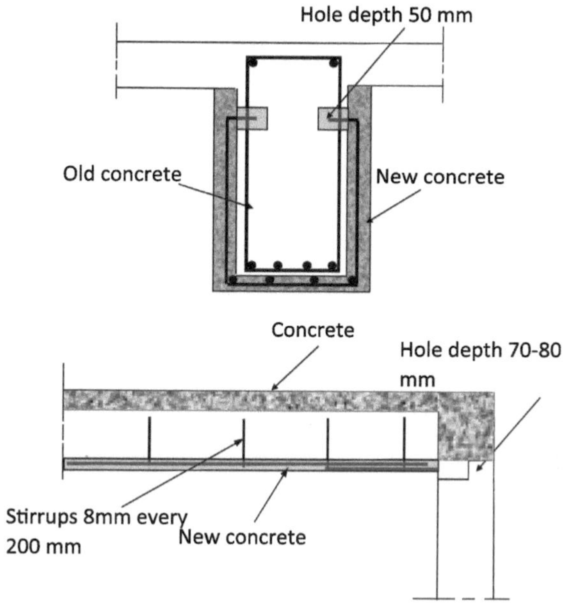

Fig. 5.28 Beam repair.

Fig. 5.29 Shotcrete application.

As for column repair, the addition of steel as shown in Fig. 5.30 and for the first floor will be as shown in Fig. 5.30, which shows how the casting on-site is done, but note that the minimum distance allowed to cast concrete easily is about 100–120 mm from each side.

Fig. 5.30 Adding steel bars for the column.

The dowel is fixed on the first floor as in Fig. 5.31, which does not require drilling holes in the base, but making the legs to steel bars to rest on the foundation. The steel bars will be distributed around the column circumference to compensate for the reduction of steel bars cross-sectional area due to corrosion.

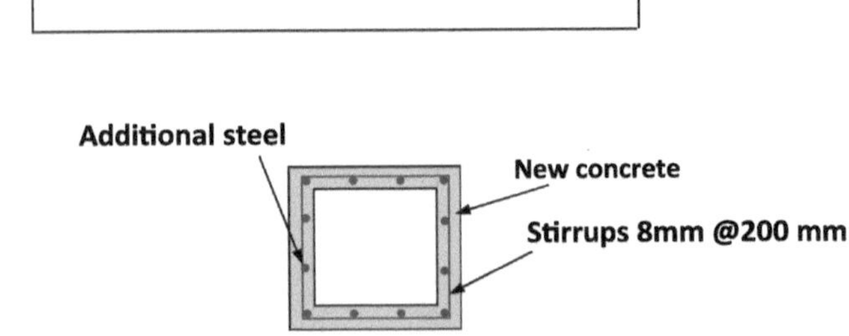

Fig. 5.31 Repair concrete column in first floor.

The old steel bars are coated by epoxies after their surface is cleaned completely, especially from the effects of chlorides. Sometimes the approach is to use slurry, which is a mixture of rich cement and water, instead of epoxy, as it will protect steel from corrosion due to the alkaline coating around the steel bars.

The cement slurry may dry quickly, so that it will not be effective in repairs that require the installation of form after painting the bars, but it works well in cases where the time between painting steel and pouring cement mortar is short; this period should be less than 15 min, as stated by U.S Army[8] manual in 1995.

5.4.7 New concrete properties

There are some mixtures available on the market that have been mixed in a certain way so that they can be used easily in the repairs of small areas, as this mixture is highly expensive.

In the case of repair of large surface areas, the mixture preparation and mixing will be on-site for less cost, but this must be carried out by experts.

In addition to those properties of concrete characteristics, using a pump for pouring or shotcrete needs a special design mix.

In the United States and Canada, most of the contractors who work on bridge repair have their own concrete mixture, with the secret of the mixing proportions, based on the available materials in their local market.

The ready-mix concrete factories can provide a guarantee in the case of corrosion due to carbonation, but if it is due to chlorides they do not provide a guarantee, as they are afraid of the presence of chloride after the repair process.

Note that most manufacturers of materials used in mixing are field execution contractors, but when they supply materials only, they will provide all the information and technical recommendations for the execution, with performance rates to calculate the required amount of the materials and appropriate method of operation.

Worldwide, all companies operating in construction chemicals have a competent technical staff to assist on-site or to supervise, to avoid any errors and to define the responsibility in case of any defects.

There are two types of materials used as a new mortar for repair: polymer mortar and cement mortar enhanced by polymer. Both of them are described in the following sections.

A.Polymer mortar

This mortar was used in the past as it provides the required workability and good protection from corrosion as a physical barrier. The polymer mortar provides a good bond with the old concrete, so in this case there is no need to use epoxy to achieve a bond between old concrete and new concrete. This mortar has a high compression strength of around 50 to 100 MPa and high tensile strength also.

Despite the many advantages of this mortar, problems have arisen due to the difference in the properties of this mortar and the existing concrete, as the polymers have a coefficient of thermal expansion equal to 65×10^{-6}/°C corresponding to the concrete thermal expansion coefficient equal to 12×10^{-6}/°C. Moreover, the modulus of elasticity for mortar is much less than that of concrete. These differences in the properties lead to the presence of cracks as a result of generating internal stresses, but this can be enhanced by using aggregate with sieve grading to make the proportion of polymers fewer, which reduces the difference in the natural properties.

There are many tests to define this mortar and they are stated in ASTM C 884-92 Standard Test Method for PCBs Comparability between Concrete and Epoxy-Resin Overlay.

B.Cement mortar

The benefit from the polymer mortar is that it provides a physical protection to the steel bars. However, the cement mortar gives passive protection, as it increases the alkalinity around the steel reinforcement. The trend these days is to use cement mortar and adding a polymer, as its properties will match with the existing concrete to avoid the difference in thermal coefficient and at the same time it gives passive protection to steel, the basis for the protection from steel corrosion.

Ohama (1995) and ACI Committee 548 (1994) describe the use of some polymer additives to the cement mortar through liquid or powder to improve some of the properties of the cement mortar and to increase the flexural resistance and increase the elongation, reduce the water permeability, increase the bond between the old and new concrete, and increase the effectiveness of its operation.

The polymer used is identical with the cement mortar mixture and the properties that control the polymers in the cement mortar are stated in ASTM C 1059-91 Specification for Latex Agents for Bonding Fresh to Hardended Concrete.

Silica fume can be used with the mixture, as well as super plasticizers to improve the properties of the mortar used and reduce shrinkage.

5.4.8 Tank ring beam repair

The following figures depict a case study of the repair of a reinforced concrete ring beam supporting steel tanks containing oil, as in the case study in Chapter 4.

The ring beam structurally carries a tensile strength due to earth load affecting the beam, which causes the ring beam to be under tension. Due to corrosion of the steel reinforcement bars, there is a mean reduction of the capacity of the ring beam. So, the best solution is to increase the steel bars to overcome the reduction of the steel cross-sectional area.

Fig. 5.32 depicts the process of repair. Step 1 is to drill holes and put steel stirrups in the holes, fixed by epoxy.

Step 2 is to install the main steel bars and coat the concrete surface by epoxy to bond the existing old concrete with the new concrete.

In Step 3, the new concrete is poured with the proper mixture design improved by additives such as polymers.

The repair of the sleepers underneath the pipeline is shown in Fig. 5.33. The figure presents the adding of the steel bars and then pouring the concrete either as normal, or a shotcrete can be used.

5.4.9 Objective of strengthening concrete members

Generally, the goal of strengthening the structure is the restoration of the ability of the concrete member to withstand loads, either to provide or increase the ability to carry loads more than the designed-for loads, or to reduce deflection that occurs when the concrete member is overloaded. The strengthening process can proceed as a result of the presence of cracks and should stop the growth of cracks or reduce the crack width in the concrete member.

Therefore, strengthening is required in the following cases:

- An error in the design or execution that causes a reduction in the steel cross-sectional area or a decrease in concrete section dimensions.
- A deterioration in the state of the structure that has led to the weakness of the resistance; an explicit example of this is corrosion in steel reinforcement.

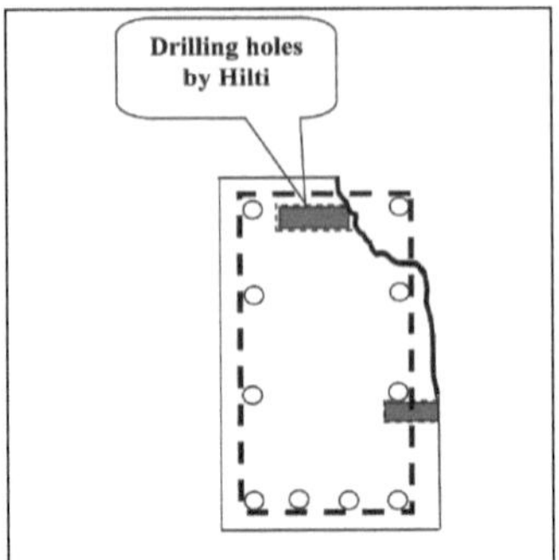

Step 1: To drill holes

Step 2: Insert stirrups and add steel bars

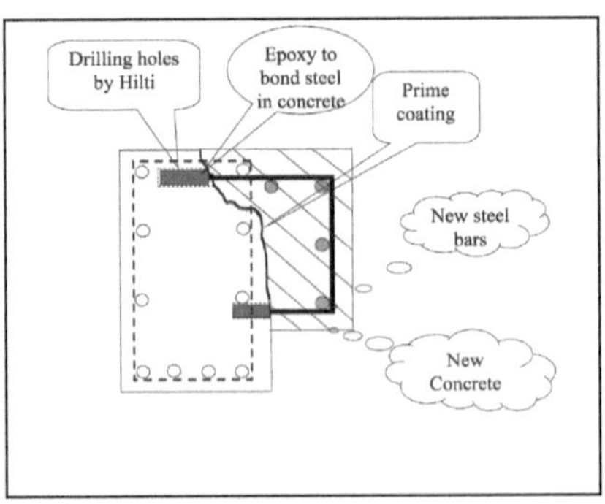

Step 3

Fig. 5.32 Tank ring beam repair steps.

Fig. 5.33 Adding steel for pipe support.

- An increase in loads; as an example an increase the number of stories or a change in the building function such as a change from residential building to office building.
- A change in the structural system such as removing walls.
- Making holes in the slab, which reduces the slab strength.
- Some special structures need strengthening, such as bridges if the flow of traffic is increasing with time or if the flow of traffic and number of heavier trucks are greater than the design so strengthening the bridge is needed to accommodate the overloaded and new operations mode.
- The load may change with time such as a building in an earthquake zone that changes with time; for example, in Egypt the earthquake map was changed and consequently some buildings needed strengthening to match with the specifications.
- The structure is exposed to high temperature due to fire.
- An industrial structure needs a large machine to be installed.
- Also, in an industrial building the machines are mature with time or they change mode of operation, which results in higher vibration so strengthening is needed to the structure.

5.4.10 Slab on grade repair

A slab on grade usually exists in industrial plants in general, such as the floor for a workshop. Usually the cracks will appear as shown in Fig. 5.34. If it is

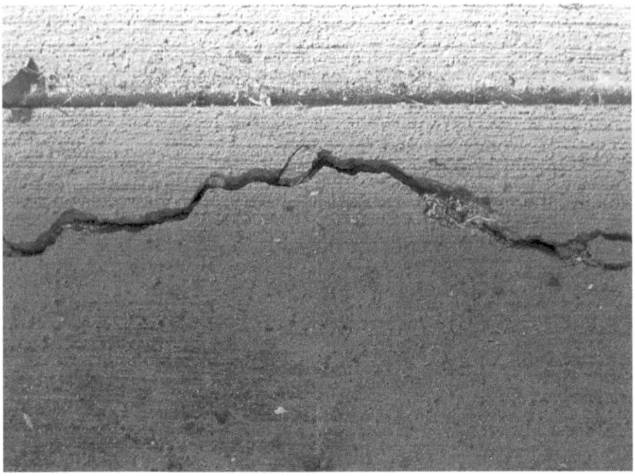

Fig. 5.34 Cracks for slab on grade.

static and there is no increase in the crack width, the most common method is opening the crack, cleaning it, and then filling it with epoxy, as shown in Fig. 5.35A, B and C.

5.4.11 Strengthening concrete by steel sections

Using steel sections on a large scale to strengthen a reinforced concrete structure has several advantages. The method of strengthening the concrete member can vary depending on the member that needs to be strengthened. The strengthening process is calculated based on the reduction in strength in the concrete section and this reduction is compensated for by using a steel section.

If a concrete beam needs strengthening due to a reduction of steel, reinforcement can be used in a U-shape, as in Fig. 5.36, when a big change from an architectural view is not needed and there is only a minor reduction in member capacity.

On the other hand, in the case of a significant reduction in concrete member capacity and a need to increase the depth significantly, the I-beam section as in Fig. 5.37 is used. In most cases mechanical bolts or epoxy adhesive are used to connect the steel section to the concrete.

Another older method is to strengthen beams by adding in the bottom of the beam steel plate in the ends and one in the middle with a length around 300 to 500 mm and width around 100 mm, with thickness defined by the calculation to compensate for the reduction in inertia or to increase its

Fig. 5.35 (A) Step 1: Crack opening. (B) Clean opening from debris and dust. (C) Fill with epoxy.

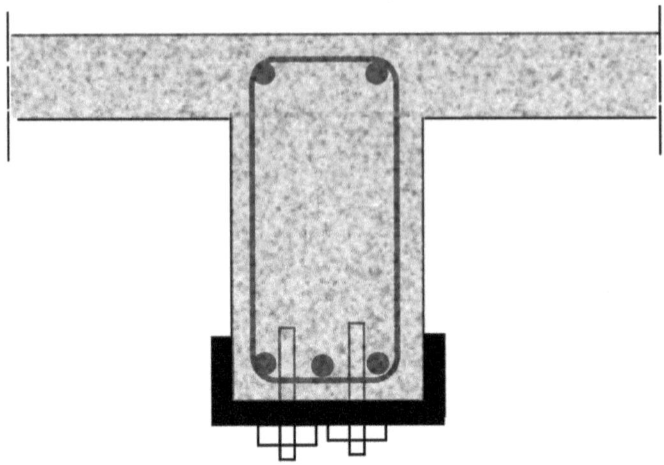

Fig. 5.36 Beam strengthening by using channel section.

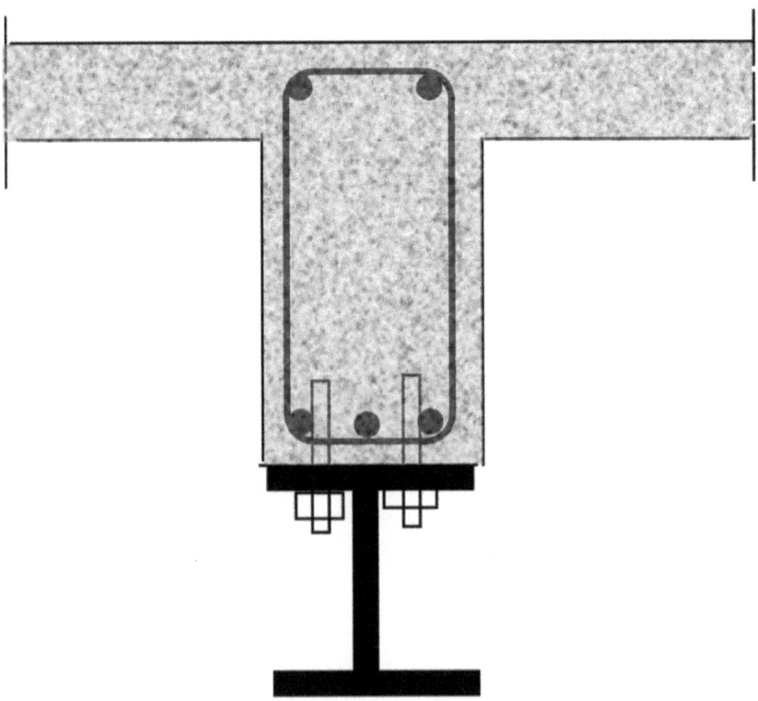

Fig. 5.37 Beam strengthening by using I-beam section.

strength to add more load. These three plates are fixed by mechanical bolts or through chemical substances such as epoxy. The purpose of adding this steel sheet is to increase the moment of inertia of the beam, thereby increasing its ability to withstand stress more than the design flexural stresses.

If there is a shear problem, such as the presence of cracks due to a shortage in steel reinforcement, the section can be strengthened to carry the shear load, as shown in Fig. 5.38.

The presence of cracks on the upper surface of the slab in the beam direction means that there is a shortage in the upper reinforcement of the slab and it can be strengthened by adding steel plates on the upper surface of the beam and fixing these plates by bolts with nuts, which will be fixed from the lower direction of the slab as shown in the figure.

Strengthening a flat slab is performed as shown in Fig. 5.39, with the steel angles fixed between slab and columns.

Short cantilevers are used in the frames in factories or bridges; the main high stresses are due to shear stresses and the other stresses are due to flexural stress. Strengthening the cantilever is as shown in Fig. 5.40, through adding at its two sides a steel sheet and applying compression force with bolts, thereby reducing the likelihood of a collapse due to shear.

5.4.12 Fiber reinforced polymer (FRP)

Nowadays, for many reasons, there is a trend toward using FRP systems for strengthening and repairing concrete structures. Using FRP, which is fibers in resins, is still expensive compared with traditional methods of repair, such

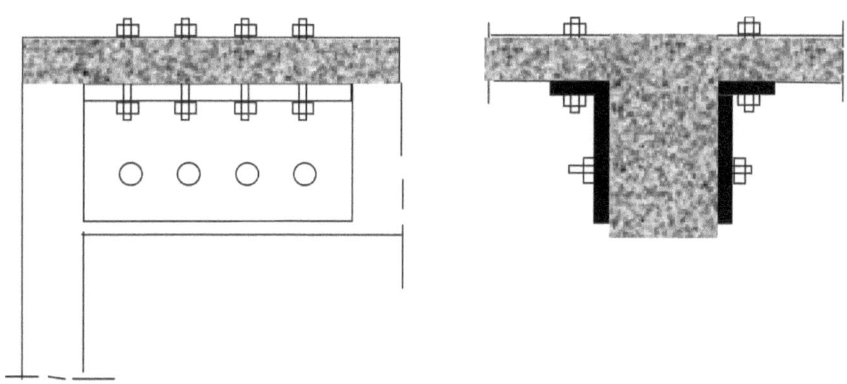

Fig. 5.38 Beam strengthen in direction of shear.

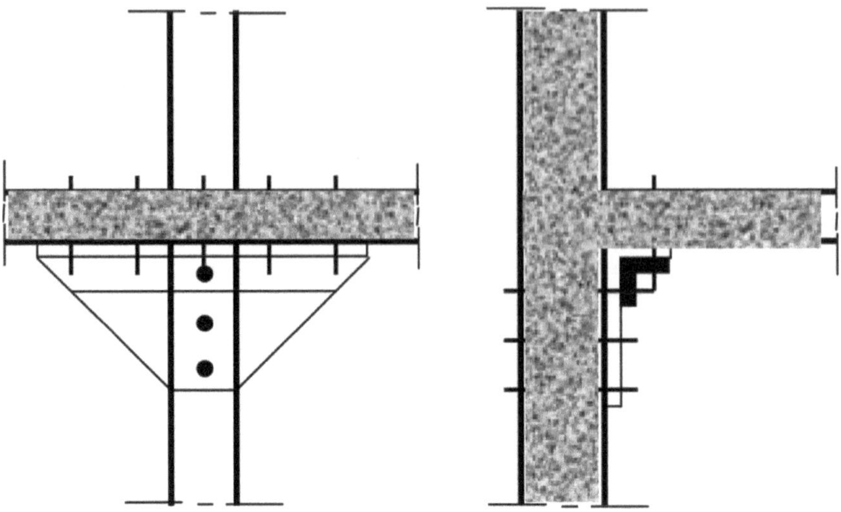

Fig. 5.39 Strengthening the flat slab.

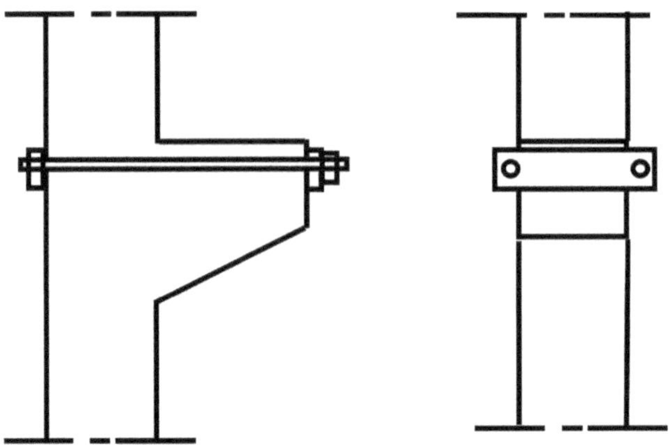

Fig. 5.40 Strengthening short cantilevers.

as by using concrete or using steel sections. However, the labor and equipment costs to install FRP systems are often lower.

In case of limited access areas such as a concrete slab shielded by pipes and conduit, using FRP systems can be the best choice rather than using the traditional strengthening methods as discussed earlier. Also, FRP systems will have lower life-cycle costs than conventional strengthening techniques because the FRP system is not affected by corrosion.

FRP is currently used in a wide range of applications and it is considered a very successful alternative to steel sections used in strengthening due to the distinctive characteristics. For example, it does not corrode over time and can be used in harsh environments. By using FRP it will be not affect of corrosion so it will be durable along its life time and also resist the heat, electricity and chemical by selecting the right type.

Different kinds of FRP are available on the market, and the most common type used is carbon fiber reinforced polymer (CFRP reinforced), as it is appropriate for practical applications and because of its unique properties in terms of resistance and the resistance with time, as well as resistance to stress.

5.4.12.1 Carbon fiber reinforced polymer (CFRP)

CFRPs are available on the market and are growing very rapidly. The available dimensions for most manufacturers of CFRP are commonly 1–1.5 mm thickness and with width of 50–150 mm. All these types consist of 60%–70% of carbon fiber in one direction, with a diameter of about 10 μm, embedded in epoxy resin.

The mechanical properties are different from one type to another, but generally the modulus of elasticity has a value ranging between 165 MPa to 300 MPa and the tensile strength has a value range of 2800 MPa to a lower value of 1300 MPa; we find that the lower value of the modulus of elasticity corresponds to the maximum tensile strength value.

The carbon fiber sheet is as shown in Fig. 5.41A and B; this sheet is embedded in resin of epoxy.

The important property is the density. The density of carbon fiber is about $1.6\,t/m^3$, while the density of steel is $7.8\,t/m^3$; therefore there is a big difference since the weight is about 10 times less than the weight of steel and this is what makes CFRP so much easier to install and repair.

The adhesive materials used with the sheets of carbon fiber resin are a mixture of high resistance and with the filling quartz and its tensile strength is about 30 MPa, up from the concrete tensile strength to about 10 times. In addition, the rate of shrinkage and creep is small and it bears high temperatures and exposure to chemical well.

The mechanical properties of carbon fiber sheets in the longitudinal direction are often controlled by the fiber and its behavior is linear elastic until collapse.

5.4.12.2 Application on-site

The secret of successful strengthening by CFRP is to have an excellent application. Before starting to work on the strengthening of a concrete structure,

Fig. 5.41 (A) Carbon fiber textile flexible sheet. (B) Carbon fiber rigid sheet.

it is necessary to work with a full examination of all the circumstances surrounding the structure in order to determine that it is necessary to strengthen the structure. In most cases, the materials manufacturer has a list of designer offices and applicators. These applicators are trained by them and they should be on their list, as they will guarantee the materials if the application is done in an appropriate manner, to protect their professional reputation.

If strengthening is needed because of steel corrosion or due to chemical attack, it is important and necessary for the success of the process of strengthening to remove rust from the steel bars and any chemical substances that have attacked the concrete before applying the strengthening procedure.

The concrete surface must be settled through sand blasting for leveling the surface, eliminating any gaps or delaminated concrete after applying the sand blast.

The next step is to collect all the damaged concrete and sand by air suction, air pressure, or water pressure to clean the surface.

It is mandatory before fixing the CFRP that the surface is leveled and clean and has a concrete strength the same as the member. In case of any cracks or delaminated concrete, the surface must be repaired before fixing the CFRP.

Prepare the epoxy that provides the bond between the segments of CFRP and the concrete surface. It is crucial for a successful strengthening, as this bond strength is mainly responsible for transferring the stresses from the concrete to the CFRP.

Based on this, the preparation of the epoxy mix and the percentage of mixture components must be precise and should be under supervision of the consultant engineer. The contractor should use workers with long experience in application, transportation, handling, and storage from both technical and safety point of views.

These epoxy materials are usually a blend between (a) and (b), which will be mixed by an electric mixer for a period of about 3 min, until the mixture homogeneity is clear. When there is a turbulence in homogeneity at the bottom and sides of the container containing a mixture, it is required to do slow mixing. In general, the highest allowed is about 500 rpm.

The epoxy which will be applied to the concrete surface is about 0.5–2 mm thick and must be applied strictly and accurately, making sure that all the gaps have been filled and there are no air voids.

The repair of a beam due to over shear stress or reduction in shear stress capacity is as shown in Fig. 5.42.

To increase the flexural strength of a concrete beam, the procedure is the same as shown in Fig. 5.43, by adding steel plates on the bottom of the beam. In this case CFRP is applied, as shown in Fig. 5.43. It can be shown that it will be simpler than using steel plates and also there is no effect in decoration,

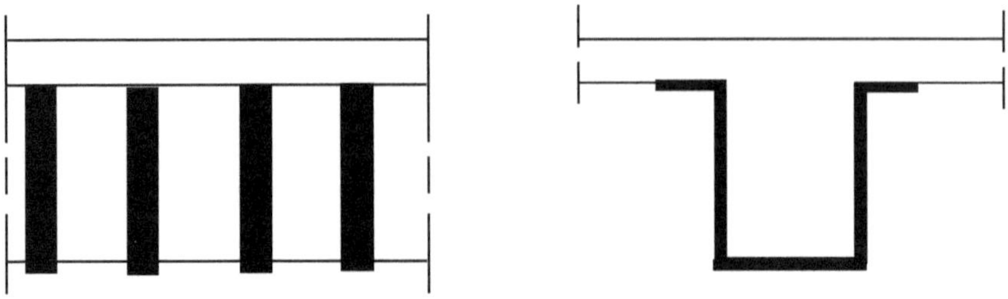

Fig. 5.42 Strengthening beam for shear failure.

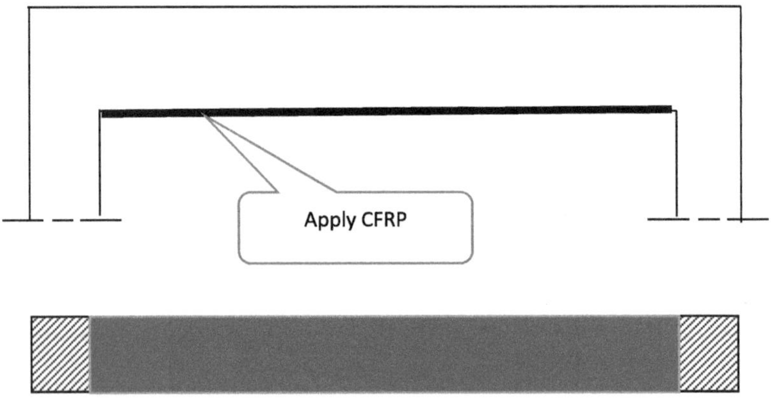

Fig. 5.43 Steel plate to strengthen concrete beam.

as its thickness is about 2 mm, which is less than the steel, which can reach 20 mm.

A column concrete repair using CFRP is shown in Fig. 5.44. In this case a sheet of CFRP is wrapped around the concrete column and it is preferred to be a rounded column to increase confinement and consequently increase the column capacity.

5.4.12.3 General precautions

Note that the safety of a structure depends mainly on the quality of the repair process and procedure during the execution phase and design or planning. As mentioned earlier, before starting the repair process it is mandatory to establish temporary support for the building and the location, depending mainly on the structure system characteristics and the load path of the existing condition.

Fig. 5.44 Column repair.

Also, determining the method of the breakers of the defective concrete and the method of removal, as well as choosing the materials to be used in the repair, must all be in full conformity with the state of the structure and the nature of the site of the structure or the location of the retrofit members within the overall structural system. All of these elements require special expertise. Without these experiences in the repair process, the possibility of risk in terms of safety of the structure is very high. In addition to that if the repair is not effective it will be a waste of money.

From an economic point of view, it must be remembered that the repair process in general costs a lot of money due to using special materials and competent resources, in addition to the required precautions that should be followed during execution, or in terms of training workers to carry out the repair process.

The repair process requires attention to the safety of the building, to protect the health and safety of the workers doing the repair, as it often uses epoxy materials or polymers that can cause breathing and skin problems over time. In addition, other tools in use require cautious consideration for health and safety.

Unfortunately, one tends to find that HSE precautions and rules are well observed in the richer countries and mainly in the oil and gas business, while not receiving the same degree of attention in the developing countries and businesses other than petroleum investments.

References

Abdullah, E., Pegah, Z., Arman, Z., 2019. An overview of marine pipeline repair methods. In: 8th International Offshore Industries Conference June 2019, Tehran. Sharif University of Technology.

ACI Committee 548, 1994. State of the Art Report on Polmer Modified Concrete. ACI Manual of Concrete Practice. American Concrete Institute.

Antaki, G.A., 2003. Piping and Pipeline Engineering-Design, Construction, Integrity, and Repair.

Ohama, Y., 1995. Handbook of Polymer-Modified Concrete and Mortars. Noyes Publication, USA.

Vorster, M., Merrigan, J.P., Lewis, R.W., Weyers, R.E., 1992. Techniques for Concrete Removal and Bar Cleaning on Bridge Rehabilitation Projects, SHRP-S-336. National Research Council, Washington, DC.

Further reading

Abdul-Wahab, H.M., 1989. Strength of reinforced concrete corbles with fipers. ACI Struct. J. 86 (1), 60–66.

ACI Committee 506, 1994. Proposed revision of: Specification for materials, proportioning and application of shotcrete. ACI Mater. J. 91 (1), 108–115.

American Concrete Institute (ACI) Committee 440, 2000. Guide for the Design and Construction of Externally Bonded FRP Systems for Strengthening Concrete Structures. September.

Department of the Army U.S. Army Corps of Engineers Washington, 1995. Engineering and Repair of Concrete Structures, DC Manual No. 1110-2-2002.

Mufi, A., On Ofei, M., Benmok, B., Banthica, N., Boulfiza, M., Newhook, J., Bakht, B., Tedros, G., Brett, P., 2005. Durability of GFRP reinforced concrete in field structure. In: 7th International Symposium on FRP for Reinforced Concrete structure, (FRPRC5–7), USA, November 7, 2005.

RILEM Committee 124-SRC, 1994. Draft recommendation for repair strategies for concrete structures damaged by steel corrosion. Mater. Struct. 27 (171), 415–436.

CHAPTER 6

Proactive approach to integrity

6.1 Introduction

The main concept of integrity management is to be proactive in taking remedial action, before an event occurs that causes the action to be converted to a corrective action. When this happens, it will be out of the integrity management standard.

The major actions are to select materials and to establish a maintenance system for the facilities during their lifetime. The selection of the materials and system depends on many factors, such as any experience with problems in the past, the mode of operations, the maintainability of the system selected, and the economics of choosing between the alternatives.

It is crucial to include the integrity engineer in any new projects to provide input into selecting the best materials and protection system specific to the selection phase and during the front end engineering design (FEED).

6.2 Onshore structure protection

The minimum requirements in most of the codes are often insufficient to guarantee durability of a reinforced concrete structure during its lifetime in a harsh environment, such as marine structures in splash zones, bridges, and parking structures where deicing by salts is a factor, or in the case of a normal building with fewer quality-control procedures.

In addition, some of the newer structures, for example, hotels, commercial buildings, and others, constructed in marine areas within 200 m of the shore experience corrosion problems due to airborne chlorides.

Most oil and gas plants are located near the shore to receive and send fluids to and from the offshore facilities. Some are located in high temperature climates such as the Middle East, Singapore, Hong Kong, and South Florida, which also need special protection methods.

As discussed in Chapter 1, there are many factors that affect corrosion. However, with good quality control that follows the codes and standards precautions for concrete cover thickness, water-cement ratio (W/C), and

Asset Integrity Management for Offshore and Onshore Structures
https://doi.org/10.1016/B978-0-12-824540-8.00006-5

minimum cement content based on the environmental conditions a structure can be maintained for a reasonable time, given good construction quality control.

This is considered one of the main proactive approaches to maintain the structure integrity within its lifetime. The other approach is to use other assistance methods to be proactive and in general this approach is based on time: extend the time until chloride or carbonation effects occur, or until sulfate reaches the steel reinforcements of the foundation.

The time for corrosion to start can be predicted by the choice of protection methods. This concept applies to offshore and onshore structures, pipelines, tanks, or any production facilities. From this prediction, the required inspection time can be arranged and also the economic analysis to choose between alternatives can be done. The other factor is the characteristics of the structure. For example, in Dubai there is a tunnel that is critical; they applied a cathodic protection system that was very expensive, but if repairs are needed to the tunnel, the traffic would have to stop and this would cause a dramatic problem. On the other hand, in oil and gas plants, when a repair starts, it will affect production: for example, whether there is a standby for the unit or not affects the choice of the best protection method for the foundation under the unit.

The corrosion process for steel bars in concrete is discussed in detail in previous chapters, and also the different causes of corrosion of steel bars in concrete are discussed.

This chapter presents the most up-to-date techniques used to protect steel reinforcements in different applications and under different environmental circumstances. Recently, many research and development projects have been undertaken to provide economical and effective methods to protect steel reinforcements.

The method selected depends on protecting the steel reinforcements by performing design and construction using a good quality control system and taking all precautions in design and construction to avoid corrosion, based on the codes and specifications. These specifications can vary with different weather conditions surrounding the structure as well as the structure function, such as offshore, underground, or surface structures.

A second way of protecting the steel bars is by the use of additional protective techniques such as using galvanized steel bars, epoxy coatings, stainless steel, external membranes to prevent water permeability, additives such as cathode or anode inhibitors, or by use of advanced methods such as the cathodic protection (CP) method.

All these methods have advantages and disadvantages and the most suitable method should be selected during the front end engineering design (FEED). Thus it is recommended that the integrity team should be involved in the project in the early stages, since the selected method depends on the criticality of the structure during operations and also the most suitable method should be determined from a maintenance point of view.

Most of the engineering firms and project engineering teams focus on following the codes and standards and determining what is available on the market. However, the choice of the best method and its efficiency on the site is mainly the responsibility of the integrity team, since they have the experience to decide on the optimum options for the structure from the operations and maintenance points of view.

The methods of protection with their advantages and disadvantages are presented in the next section.

6.2.1 Corrosion inhibitor

There are basically two types of corrosion inhibitors, anodic inhibitors and cathodic inhibitors.

The first type of corrosion protection is the anodic inhibitor, which depends on a passive protection layer for the steel reinforcement. The second type is the cathodic inhibitor; its protection is based on preventing the propagation of oxygen in the concrete. It has been observed that the most effective protection is the anodic inhibitor over the cathodic inhibitor, so the anodic is commonly used in practice.

Now we will explain each type of protection, to clarify the advantages and disadvantages of each.

6.2.2 Anodic inhibitors

The most common anodic inhibitor is calcium nitrate. It is compatible with the process of pouring concrete on-site, having no impact on the properties of the concrete in either the fresh or the hardening state.

There are two other types, sodium nitrate and potassium nitrate. They are highly efficient in the prevention of corrosion, but they are not used because concrete has aggregates with high alkalinity and these chemicals react with cement and cause extensive damage to the concrete.

Calcium nitrate has been used widely since the mid-1970s and has been found to accelerate the concrete setting time. Therefore retarders are added to the concrete mix at the mixing plant to avoid this setting time acceleration.

Table 6.1 The required quantity of calcium nitrate to protect steel bars from chlorides.

Calcium nitrate amount (kg/m^3 and 30% solution)	Amount of chloride ions on the steel (kg/m^3)
10	3.6
15	5.9
20	7.7
25	8.9
30	9.5

This chemical will inhibit corrosion and provide stabilization of the passive layer, which helps to prevent damage when chloride ions are present at the steel surface.

Anodic materials are more efficient for concrete directly exposed to chlorides, such as contact with seawater, where the corrosion inhibitor reacts with chlorides and therefore increases the necessary amount of chloride concentration required to cause corrosion, based on test results, as indicated in Table 6.1.

The amount of calcium nitrate to add depends on the amount of chlorides the concrete is exposed to. This can be determined in practice or through previous experience of knowing the quantity of chlorides facing the structure.

6.2.3 Cathodic inhibitor

A cathodic inhibitor is added to the concrete during mixing. An advanced technique involves painting the concrete surface. The painting then propagates into the concrete porosity until reaching the steel bars, where it reduces the quantity of the oxygen that has propagated into the concrete by reacting with it; as discussed in Chapter 1, oxygen is the main driver of the corrosion process.

Many tests have been performed involving corrosion inhibitors and these materials follow ASTM G109-92. In reviewing these tests, it can be seen that their target is to define the effect of the additives on the corrosion of steel reinforcement embedded in concrete. In general, it is found that a corrosion inhibitor increases the protection of the steel bars more than control of concrete quality only. But it is found that the effect of cathodic inhibitors is less than the effect of anodic inhibitors.

From these tests it was found that to obtain higher efficiency from the cathode protection, it was essential to add a greater amount in the concrete mix. In addition, cathodic inhibitors like Ameen, phosphates, and zinc strongly retard the concrete setting time when a large quantity is used, so the setting time should be considered.

In summary, the anodic inhibitor is generally more effective than the cathodic inhibitor, so the anodic inhibitor is more practical in use.

6.2.4 Coating of steel bars by epoxy

Beginning in 1970, studies showed the importance of using epoxy to coat steel reinforcement bars in bridges and also marine structures, as certain types of epoxies were able to protect steel from corrosion. This method provided good results specifically in marine structures where steel was exposed to seawater, as shown in a study carried out by the Federal Highways Association (FHWA), which evaluated the use of epoxy coating of steel bars exposed to chloride attack.

Pike et al. (1972), Cairns (1992), and Satake et al. (1983) also showed the advantages and the importance of the use of coating steel bars. However, some precautions should be followed during application of the coating, such as avoiding any friction between the bars that would remove the coating from the bar surface.

With an epoxy coating, it is difficult to use methods for measuring the corrosion rate, such as polarization or half cell, so it can be difficult to predict the steel corrosion performance or measure the corrosion rate of the steel bars.

The coating of steel bars should comply with ASTM A 775M/77M-93. This standard has some limitations and guidelines that apply to acceptance of the coating:

- The coating thickness is in the range of 5–12 mils where (1 mil=0.001 in.).
- Avoid formation of cracks in the epoxy coating when bending the coated bar around a standard mandrel.
- The maximum number of pinhole defects is 6/m.
- The damaged area of the coating of the steel bars is less than 2%.

These issues usually occur during transportation, when putting the steel on the wood or steel shuttering, and during the preparation for pouring the concrete. Some precautions must be taken during these phases to avoid cracks in the paint, as described in Andrade et al. (1994) and Cairns

(1992), who defined methods of storage and steel reinforcement bending, carrying the steel, and pouring concrete.

By coating the steel bars, the bond between the concrete and steel will be reduced; therefore, it is required to increase the length of the steel bars to compensate for the reduction in bond strength. As defined in ACI 318, the increase of the development length is about 20%–50%. It is also stated that the development length is 50% more when the concrete cover is less than three times the steel bar diameter or the distance between the steel bars is less than six times the bar diameter. In other circumstances the development length can increase by 20%.

It is worth mentioning that the Egyptian code of practice (ECP) did not take painting coated steel bars into account. A study carried out by El-Reedy et al. (1995) found that the equation given in the ECP for the development length calculation has a higher factor of safety, so it can be applied for coating steel bars with epoxy without increasing the development length.

Based on ACI, the coating thickness should not be more than 300 μm or 12 mils. Moreover, it is prohibited to paint mild steel at all, since the bond strength is due to friction, and with coating, the bond strength would be very much reduced. Also, some research studies found that when painting steel reinforcement in concrete slabs at 14 mils thickness, too many cracks were observed that caused separation between the steel bars and the concrete.

A comparison was done between steel bars without and with epoxy coating. Samples were exposed to tap water and salt water containing sodium chloride and sodium sulfate. The comparison between them is shown in Table 6.2; it was found that in every case the coated steel bars had a corrosion rate less than 10 times that of the uncoated steel bars.

The epoxy coat provides physical protection and it is considered a relatively cheap method, so it has been used worldwide in harsh environmental conditions.

It is very important to know that use of epoxy coating on the steel bars does not allow for violations of codes and standard requirements, failure to

Table 6.2 Corrosion rate for coated and uncoated steel bars.

	Corrosion rate (mm/years)	
Case	**Tap water**	**NaCl 1% + Na_2So_4 0.5%**
Normal steel bars	0.068	0.1
Coated steel bars	0.007	0.013

follow the quality control procedures during construction, or failure to apply the required conditions of concrete cover, w/c ratio, and cement content.

Note that it is preferred to use steel bars that are coated by the steel manufacturer, as it is difficult to maintain the coating thickness within the limits if it is done on-site.

6.2.5 Galvanized steel bars

Galvanized steel bars are used widely in the United States in reinforced concrete structures. In addition, the Federal Highway Association (FHWA) states that galvanized steel can reach an age of 15 years under the influence of attacking chlorides if high-quality concrete is used, according to the research of Andrade et al. (1994). Galvanized bars are used effectively in structures exposed to carbonation. It is very important to avoid contact of galvanized steel bars with nongalvanized steel bars, as this accelerates depletion of the galvanizing.

In brief, the process of galvanization involves immersing the steel rod in a zinc solution at a temperature of 450 °C and then cooling it. Hence the zinc covers the steel bar. This cover consists of four layers, the outer layer being a layer of pure zinc and the other layers being mixed between zinc and steel.

With zinc, as with most metals, corrosion will happen over time. The rate of corrosion under different weather factors can be calculated from the corrosion on the zinc layer and the time required.

The galvanized coating must be tested after bending the steel bars. Take care with the use of welding in manufacturing. The maximum zinc cover is around 200 μm. This follows the ASTM A767/A767A M-90 standard.

It is found that zinc stability mainly depends on the stability of the pH value of the concrete. A pH value of 13.3 is the value at which the passive protection layer is formed, but when the value of pH increases to more than 13.3, the zinc will melt, until it vanishes completely.

Therefore we find that the process of galvanization is totally dependent on the value of pH in the concrete pores and relies mainly on the alkalinity of Portland cement. Therefore a reduction in the zinc layer of 2 μm occurs if the pH is equal to 12.6, and if pH is equal to 13.2, the reduction in the galvanized layer is 18 μm, which occurs before the passive protection layer is present.

According to laboratory tests by Building Research in the UK in 1969 of different types of Portland cement with different alkalinity, when the thickness of the cover was 60 μm, the lifetime was about 200 years in the case of a lower cement alkalinity, but was 11 years with high alkalinity of cement.

Therefore, it was found that the thickness of the cover must be more than 200 μm. The ASTM 767/A767 M-90 standard is especially for galvanized steel bars and it identifies two types of galvanization, Class Type I and Class Type II, which have a cover thickness to produce a weight per surface area of more than 1070 and 610 g/m^2, respectively, which is the equivalent of 85 μm and 150 μm thicknesses, respectively.

Per Building Research UK in 1969, the maximum thickness of the cover of zinc is about 200 μm, as an increase in cover thickness will reduce the bond between the steel and concrete. As stated before, the pH is between 8 and 12.6. Note that zinc is more stable in this case than in the case of an increased pH value. Note that the zinc gives high efficiency with a structure exposed to carbonation, as it will reduce the pH.

However, with chloride attacks, the galvanization will not work as a corrosion prevention, but significantly reduces the rate of corrosion. A study from Yeomans (1994) mentioned that galvanization can increase the value of chlorides which cause corrosion by 150%–200%; this will increase the time to cause corrosion by four times.

To examine the advantages and limits of galvanization, the following are some of the important points that should be noted when using galvanization, according to the specifications of the Institute of Equatorial Concrete in 1994:

- Galvanization increases the protection from steel corrosion, but does not compensate for the use of concrete with a good appropriate concrete cover.
- It is prohibited to put galvanized steel with nongalvanized steel, as this will cause corrosion quickly to the galvanic layer.
- The galvanic layer should be examined after bending the steel reinforcement and fabrication and it is recommended to increase the bending diameter.
- Some precautions should be taken when using welding with galvanized steel.

6.2.6 Stainless steel

Stainless steel is rarely used but it is available on the market. Some special applications use reinforcing steel bars made from stainless steel material in order to avoid corrosion, but it is costly. Thus normal steel bars are available on the market coated with a layer of stainless steel with a thickness from 1 to 2 mm. These are less costly than the stainless steel bars but are still higher cost than steel bars.

The same precaution mentioned for galvanized steel bars also applies to stainless steel. Also, avoid putting stainless steel bars or steel bars coated with stainless steel adjacent to or contacting carbon steel bars, as this will quickly lead to the start of the corrosion process.

Miller (1994) stated that the design of some balconies used reinforcing stainless steel adjacent to normal steel and corrosion appeared rapidly. Another precaution is that stainless steel is not a good choice for welding; but if it is necessary to do welding, it should be done using tungsten inert gas and the welding wire should be stainless steel.

The cost of stainless steel is about 8–10 times the cost of the normal steel commonly used. The specifications for steel coated by stainless steel are now being prepared and discussed by ASTM under the title "Standard Specification for Deformed and Plain Clad Stainless Steel Carbon Steel Bars for Concrete Reinforcement," which provides specifications for the installation of steel and follows specifications A 480/480M, for types 304, 316, or 316L. They defined three levels of the yield stress: 300, 420, and 520 MPa. In the commercial market, there is reinforcing steel coated by stainless steel with a yield strength of about 500 MPa with maximum tensile stress of about 700 MPa; the available diameters are 16, 19, 22, 25, and 32 mm, and now some factories have increased the diameter to 40 mm.

6.2.7 Fiber reinforcement bars

Over the last decade, there have been many research studies, both theoretical and practical, on the replacement of steel reinforcement bars by fiber reinforced polymer (FRP), as this material is not affected by corrosion and thus is very economical during the structure lifetime. FRP bars are being developed and manufactured, to use as an alternative to the steel bars in concrete structures.

As seen in Fig. 6.1, FRP is manufactured with protrusions on the surface, which gives the ability to bond with the concrete. Currently, it is manufactured and used on a small scale as it is still under experimentation in research and on the market. There are many research studies being conducted to study the performance of these bars over time.

As a result of limited production, the cost is still very high, but on the other hand, the maintenance is very low or zero. Moreover, it reduces the dead load on the structure, which decreases the total cost of the structure.

Some structures in Canada were constructed using GFRP bars, including a marine structure at harbor in Canada. This structure is a thick wall of precast concrete designed with a maximum compressive strength of

Fig. 6.1 Shape of reinforcement bar from GFRP.

450 kg/cm^2. This structure is exposed to temperatures from 35 to −35 °C. Also in Quebec, Canada a bridge was constructed using GFRP bars that was exposed to salt used to dissolve ice. Other such bridges were built in Antherino and Vancouver Island.

In the Vancouver bridges they performed tests on all structures with FRP exposed to different weather conditions. The tests were conducted through sampling cores of reinforced concrete and using an X-ray machine as well. These structures were aged from 5 to 8 years.

The results of that research showed no impact of various factors on the GFRP and the cycles of wet and dry did not affect the GFRP bars. Microscopy examination found a strong bond between the bars manufactured from GFRP and concrete.

The mechanical properties include a maximum strength of about 600 MPa and a maximum bond strength of 11.8 MPa, but the modulus of elasticity was less than the steel modulus of elasticity by about five times. According to ACI 440, it is suggested that the tensile stress must be not less than 20% of the maximum tensile strength, to overcome the problems that might arise from creep.

6.2.8 Protecting the concrete surface

The concrete surface should be protected from permeability to water and propagation of chloride, such as in offshore structures exposed to sea waves or bridge decks on which salt is used to melt ice. In addition, prevention of permeability protects the concrete surface from carbonation and prevents propagation of carbon dioxide. Other factors that increase the corrosion rate, such as humidity and oxygen, should also be prevented from propagating inside the concrete.

There are two common methods used to protect the concrete surface: liquid materials applied by spraying or painting with a brush, or sheets and membranes made of rubber/plastic or textiles immersed in bitumen, more commonly used in smaller works.

These methods can encounter the same problems as epoxy painting of reinforcement steel, as discussed earlier: thus it is crucial to avoid any damage or erosion to this layer due to external mechanical factors, such as tears, exposure to heat sources, or other factors.

It is important to follow the required specifications during execution. Some specifications define the execution methodology, in addition to the supplier recommendations.

The function of all these surface protection methods is the same: to prevent water propagating inside the concrete through concrete voids.

6.2.8.1 Sealers and membranes

Sealers and membranes are used to provide protection to reinforced concrete structures exposed to water, severe chemical attack, or other situations. This is a common method used in most buildings; as a rule of thumb, it is used for the foundation, in the floor of the w/c, and on the roof of the building. The objective is to protect from water or oxygen propagating inside the concrete in addition to sulfate, chloride and carbonation.

In general, membranes and sealers will provide protection by:

(i) Eliminating or slowing down the chlorides and carbonation propagation inside concrete, to maintain steel passivity.

(ii) Reducing or eliminating the moisture moving inside the concrete, to keep it dry and slow corrosion reactions.

The sealers and membranes can be classified into three categories based on the means of protection and different chemical composition, with different commercial names.

The three categories are described in the following paragraphs.

Coating and sealers

These consist of a continuous film applied on the concrete surface having a thickness in the range of 100–300 μm. The film is composed of a binder and fillers, with the performance based on the compounding of the fillers.

Therefore compositions having a similar binder may be quite different in their performance.

The coating or sealers are available in the commercial market in a liquid form, which can be brushed or sprayed on the concrete surface, similar to the painting process.

The successful performance of coating and sealers depends not only on the quality of the materials used, but also on the application. The main step in the application is to ensure the concrete surface is clean and sound. Therefore any weak and cracked concrete should be removed, holes should be filled, and, if necessary, apply a leveling coat.

The surface should be dry if the applied membrane is of polymeric composition. Silane treatments can be done in wet concrete, to facilitate better penetration of the liquid compound into the pores. The coating is normally applied in two layers to obtain a continuous film without pinholes.

The service life of the membranes or sealers is from 15 to 20 years. Therefore continuous maintenance and follow up is necessary, including additional treatments.

Pore lining

The pore lining is based on processes by which the surface of the pores in the concrete are lined with materials that reduce the surface energy, to make the concrete water repellent. Silicone compounds are most frequently used for this purpose. Silicone resin can be dissolved in organic liquid which, after evaporation, deposits a film of the resin on the pore surface.

Pore blocking

The pore blocking sealer is a specific material that propagates into the concrete pores and then reacts with some of the concrete's constituents. The resulting products from the reactions are insoluble and are deposited in the pores to block them. The most common materials for this type of sealers are liquid silicates and liquid silicofluorides.

From experience, using membranes to prevent water permeability has a lifetime of around 15 years. After that, they must be replaced; this time frame can be as long as 25 years but only with good materials and excellent application technique.

Being proactive is very expensive, because the floor tiles and plumbing accessories must be demolished in order to start from scratch. Thus in residential buildings the maintenance will be mainly corrective action, meaning run to fail, which is different from the integrity management concept. In any mature reinforced concrete structure, 80% of the problems are due to the sealer and membrane. The main steel of the slab is reduced and the signs of corrosion are parallel cracks in the steel bars. The signs will be seen on the lower floor so the user of the bathroom cannot observe it and, as mentioned, it will be run to fail so the repair will be done until the concrete cover falls leading to a serious corrosion problem in the steel bars.

When using membranes, care must be taken in execution, as any tear in the membrane or bad execution through violating the technical specifications and supplier recommendations can lead to losing money, as any defect will cause permeability of water and allow corrosion to start.

There are two well-known methods of application of sealer for a foundation, swimming pool, or tanks: these methods are to isolate the surface exposed to water or isolate the surface that is not exposed directly to water. These types of sealing must be executed by professional contractors with a great deal of experience in this type of work, as working with the chemicals and dealing with the membrane requires competent workers, available only with trusted contractors. Recently, some research has found that there is a problem with using membranes with cathodic protection, as gas can accumulate in the anode and needs to escape, but is prevented by the membrane and this prevents completing the electrical circuit.

6.2.9 Cathodic protection system

Cathodic protection (CP) is the most expensive method of protection, and it is typically used to protect pipelines in the petroleum industry. It is intended for use in reinforced concrete structures, but only for particular structures due to its higher cost. Special studies, design, execution, and monitoring are required.

Cathodic protection is a very effective system but it must be applied properly. Its main function is to prevent corrosion of steel in concrete, and in existing older buildings it will stop the corrosion process, maintain the structure as it is, and stop further deterioration.

The concept of CP in a concrete structure is to use the steel bar as a cathode through use of an external anode. Electrons will transfer to the reinforcing steel bars from the anode, through the ionically conductive concrete. The current supplied should be sufficiently high so that all the local cells are inhibited and all the steel surface becomes anodic.

The external current can be supplied by connecting the steel to a metal that is higher in the electrochemical series, such as zinc or aluminum.

It serves as the anode relative to the cathodic steel. This is called the sacrificial anode, as the anode will gradually dissolve as it oxidizes and provides electrons to the cathodic steel.

Another main method is the impressed current. This CP method is based on supplying electrons to the reinforcing steel from an external electrical power source. This power source is fed into an inert material that serves as the anode and that is placed on the concrete surface. The anode is frequently called a fixed anode.

In 1824 Sir Humphrey Davey discovered the principle of the sacrificial anode. Since then, his discoveries have been used to protect the metal parts of boats completely submerged under the water from corrosion. At the beginning of the 20th century this technology was used for pipelines buried under the ground, but when it was discovered that the soil resistance to electricity, it is required to use very high cathodic protection system with constant direct current.

The CP system is used in modern structures having a higher importance level and special characteristics. The first practical application of CP in a reinforced concrete structure was in a bridge in the north of Italy on a highway in a mountainous area. Then the use of CP became more widespread with more research and the development of accompanying technology. Broom Field (1995) stated that a survey of consultants in England working in the area of cathode protection of concrete structures that were used a CP to protect around 64,000 m^2 in 24 of different structures in the United Kingdom and the Middle East.

The largest anode manufacturer supplied anodes that cover about 400,000 m^2 around the world and includes a garage for 60 cars and 400 different forms of bridges and tunnels.

Based on the risk-based maintenance approach, as discussed in Chapter 7, a CP system could be the optimum solution for some special or critical structures the failure or collapse of which would have serious consequences, or for which the maintenance or repair would be very expensive.

For existing structures with corrosion due to chlorides inside the concrete, a CP system could be the best option to keep the structure as it is. This could apply to foundations under critical equipment in oil and gas, refineries, or petrochemical plants. When there is no standby and a complete demolish would require a plant shutdown, a CP system could be the best solution.

The US Federal Highway Association (FHWA) has stated that the CP method is the only method proven to stop corrosion in salt-contaminated concrete bridges, regardless of the chloride content of the concrete.

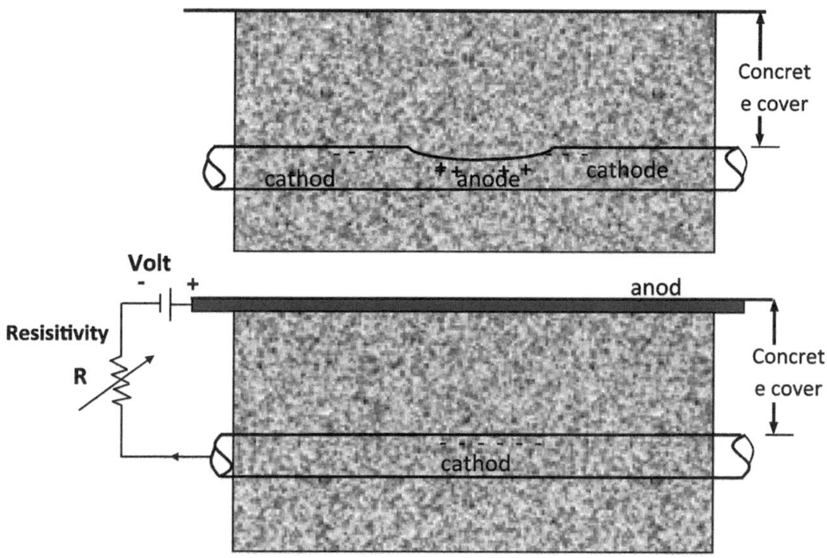

Fig. 6.2 Cathodic protection method.

Cathodic protection is specifically used to prevent the corrosion of steel reinforcements in concrete or to stop a corrosion process that has already begun. The method relies on making steel continuously cathodic by using an external anode, as presented in Fig. 6.2.

As shown in Fig. 6.2, the electrons transfer to the steel reinforcement surface when the anode is placed on the concrete structure. The concrete will be electrically conductive between the anode and steel reinforcement, so cathodic protection is provided to the steel bars. As shown in the figure, the positive is the anode that will transfer electrons to the negatively charged cathode; this method of cathodic protection is called sacrificial protection and the anode in this case is called a sacrificial anode, as described earlier.

There are two main methods for cathodic protection, the first method is using sacrificial anode and this method of protection is called sacrificial protection. In this method the anode will be from zinc or aluminum. The aluminum or zinc will transfer electron to steel to protect the steel and with time these anodes will be depleted.

In a RILEM report consideration was given to ways to make cathodic protection more valuable:

- The electrical conductivity to the steel reinforcement should be continuous.

- The concrete needs a lower electrical resistivity to be capable of transferring electrons between the anode and the steel bars.
- Avoid aggregate with high alkalinity.

6.2.9.1 CP components and design considerations

The main components of the CP system are the impressed current, anode, and the electric conductive, which in this case is the concrete. Therefore CP depends on the relative humidity of the concrete, as it has a strong effect on the electrical cables and the negative pole system on the steel reinforcement. The final component of this system is the wiring that conducts between the anode and the source of direct current, which is a rectifier. But one should consider the last important element, which is monitoring and measuring the electric current to ensure that it is continuous.

The anode is the most expensive part of the CP system for the impressed current method. The anode should be capable of resisting the chemical, mechanical, and environmental conditions that affect it during the structure's lifetime. In general, it is preferable that its lifetime is greater than that of the coating layer. The anode surface must be large enough to provide density to the electric current so as to prevent any deterioration to the anode surface.

From previous discussions, cathodic protection of a concrete structure is different from other applications, as concrete has pores and voids containing water in some parts while other parts may be dry or have water on the surface only, which is different from marine structures or structures buried in soil.

The major fact is that a concrete structure cannot contain 100% water inside its pores. The problem is clear if the concrete around the anode is completely dry, which means the voltage would need to be increased to 10–15 V instead of 1–5 V. On the other hand, dry conditions around the steel bars prevent the causes of corrosion. Practically, however, this can be seen only in summer without high relative humidity.

Source of impressed current

Most design methods take a current of about 10–20 mm amber on the surface of steel reinforcement and take into consideration the lower layer of steel reinforcement. A voltage potential of around 12–24 V is used, to ensure that the chance of delivering an electric shock to a human or animal is very small.

The source of electric current is chosen in the design stage and the choice of electric source should be enough to stop the corrosion process. A higher estimate for electric current is often assumed and a 50% safety factor is added. This increase will produce some heat that will cause some loss in the current.

The CP system designer should receive the complete drawing package with the steel reinforcement detailing in order to design a suitable CP system and define the location of the anodes as well as the connections and monitoring system.

Anodes for columns, beam, and foundations

In this type of concrete structure member, titanium mesh is used, and often the concrete is applied using shotcrete on the surface of the mesh to cover the anode. Note that extreme caution is needed in using shotcrete, as it requires better trained workers, supervision, and a contractor, and also needs special curing to guarantee good quality concrete and good adhesion between it and the old concrete surface. Some precautions must be taken during construction to ensure that the layer which is saturated 100% has a very strong bond with old concrete and the electric conductivity will be through pieces of titanium sheet welded on the mesh.

In a way, we may use the rod of titanium platinum and placed inside the concrete member of a subject inside the coke so as to reduce the impact of acid generated on the anode. One of the most important precautions during implementation is to make sure that there is no possibility of causing a short circuit between the anode and steel reinforcement. This can be prevented through the use of appropriate cover placed after the hole in the concrete is made, as shown in Fig. 6.3.

The anodes distribution should be designed so as to ensure the protection of the steel reinforcement in the structure; this method is also used in horizontal surfaces.

Conductive layer

This is one of the most important commercially successful methods. The anode is a main pole of inert metal and the secondary poles consist of layers of mortar, asphalt, and coating; this layer is electrically conductive due to the carbon particles (Fig. 6.4).

These layers are not durable enough over time to protect the titanium mesh, but this is considered the cheapest and easiest method from a construction and maintenance point of view. The layers can be painted using a reasonable painting match with the architectural design and decoration of the building. In addition, it can be used in ceilings and both vertical and horizontal surfaces. The main disadvantage is that it cannot be used on a

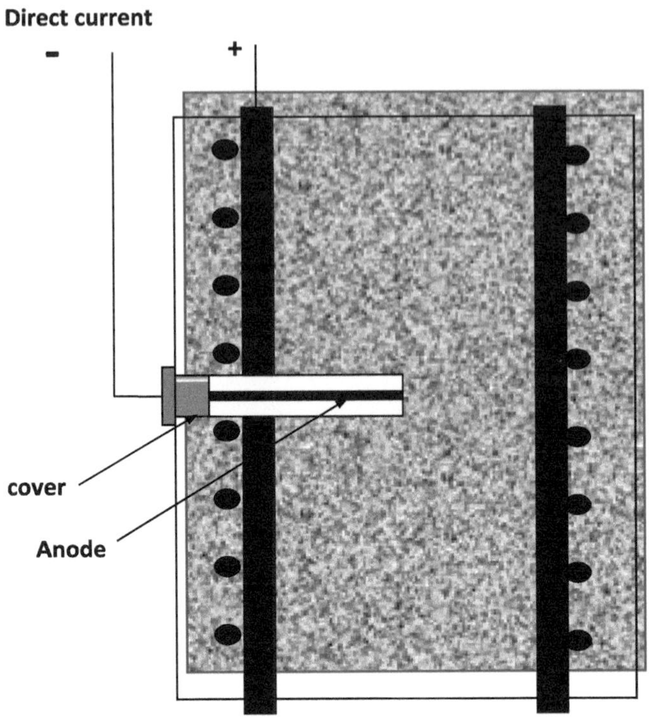

Fig. 6.3 Anodes inside walls and columns.

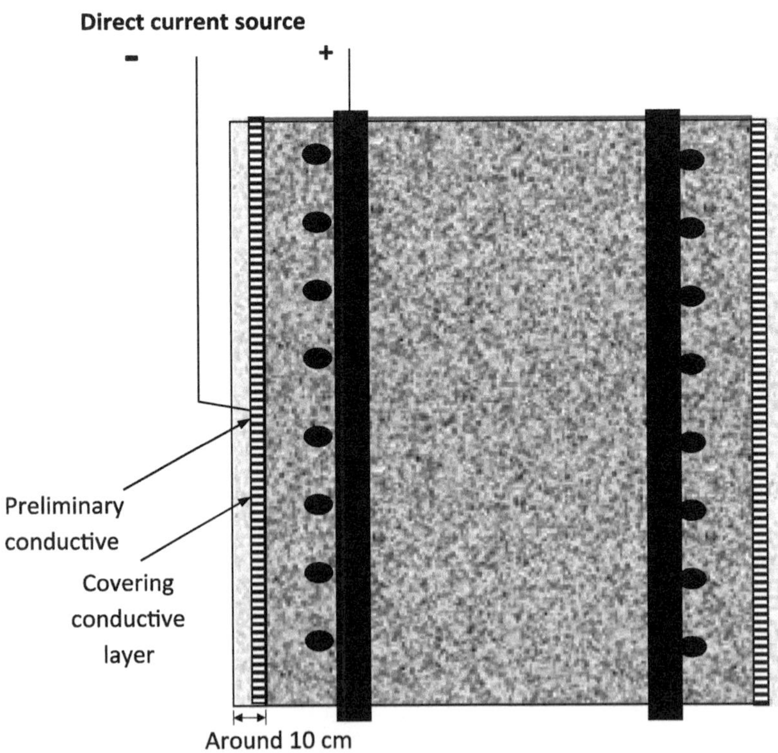

Fig. 6.4 Titanium anode mesh around the column.

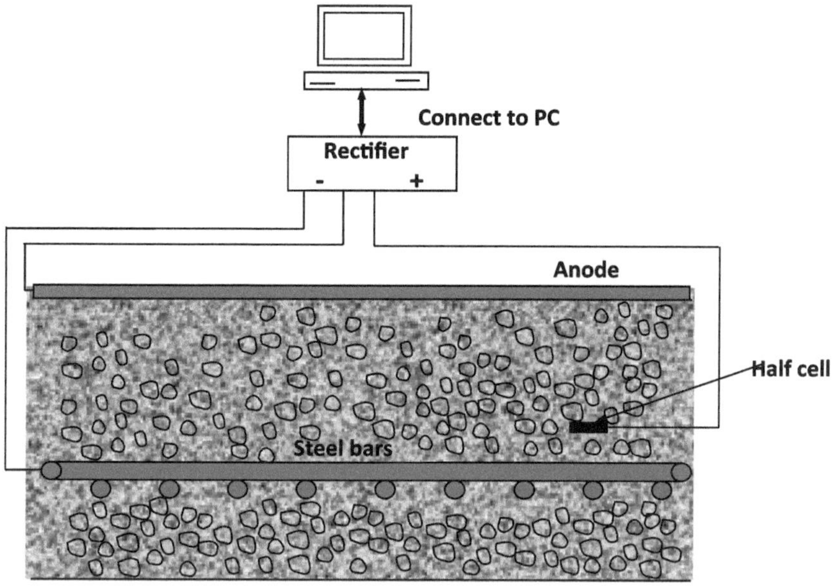

Fig. 6.5 CP monitoring system.

surface exposed to abrasion, as its capacity is low and its lifetime is about 5–10 years.

A main feature of CP systems is that they are required to be monitored regularly to guarantee that the anode transfers electrons to the steel bars continuously with no obstacles, as in Fig. 6.5. The figure shows the insertion of a half cell pole near the steel bars connected to a computer through a cable modem; the computer will process the data to monitor the measurements continuously.

The system can deliver more than one point in the structure to the computer, which has special software and hardware.

The efficiency of the cathodic protection can be tracked by use of the computer. More advanced systems can be used to control the source of the electric current. The current density and potential can be changed from a long distance away from the structure using a microprocessor.

6.2.10 Comparison between cathodic protection and others

Note that cathodic protection needs to be measured regularly or measured from a distance away from the structure, as stated previously, using the computer, to ensure the safety of the design and the proper functioning of the cathode protection. Therefore cathode protection is more costly than other types of steel reinforcement protection.

Thus CP is often used in important structures, having a special nature. This method has been used in bridges and tunnels frequently from the late 1980s and early 1990s.

It was used in an underground garage at the World Trade Center in New York, as well as in the Arab Emirates. Cathodic protection could be implemented during the implementation of civil works and therefore does not need excess labor, as the anodes are placed with the reinforcing steel.

The impact of electric current on the bond strength between steel reinforcement and concrete is considered one of the most important subjects and many researchers have studied it for several years. From a practical point of view, there has been no impact of the cathodic protection on the bond strength.

Some 20-year-old structures have been studied and there has been little impact on the bond strength, as most of bond strength is carried by the ribs that exist in the deformed steel reinforcement.

Practical experience has proven that the bond strength is increased with the presence of rust on the steel surface.

Note that with use of cathode protection, there is no need to use another protection method, such as membranes that prevent water permeability, other additives, or painting the steel bars. Thus this reduces the total cost of the project and because of this, in a deteriorating concrete structure or some of its members, CP is used at a high cost. Therefore the cost of repairing the concrete member may be higher than the cost of constructing it.

In some structures, due to the special nature of the construction, it is difficult to perform the necessary repairs so cathode protection is the best and only solution. Table 6.3 presents a summary of the protection methods, which gives the advantages and disadvantages of each method and the appropriateness of use with cathode protection at the same time.

6.3 Offshore structure protection

The main method for protection of the offshore structure is by coating its surface and this is specific in topside mainly and a cathodic protection system for the jacket. There are many types of CP. When choosing the CP system, the economic factor is very important in the decision.

Table 6.3 Advantages and disadvantages of different protection types.

Method	Advantage	Disadvantage	Compatibility with cathodic protection
Increase concrete cover	No cost increase Obtain from good design	Has limits Increased cover will increase the shrinkage crack	Yes
Not permeable concrete	Depends on design	Need additives with very good curing	Yes, but silica fume must be $<10\%$ cement weight
Penetrating sealer	Relatively not costly	Complicated practically	Yes
Membrane prevents water	Very known technology	Can be damage Difficult to fix without fault	Yes, but cathodic protection under the membrane
Epoxy coated steel bars	Known technology Little maintenance	Problem in quality control under certain weather conditions	Yes, with continuous electric current
Galvanized steel	Easy use on site	Rapid damage by contact with nongalvanized steel	Yes, need higher level of protection
Stainless steel	Excellent for corrosion prevention	High cost	Yes
Corrosion inhibitor	Calcium nitrate effect for little chloride concentration	Must know the chloride amount Effect over long time not known	Yes
Cathodic protection	Known technology Long life of titanium anode	Regularly inspect the rectifier	

There are three major types of cathodic protection system, each of which, when correctly designed, installed, and operated, can effectively protect a fixed offshore steel structure for its design life. These are:

Sacrificial: Comprising anodes cast from reactive metals (normally zinc or aluminum alloys), which are more electronegative than the structures requiring protection, and which require no external source of power.

Impressed current: Comprising anodes manufactured from materials that are essentially inert and powered by an external source of direct current.

Hybrid: Comprising a mixture of sacrificial anodes and externally powered impressed current anodes. The principal technical advantages and disadvantages of sacrificial, impressed current and hybrid systems are summarized in Table 6.4.

Table 6.4 Comparison between different CP systems.

	Sacrificial anode	Impressed current	Hybrid
Advantages	• Simple, reliable, and free from in-service operator surveillance • System installation is simple. • Permanent potential monitoring system not essential	• Flexible under widely varying operating conditions • Weight advantage for large capacity, long-life systems	• Flexible under widely varying operating conditions • Weight advantage for large capacity, long-life systems
Disadvantages	• Large weight penalty for large capacity, long-life system • Response to varying operating conditions is limited • Hydrodynamic loadings can be high	• Complex system requires high level expertise • System installation is complex, and it requires a power source • Perceived driver risk from electric shock • In service operator surveillance required • Permanent potential monitoring system essential • Vulnerable to loss of power	• Complex system requires high level of expertise • System installation is complex, and a power source is required • Perceived driver risk from electric shock • In service operator surveillance required • Permanent potential monitoring system essential

The use of the term "impressed current system" can be misleading, because for most offshore applications an impressed current system is used in combination with a small number of sacrificial anodes, forming a hybrid system. Sacrificial anodes in hybrid systems are provided on structures to ensure that adequate polarization of the critical nodes is maintained at all times, even if the power supply to the impressed current anodes fails or is switched off temporarily to permit manual underwater inspection or cleaning of the structure by divers. Some early impressed current systems were provided with inadequate sacrificial anode back-up to perform this critical task and significant corrosion damage has been reported in times of unplanned and planned impressed current shutdowns.

The same considerations apply equally to jacket structures with one important addition, namely that a power source to drive the impressed current system is generally not available until the topside power generation equipment is installed and commissioned. On large deepwater jackets in the North Sea, this may be a year or more after installation of the jacket, protection for the interim period being provided by high-current, short-life sacrificial anodes.

It is strongly suggested that designers contemplating an impressed current system for North Sea applications should provide full sacrificial back-up. The sacrificial anodes should provide full protection for a minimum of 2 years, plus an allowance for periods of possible impressed current system shutdown during subsea surveys and maintenance throughout the design life.

The obvious technical attractions of sacrificial systems, as illustrated in Table 6.4, make this type most often chosen for offshore structures. Also, for many offshore structures, sacrificial systems are the most economical option for the owner when taking into account both capital expenditure and the running costs over the design life of the structure. However, generalizing on economic advantages and disadvantages can be misleading, because they differ widely for each type and size of structure and according to the design constraints imposed by the environmental conditions prevailing at different offshore locations. For this reason, economics are not included in Table 6.4.

Faced with severe weight constraints, the designers of the Murchison and Hutton platforms in the North Sea carried out detailed assessments of alternative sacrificial and impressed current designs. These showed that, although sacrificial anodes could not be dispensed with entirely, substantial weight savings could be gained by using impressed current systems as the primary means of protection on both platforms. This was despite their vastly different structural configurations, Murchison being a deep water conventional jacket and Hutton the world's first tension leg platform.

The adoption of hybrid systems for weight-saving reasons on Murchison and Hutton is significant, for this is perceived to be their primary advantage over sacrificial systems. In the case of Hutton, the installed weight of the primary impressed current system, plus supplementary sacrificial anodes located close to the main node joints, was approximately 60 tons. An equivalent totally sacrificial system would have weighed around 250 tons. In most cases, impressed current systems are more likely to be commercially competitive for buoyant structures such as tension leg platforms and submersibles than for conventional jacket structures.

The relatively simple geometry and large, flat surfaces of buoyant structures are ideally suited for protection to be provided by a small number of high current, low voltage, flush-mounted anodes. Cables to reference electrodes and anodes can be easily and economically routed through ballast tanks and many access ways in the pontoons and columns of the hull (Table 6.5).

Table 6.5 Comparison between the requirements for ICCP and SACP.

	Impressed current CP systems ICCP	**Sacrificial anode CP systems SACP**
Power Source	External continuous power source required	Independent of any power source
Connections	Can be inadvertently misconnected resulting in reversed DC polarity, which will accelerate corrosion Isolation of anodes from the structure is essential Fewer connections required but more complex	Straightforward welded or bolted connections to structure Cannot be wrongly connected and connections are cathodically protected
Control	Simple controls—manual/automatic Automatic will maintain potentials within Pre-set limits but requires additional fixed Reference electrodes monitoring required at regular intervals	A tendency for current to be self-adjusting Dependent on initial design and if not adequate will require additional anodes

Table 6.5 Comparison between the requirements for ICCP and SACP—cont'd

	Impressed current CP systems ICCP	**Sacrificial anode CP systems SACP**
Anodes	Usually lighter and fewer numbers May affect other structures in close proximity to anodes and should be assessed for any interaction	Relatively heavy and large numbers Bulk of material may restrict water flow and induce turbulence and drag Less likely to cause interaction with neighboring structures as output is low Lifespan varies with conditions so replacement may be required at different times
Damage	Anodes lighter in construction and therefore less resistant to damage Loss of an anode can be more critical to the effectiveness of the system due to high current output for each anode	Anodes are robust and not susceptible to mechanical damage Where a system comprises a large number of anodes, the loss of a few anodes has little overall effect on the system
Maintenance	Equipment designed for long life but regular checks on electrical equipment in service required	Generally, maintenance free Renewals usually required at regular intervals unless designed for life of structure
Environment	May cause overprotection, causing coating disbondment or hydrogen-induced cracking of high tensile steels Less restricted by high resistivity conditions	Not practical for use in high resistivity conditions Limited current availability
Installation	Requires a high level of detailed design and installation expertise	Straightforward Often bulky and large numbers involved
Hazards	Diver risk from electrical shock, system needs to be switched off when diver is near anodes	Magnesium anodes can be used in potable tanks but never in areas containing hydrocarbons Aluminum and zinc anodes must never be used in potable water tanks
Cost	Generally initial cost high but service cost normally low	Initial cost dependent on design life but relatively low, periodic renewals necessary and cumulative costs high

Impressed current systems may also be cost competitive for conventional jacket structures of simple geometry, located in relatively benign offshore waters. and on which reference electrode and anode cables can be safely installed in substantial conduits routed along the outside of the structural tubular members.

But the impressed current systems are less likely to be cost competitive on large jackets of complex geometry located in hostile environmental conditions. Complex node geometries are unlikely to allow large capacity anodes to protect all surfaces adequately on account of shielding effects. Moreover, the hostile environment may demand the difficult and expensive routing of anode and reference electrode cables inside the structural tubular members in order to ensure their mechanical safety. This necessitates a large number of stress raising penetrations in the tubular members below water adjacent to the anodes and reference electrodes.

As a general rule, increasing anode operating temperatures causes a decrease in both anode ampere-hour capacity and driving potential; at temperatures exceeding 50 °C, zinc alloys experience intergranular corrosion and they should not be used at low anode current densities. On the other hand, the ampere-hour capacity of aluminum alloys tends to decrease significantly. In order to realize the performance claimed by anode manufacturers and thus to ensure the successful operation of the cathodic protection system, it is imperative that strict quality assurance and quality control of the anode manufacturing process should be achieved and maintained throughout production. The quality assurance, quality control, and tests will be discussed in later sections. It is worth mentioning that the requirements contained in Det Norske Veritas DNV (2005) are considered to set the minimum standards for offshore work, with supplementary requirements for specific project applications to be determined and specified by the designer.

6.3.1 Geometric shape

Sacrificial anodes are generally cast in three basic geometric shapes: long slender stand-off type, flat plate flush-mounted type, and bracelet type. Typical examples of the first two basic geometries are shown in Figs. 6.6 and 6.7, respectively.

The most common anode shape used for offshore structures is the long slender type of trapezoidal or circular cross section. The principal advantages of this anode geometry are high current output and good current distribution for a given mass, simple fabrication and casting requirements.

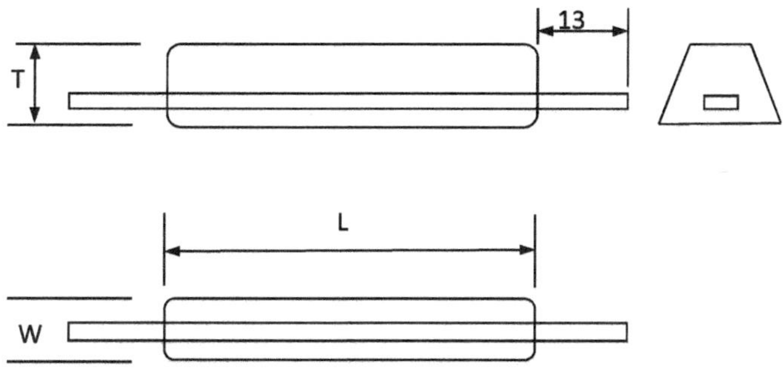

40 mm wide x 6mm thick flat steel core with 2700 mm

Fig. 6.6 Flushed mounted anode.

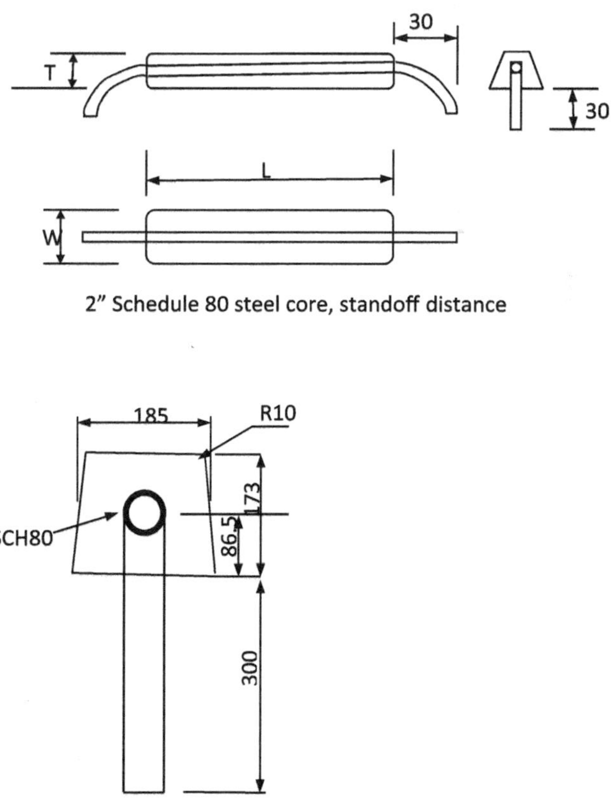

Fig. 6.7 Standoff anode type.

Flat plate anodes are generally best suited in complex fabrications where space limitations prevent the use of larger stand-off anodes and/or cathode current densities are low. Examples are heavily reinforced mud mats and large flat plate painted surfaces; the designer should determine if the chosen anode shapes can be more economically chosen from a manufacturer's standard units or whether, because of the large number required for a new structure, a preferred design could raise costs. Anode manufacturers offer a large variety of standard anode and insert core types, the choice of steel insert being usually between bar tube or rod, in either straight lengths or prefabricated, weld jointed shapes.

Stand-off and flush-mounted anodes may further be divided into "short" and "long," based on the length-to-width ratio, and they are presented in Figs. 6.6 and 6.7, respectively. The anode type is a main factor in anode resistance and utilization factor, as we will illustrate later.

The slender stand-off type is typically cast on a tubular insert and used for relatively large anodes on, for example, platform substructures and subsea templates. The current output, I_a (A), in relation to net anode mass, M_a (kg), is high, as is the utilization factor (u).

The modern type places the anode in a steel frame called a sled and connects it with a special clamp to the steel structure. This system is very easy and less costly in the case of retrofitting the existing structure.

Stand-off anodes can be manufactured and obtained up to a net anode mass of several 100 kg. In surface waters, drag forces exerted by sea currents are significant.

Bracelet anodes are used primarily for pipelines but now they are used in the upper zone for the platform legs.

It is mandatory to have a suitable coating system to apply on the surface that is facing the object that is required to be protected for all flush-mounted anodes, as this procedure assists to avoid build-up of anode corrosion products that will cause anode fastening devices to be distorted and then fracture.

During the design conceptual phase for CP systems, the type of anodes and any special requirements for anode fastening will be defined, considering the applied forces during installation such as piling operations and the effect of wave forces during the structure lifetime. For stand-off type anodes, special precautions may be necessary during anode design and distribution of anodes to avoid impeding subsea operations.

The insert should be structurally suitable for the anode weight and for the forces it is likely to encounter during its lifetime, including impact, storm damage, wave action, and possibly ice. The insert should normally be made from materials that can be welded to the structural steel. The typical grades of steel BS4360 grades 40A, 43A or 50C, or API 5L grades B, X42 or X52 are usually used.

If anode inserts are fabricated by welding, the latter has to be in accordance with a recognized, quality-controlled standard. Inserts should be prepared by abrasive cleaning to a minimum standard of SIS 05–59-001967 Sa 2.5. Zinc anode inserts are normally zinc coated to BS1729 or to BS1706. NACE recommended practice RP03-87 may be followed.

Aluminum anode steel insert specifications are similar to those for zinc, except that the surface must not be zinc coated nor galvanized after cleaning.

Bracelet anodes are the most commonly used type for protection of submarine pipelines, for which their wrap-around construction is ideally suited. They are rarely used on new offshore platform constructions, because of their low current output to mass ratio compared with long slender anodes. However, bracelet anodes do lend themselves to retrofitting on existing structures to supplement or replace the original failed, deficient, or end-of-life cathodic protection systems.

6.4 Steel structure coatings and corrosion protection

The offshore platform steel is subject to different types of corrosion, such as atmospheric corrosion, splash zone corrosion, crevice corrosion, and others.

The corrosion rates found for uncoated steel in seawater are different. In the splash zone the rate is about 0.15 mm/year; for submerged zones it is around 0.07 mm/year but in cold water with fast running tides carrying silt or other abrasive materials, it can reach 0.3 mm/year. Some studies provide a value for the corrosion rate of 0.127 mm/year for uncoated steel in seawater.

The application procedure for painting is very exacting, so it will be done in the shop under high quality control supervision. The application should follow the standards regarding humidity conditions and protection in case of extreme weather. If some work is mandatory in the field, it should be done when the surfaces are dry and with the required temperature for the painting procedure.

Based on the studies of good coatings, corrosion will start within 10–20 years. The DNV rules require that the coating method statement include:

1. Provision of coating yard description and the general application conditions.
2. Surface preparation procedure and equipment.

3. Temperature and relative humidity for the location.
4. Method statement for application, including the time limit between surface preparation and starting the first coat.
5. Description for single coat of the minimum and maximum limit of dry film thickness.
6. The total coating thickness and the coat numbers.
7. Repair procedure for damaged coating.
8. Provision of the inspection methods, such as adhesion testing and holiday detection.

The coating manufacturer will provide the limits of air humidity and temperature that should be followed in the surface preparation and application.

It has been found that organic coatings are the most effective over metallized zinc, such as vinyl mastic on urethane in temperate zones over zinc, or zinc silicate and phenolic over zinc primer in the arctic and subarctic.

For protection coating, the US Army Corps of Engineers recommends shot blasting, a single coat of zinc primer 0.625–0.1 mm thick, followed by two coats of zinc rich vinyl immersion coating to 0.175 mm. For subsea and in the splash zone, they apply epoxy shop primer, followed by a top coat epoxy to 1.0 mm.

Corrosion protection for subsea steel members is provided by sacrificial anodes or impressed current cathodic protection. The nodes design and installation should be protected from damage during all installation activities from transportation until driving the pile.

It is important to have a good electrical connection between sacrificial anodes and the steel structure. The impressed current is more effective, but it needs continued monitoring, as if there is a shutdown of electric current, corrosion will be accelerated.

Coatings on subsea members can reduce requirements for cathodic protection, as long as the coating has adequate resistance to cathodic disbondment. Zinc-based and aluminum-based alloys have been applied by thermal spray. In the splash zone, Monel wrap, copper nickel, austenitic stainless steel, and other methods can be applied for added protection.

For example, at least one offshore operator has successfully employed thermal sprayed aluminum coatings for critical high strength steel components on a major North Sea structure, as an alternative to a conventional sacrificial anode system. The benefit in using this galvanic coating is that it provides an essentially uniform protected potential of −0.800 to −90.0 mV Ag/AgCl over the entire coated surface area (Tables 6.6–6.8).

Table 6.6 Guide to coating breakdown criteria for COP design.

	Coating breakdown (% of area)		
Year	**Initial**	**Mean**	**Final**
10	2	7	10
20	2	15	30
30	2	25	60
40	2	40	90

Table 6.7 Guidance on the minimum design current densities for cathodic protection for bare steel.

	Current densities (mA/m^2)		
Location	**Initial**	**Mean**	**Final**
North Sea	180	90	120
Arabian Gulf	130	70	90
India	130	70	90
GoM	110	60	80
Indonesia	110	60	80

Table 6.8 Potential limit for cathodic protection of steel.

	Positive limit	**Negative limit**
Aerated seawater	−800 mV Ag/Ag +250 mV Zn	−1050 mV 0.0 mV
Anaerobic condition	−900 mV Ag/AgCl +150 mV Zn	−1050 mV 0.0 mV

Carbon steel corrosion in seawater is mainly a function of the availability of oxygen to the metal surface. The range of the carbon steel corrosion rate is from 100 to 200 mm/year; this range depends on the variability of oxygen level and temperature (Fig. 6.8).

The increase of mass flow of oxygen on the surface depends on the flow rate, as it can increase 100 times if the flow increases from 0 to 40 m/s.

It was found that carbon steel corrosion increases by approximately 50% from the winter session with average temperature 7 °C and summer with a temperature range of 27–29 °C. The higher temperature increases reaction

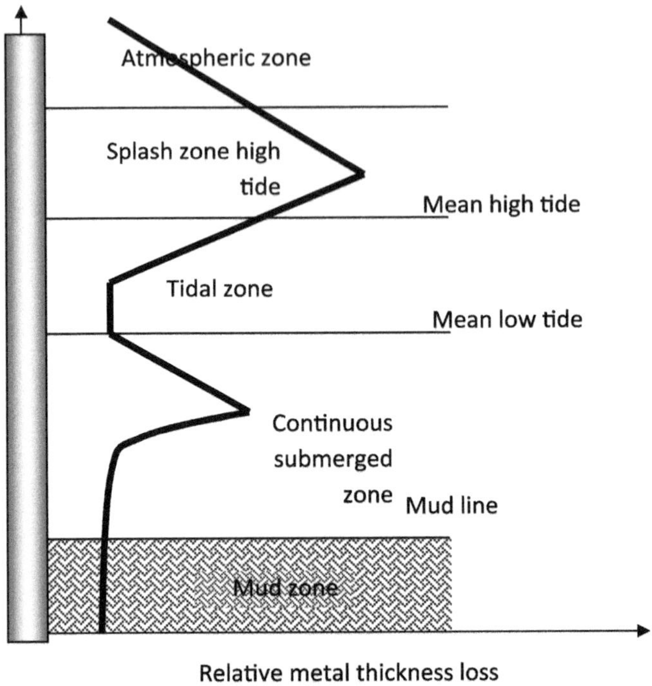

Fig. 6.8 Corrosion profile of steel piling after 5 years exposure (humble).

rate and some research on steel in potable waters showed that the temperature effect is more important and that steel corrosion increases with temperature.

A part of ISO 12944 defines the classification of the principal environments to which steel structures are exposed, and the corrosivity of these environments. In addition, it provides some data about corrosion stresses that can increase the corrosion rate and provides the corrosion rate based on the environmental classifications.

6.5 Corrosion stresses due to the atmosphere, water, and soil

Atmospheric corrosion is a process that happens in a moisture film on the metal surface. The moisture cannot be seen by eye. The corrosion rate can increase due to an increase in relative humidity; from the occurrence of condensation, when the surface temperature is at or below the dew point; or from an increase in the pollution amount in the atmosphere, in which the corrosive pollutants can react with the steel and may form deposits on the surface.

Experience has shown that significant corrosion is likely to take place if the relative humidity is above 80% and the temperature above 0 °C.

The atmospheric humidity and air temperature in a particular region of the world depends on the climate prevailing in that part of the world. If structures are exposed to the open air, under effects of rain, sun, and pollutants in the form of gases or aerosols, corrosion is affected. A high corrosion rate can be caused by poor ventilation, high humidity, or condensation.

Special care shall be taken when considering structures that are partly immersed in water or partly buried in soil. Corrosion under such conditions is often restricted to a small part of the structure where the corrosion rate can be high.

Oxygen content of the water directly influences corrosion as do the type and quantity of dissolved substances, in addition to the water temperature. Animal or vegetable growth can also accelerate corrosion.

Three different zones for immersion in water can be defined. The underwater zone is permanently exposed to water. Fluctuating water levels define the intermediate zone, where the water level changes due to natural or artificial effects; in this zone the corrosion increases due to the effects of both water and the atmosphere. The splash zone is a very critical area and is defined as the area wetted by waves; here the water spray effect causes exceptionally high corrosion stresses.

Corrosion in soil is dependent on the mineral content of the soil and its nature, and on the organic matter present, as well as water and the oxygen content. The corrosivity of soil is strongly influenced by the degree of aeration. The oxygen content will vary, and corrosion cells may be formed. Different types of soil and differences in soil parameters are not considered as classification criteria in this part of ISO 12944.

For the selection of a protective paint system, special stresses to which a structure is subjected and special situations where the structure is located will also be taken into account. Both the design as well as the use of the structure may lead to corrosion stresses not taken into consideration in the classification system.

6.5.1 Classification of environments

For the purposes of ISO 12944, atmospheric environments are classified into six atmospheric-corrosivity categories. To determine corrosivity categories, the exposure of standard specimens is strongly recommended. Table 6.9

Table 6.9 Atmospheric corrosivity categories and examples of typical environments.

	Mass loss per unit surface/thickness loss (after first year of exposure)				Example of typical environments in a temperate climate	
	Low carbon steel		Zinc			
Corrosivity category	Mass loss (g/m^2)	Thickness loss (μm)	Mass loss (g/m^2)	Thickness loss (μm)	Exterior	Interior
C1 very low	≤10	≤1.3	≤0.7	≤0.1		Heated buildings with clean atmospheres, e.g., offices, shops, schools, hotels
C2 low	>10 to 200	>1.3 to 25	>0.7 to5	>0.1 to 0.7	Atmospheres with low level of pollution, mostly rural areas	Unheated buildings where condensations may occur, e.g., depots, sports halls
C3 medium	>200 to 400	>25 to 50	>5 to 15	>0.7 to 2.1	Urban and industrial atmospheres, moderate sulfur dioxide pollution. Coastal area with low salinity	Production room with high humidity with some air pollution, e.g., food processing plant
C4 high	>400 to 650	>50 to 80	>15 to 30	>2.1 to 4.2	Industrial area and coastal area with low salinity	Chemical yards, swimming pools, coastal ships, and boatyards
C5 (Harsh Onshore)	>650 to 1500	>80 to 200	>30 to 60	>4.2 to 8.4	Industrial areas with high humidity and aggressive atmosphere	Buildings or area with almost permanent condensation and with high pollution
CX extreme (offshore)	>650 to 1500	>80 to 200	>30 to 60	>4.2 to 8.4	Environments include extreme industrial areas, offshore areas, and salt spray	

presents the categories of corrosivity as a function of the mass or thickness loss of such standard specimens made of low-carbon steel and/or zinc after exposure for 1 year. Details of standard specimens and the treatment of the specimens prior to and after exposure are as in ISO 9226. Extrapolation of the mass or thickness losses to 1 year from shorter exposure times, or back-extrapolation from longer times, will not give reliable results and is therefore not permitted. The mass or thickness losses obtained for steel and zinc specimens may sometimes give different categories. In such cases, the higher corrosivity category should be taken. If it is not possible to expose standard specimens in the actual environment of interest, the corrosivity category may be estimated by simply considering the examples of typical environments given in Table 6.9. The examples listed are informative but may occasionally be misleading. Only the actual measurement of mass or thickness loss will give the correct classification. Note that the corrosivity categories can also be estimated by considering the combined effect of the following environmental factors: yearly periods of wetness, yearly mean concentration of sulfur dioxide, and yearly mean deposition of chloride as in ISO 9223.

6.5.1.1 Categories for water and soil

For structures immersed in water or buried in soil, corrosion is normally local in nature and corrosivity categories are difficult to define. However, for the purpose of this International Standard, various environments can be described. In Table 6.7, three different environments are given together with their designations. Note that in many such situations, cathodic protection is involved and this should be kept in mind.

Note that in Table 6.10 the loss values for the corrosivity immersion categories are identical to those given in ISO 9223. And in coastal areas in hot humid zones, the mass or thickness losses can exceed the limits of the category in the case of marine structures in category (C5-M). Special

Table 6.10 Categories for immersion in water and soil.

Category	Environment	Examples of environment and structures
Im1	Fresh water	River installations, hydroelectric power plants
Im2	Sea or brackish water	Harbor areas with structures like sluice gates, locks, jetties, offshore structures
Im3	Soil	Buried tanks, steel piles, steel pipes

Table 6.11 Calculated time of wetness and selected characteristics of various types of climate (as in ISO9223:1992).

	Mean value of the annual extreme values (°C)			
Type of climate	**Low temperature**	**High temperature**	**Highest temperature with RH > 95%**	**Calculated time for wetness at RH > 80% and temperature > 0 °C (h/year)**
Very cold	−65 °C	32 °C	20 °C	0–100
Cold	−50 °C	32 °C	20 °C	150–2500
Cold temperature	−33 °C	34 °C	23 °C	2500–4200
Warm temperature	−20 °C	35 °C	25 °C	
Warm dry	−20 °C	40 °C	27 °C	10–1600
Mild warm dry	−5 °C	40 °C	27 °C	
Extremely warm dry	3 °C	55 °C	28 °C	
Warm damp	5 °C	40 °C	31 °C	4200–6000
Warm damp, equable	13 °C	35 °C	33 °C	

precautions must therefore be taken when selecting protective paint systems for structures in such areas.

In most cases the general conclusions are that the corrosion rate is driven by climate conditions. In a cold climate or a dry climate, the corrosion rate will be lower than in a high temperate climate; it will be greatest in a hot, humid climate and in a marine climate, although considerable local differences can occur. The main factor is the time length which the structure is exposed to high humidity, which is called "time of wetness" (Table 6.11).

6.5.2 Mechanical, temperature, and combined stresses

Erosion, or abrasive stresses, may occur due to particles, such as sand being entrained by the wind. Surfaces subject to abrasion are considered to be exposed to moderate or severe mechanical stresses. On the other hand, in water, mechanical stresses may be produced by boulder movement, the abrasive action of sand, wave action, and others.

Mechanical stresses can be divided into three classes: weak, moderate, and severe. The criteria are as follows:

(a) *Weak*: The mechanical stresses occur very rarely; they can occur due to light debris or sand, with small quantities moving slowly with water.

(b) *Moderate*: The mechanical stresses occur due to solid debris, sand, gravel, shingle, or ice moving in moderate quantities, with water that is flowing at moderate speed.

(c) *Severe*: High mechanical stresses occur as a result of solid debris, sand, gravel, shingle, or ice entrained in large quantities that are moving very fast with the water.

The stresses accumulate due to medium or high temperatures. In most standards, special DNV guidelines define medium temperatures as those between +60 °C and +150 °C, and high temperatures as those between +150 °C and +400 °C. Temperatures of this magnitude only occur under special conditions during construction or operations. The higher temperature is seen usually in the case of fire; the medium temperatures are only encountered in some areas, such as in the Arabian Gulf, where the temperature can reach 40–60 °C, which is near the medium range.

Where combined stresses affect the steel, corrosion will increase. In general, corrosion may develop more quickly on surfaces exposed simultaneously to mechanical and chemical stresses. This applies to steel structures near salt water; in addition, there are mechanical stresses due to impacts, as boat landings.

6.6 General CP design considerations

Part of the platform conceptual design is to define the cathodic protection and define the compatibility of CP with metallic materials of the structure and its applied coating system. This section presents guidelines for conceptual and detail engineering for designing the cathodic protection system.

The corrosion allowance should be taken into consideration during the design and cannot be ignored based on the CP system.

6.6.1 Environmental parameters affecting CP

The major seawater parameters affecting CP in situ are:

- The amount of dissolved oxygen
- Sea current velocity
- Temperature
- Marine growth

- Salinity
- Carbonate content
- pH variation in seawater
- Seabed temperature
- Seabed bacterial growth
- Salinity of seabed
- Sediment coarseness

Depending on the geographic location, water depth, and season, the preceding parameters can vary. There is no directional equation between the seawater environmental parameters and cathodic current demands to achieve and to maintain CP.

To rationalize CP design for offshore structures, default design current densities, I_c (A/m^2), are defined based on climatic conditions in the regions, the water depth, and mean seawater surface temperature. The ambient seawater temperature and salinity determine the specific seawater resistivity, ρ (ohm m), which is used to calculate the anode resistance, R (ohm), a controlling factor for the current output from an anode.

6.6.2 Design criteria

The cathodic protection system for the substructure will be designed in accordance with Det Norske Veritas (DNV) (1993) and the requirements of this specification. The design criteria should also contain the conditions and the assumptions that were taken into consideration.

A retrofit CP design considerations may be limited only to the platform steel structure, pile guides, piles, and also any conductors, risers, or any appurtenances. When focusing on the platform, the design does not include pipelines or associated subsea structures that might be in close proximity and electrically connected to the platform, other than making an allowance for some debris on the seabed.

The design life of sacrificial anode cathodic protection (SACP) systems should be defined based on the owner requirements, taking into consideration the lifetime of the platform; this case study is over 25 years.

The retrofit CP design is based on using the conventional and modified CP criteria with a target of maintaining external structure potentials in the range from -800 to -1050 mV with respect to the Ag/AgCl reference electrode for seawater, throughout the remaining service life of the platform. A seawater resistivity of 20 Ω cm and an average seawater temperature of 22 °C have been used in design calculations.

The platforms were originally fitted with aluminum flush-fit, long slender type anodes more than 30 years ago, so the anodes are now either totally consumed or almost at end-of-life. As these types of anodes have a theoretical utilization factor of 85%, the anode retrofit design will not consider an allowance for any remaining anode alloy.

For retrofits, the designer should check that the original coating system applied to the splash zone is assumed to be comparable to today's standards, equivalent to a DNV-Cat III standard coating. For the purposes of the CP retrofit design, an initial coating breakdown factor of 2% has been used and 1% each year thereafter. This is slightly less than DNV recommends, but is comparable with other offshore operators as being acceptable practice. Based on all platforms being originally installed for 30 years, the coating breakdown factors used in the calculations for 25 years consider that the initial coating breakdown failure (CBF) is 35%, mean (CBF) is 48%, and the final CBF is equal to 60%.

Based on Det Norske Veritas DNV (2005), the cathodic protection design guideline provides the following current densities that can be used in designing offshore structure CP systems; for wetted areas 150, 70, and 100 mA/m^2 for initial, mean, and final, respectively. For buried areas 20 mA/m^2 will be taken for initial, mean, and final.

It is important to highlight that these current densities are used for new uncoated steel structures in tropical waters (>20 °C) with a water depth to a maximum of 30 m. From a practical point of view, these criteria are conservative for retrofit designs for platform steel structures remaining partially protected, and design using these figures would lead to having a large mass of anode alloy and a substantial number of anode sleds.

DNV and other international standard organizations have stated that the design parameters may be varied depending on local conditions and from operations experience.

Case studies have been carried out assessing the original CP design of SACP systems fitted to seven of the specified platforms. The assessment calculated the average current densities for the platforms based on potential levels over the last 25 years and compared these to total anode alloy weight loss over the same period. To obtain the most accurate current density figures, the total surface areas for each platform have been added together, as have the number of anodes to obtain the total mass of anode alloy consumed over 25 years.

The assessment concludes that, based on potential data, the average current density has been 24 mA/m^2, whereas anode weight loss over the same period gives an average current density of 23 mA/m^2. Based on the findings of the assessment, in some platforms the recommendations for wetted area 35 mA/m^2 and for buried area 10 mA/m^2 are the same for initial, mean, and final; this is to reduce current density criteria for platform retrofit CP designs, which is related to the design office experience.

6.6.3 Protective potentials

The Ag/AgCl/seawater is the reference electrode, so a potential of −0.80 V is accepted as the design protective potential Ec° (V) for carbon and low-alloy steels.

For seawater sediments in anaerobic environments, the design protective potential of −0.90 V will be applied.

The CP potential reading along its lifetime should be in the range of −0.90 to −1.05 V and at the end of the service life, the potential increases rapidly towards −0.80 V, and eventually to be underprotection, which means fewer negative values.

The term overprotection is only applicable to protection potentials more negative than −1.15 V. Such potentials will not apply for CP with galvanic anodes based on Al or Zn.

6.6.4 Detrimental effects of CP

A side effect of CP is that it will produce hydroxyl ions and hydrogen at the surface of the protected object. As a result, there will be an electrochemical reduction process in the interface between the coating and metal that causes a coating disbonding, and this phenomenon is called cathodic disbonding. The disbonding will be accelerated in the case of hot fluids. Some coating systems with a base of epoxy or polyurethane have good resistance to the cathodic disbonding effect.

ASTM G8 provides a standard testing of marine coatings' resistance to cathodic disbanding. It is found that the cathodic protection will generate atomic hydrogen at the metal surface in the case of aluminum or zinc base anodes with potential range −0.80 to −1.10 V with reference Ag/AgCl at seawater, considering that the production of hydrogen increases exponentially towards the negative potential limit.

Another effect of the hydrogen atoms is that they can interact with the microstructure of components subject to high stresses, causing initiation and growth of hydrogen-related cracks, which is called hydrogen-induced stress cracking.

A consequence of CP application is that a calcareous layer primarily consisting of calcium carbonate will form on bare metal surfaces. The thickness is typically around 0.1 mm, but thicker deposits may occur. The main effect of the calcareous layer is that it reduces the current demand for maintenance of CP, which is good.

CP with copper-base alloys will eliminate the antifouling effect in seawater.

6.6.5 Galvanic anode materials

For offshore structures, galvanic anodes are based on either aluminum or zinc. The generic type of anode material, which is aluminum or zinc based, is typically selected and specified in the conceptual and detail CP design.

The higher electrochemical capacity ε (A h/kg) is in aluminum-based anodes. On the other hand, the zinc-based anodes have sometimes been considered more reliable with respect to electrochemical performance for some applications in marine sediments or internal compartments in ships, such as FPSO, with high bacterial activity, both environments representing anaerobic conditions.

Based on practical experience, ferritic and ferritic-pearlitic structural steels with specified minimum yield strength (SMYS) up to at least 500 MPa have proven compatibility with marine CP systems. However, laboratory testing has demonstrated susceptibility to hydrogen-induced stress cracking during extreme conditions of yielding.

A postweld heat treatment is very important to reduce the heat-affected zone hardness and the residual stresses due to the welding process, as welding will change the steel formation. Thus there are recommendations for ferritic steels to be applied for hardness limits and design measures.

Design precautions should include measures to avoid local plastic yielding and the use of coating systems qualified for, e.g., resistance to disbondment by mechanical and physical or chemical effects.

Anodes will have a closed-circuit potential of −1.10 V (or more negative) to a copper/copper sulfate electrode and will have a minimum efficiency of 80% (maximum consumption rate of 3.68 kg per ampere-year) or such higher value as may be apparent. Each foundry pour will be tested for chemical composition and closed-circuit potential. A suitable sample will be retained from each pour for performing a rate of consumption test. A minimum of three tests will be performed for rate of consumption during the production run. In the event that any test indicates a consumption rate of more than this maximum, tests will be performed on additional samples to confirm the adequacy of the anode composition.

The active anode material will be a proven aluminum-zinc-indium based alloy suitable for long-term continuous service in seawater, saline mud, or alternating seawater and saline mud environments.

Whoever delivers the anode should propose an alloy offering the following minimum electrochemical performance characteristics, documented by long-term performance test reports:

Prior to manufacture, the supplier will demonstrate the ability of the proposed alloy to satisfy these requirements; minimum anode capacities and working potentials will be determined at operating anode current density in the range of 0.25–1.0 A/m^2 for saline mud and 0.25–4.0 A/m^2 for free seawater.

The composition is only the limits of ranges. The supplier will submit the full chemical analysis details and the specific gravity value of the proposed alloy to the client for approval and will demonstrate that the specification requirements can be satisfied at the compositional limits.

6.6.6 CP design parameters

This section describes design parameters to be used for conceptual and detailed design of galvanic anode CP systems and gives guidance on the selection of such parameters. With the exception of the design life and possible also coating breakdown factors, the actual design values to be applied for a specific project are normally selected by the designer, based on environmental and other parameters identified in the project design basis. The recommended design values represent a conservative approach. Applying these values will provide a service life that is more than the CP system design life.

The offshore structure CP designer may specify less or, in certain cases, more conservative design data, based on special considerations or expert judgment based on experience.

All electrochemical potentials associated with CP in this section refer to the Ag/AgCl/seawater reference electrode. By comparing the 1993 revision of this DNV RP with the major changes of the 2004 revision, the number of depth zones for design current densities has been extended from 2 to 4, while the number of "coating categories" is reduced from 4 to 3.

6.6.6.1 Design life

The owner will usually specify the required lifetime for the CP system. The design life will further take into account any period of time when the CP system will be active prior to operation of the protection object.

It is very important to take into consideration, when defining the CP lifetime, that the anode retrofit for sacrificial anodes in CP systems for fixed offshore structures is very costly and sometimes impractical. In some cases, the planned retrofitting of sacrificial anodes may be an economically better option than the installation of very large new anodes. This alternative should then be planned, to define the requirements for retrofitting during the preliminary design.

6.6.6.2 Design current densities

The current density, I_c, is the cathodic protection current per unit surface area (in A/m^2). The "initial" and "final" design current densities, I_{ci} (initial) and I_{cf} (final), respectively, provide the anticipated cathodic current density demand measurement to achieve cathodic protection of a steel surface during a short period of time. Knowing the initial and final current demands is a first step to calculating and defining the number and sizing of anodes. The effect of any coating on current demand is applying a factor called the coating breakdown factor.

The initial design current density is the cathodic current density required to carry out polarization of an initial structural steel surface with some rusting and mill scale.

The initial design cathodic current density should be higher than the current density of the final design due to the possibility during the initial phase of forming a calcareous scale and marine fouling layer, which reduces the subsequent current demand as the polarization resistance is reduced.

In addition, sufficient initial design current density enables rapid formation of protective calcareous scale and hence efficient polarization. It considers the current density required to repolarize a structure if such layers are partly damaged, due to periodic performance of marine growth removal.

The accurate final design current density and hence CP polarizing capacity will guarantee that the structure protection remains polarized to a potential in the range of -0.95 to $-1.05\,V$ during the whole design life.

The initial and final current densities are used to calculate the required number of anodes for a specific type to achieve sufficient polarizing capacity, using Ohm's Law and assuming the following:

- The anode potential is in accordance with the design closed circuit potential.
- The protection object potential, which means the cathode, is at the design protective potential for C-steel and low-alloy steel, taken as equal to −0.80 V.

Based on the preceding relationship, the anode current, and hence the cathodic current density, decreases linearly when the cathode is polarized towards the closed-circuit anode potential, reducing the galvanic cell driving voltage.

The total CP current for a CP unit, I_t (A), is calculated from this equation:

$$I_t = \frac{E_c - E_a}{R} \tag{6.1}$$

where R (ohm) is the total anode resistance, E_c (V) is the global protection potential, and E_a (V) is the actual anode potential as a closed circuit.

The definition of mean (or maintenance) design current density, I_{cm} (A/m^2), is a measure of the anticipated cathodic current density once the CP system has attained its steady-state protection potential. The steady-state potential is about 0.15–0.20 V more negative than the design protective potential.

The decrease in cathode potential, which means cathodic polarization, will reduce the anode current. Based on this, the average design current density will be about 50% of the initial and final design current density. The initial polarization period preceding the steady-state condition is normally short with respect to the design life; therefore the current density for the time-weighted cathodic is very close to the steady-state cathodic value.

Cathodic current densities to achieve and maintain CP are dependent on factors such as the geographical location, water depth, and others.

The initial, final, and average design current densities recommended values are shown in Tables 6.11 and 6.12, respectively, based on climatic regions and water depth. These current density design factors are conservative for harsh weather conditions, including waves and sea currents, but do not consider calcareous layer erosive effects due to silt or ice. In addition, they assume that the seawater at the surface is saturated with air, which means 0.2 bar oxygen partial pressure.

Table 6.12 Performance properties of the alloy.

Environment	Nominal resistivity (ohm cm)	Anode temp. (°C)	Min^{m} alloy capacity (A h/kg)	Working potential (mV w.r.t.) Ag/AgCl/seawater
Saline mud	100	40	1500	−1050
Saline mud	200	25	1865	−1050
Seawater	25	25	2000	−1070

The effect of seawater temperature and depth on the properties of a calcareous scale formed by cathodic protection and of the content of dissolved oxygen on the CP is presented in Tables 6.11 and 6.12. Oxygen is dissolved in the surface layer due to dissolution from air and photo synthesis, and due to the fact that at a large depth in a tropical region the oxygen content is likely to be substantially lower than in temperate or arctic surface waters of the same ambient seawater temperature. The higher design current densities in the uppermost zone are due to the wave forces and marine growth on degradation of calcareous scales and convective mass transfer of oxygen.

In some regions, the organic material decomposition may reduce and ultimately consume all oxygen in the seawater. No such reduction in oxygen content is accounted for in Tables 6.13 and 6.14.

For bare steel surfaces buried in sediments, a design current density (initial/final and average) is considered equal to 0.020 A/m^2, whatever the geographical location and depth.

In the uppermost layer of seabed sediment, bacterial activity may be the primary factor to define the CP current demand.

The additional CP current density is to account for increased convective and diffusive mass transfer of oxygen induced by heat transfer.

Table 6.13 For seawater exposed to steel surfaces, recommended initial and final design current densities (A/m^2) as a function of depth and surface water temperature.

	Tropical (>20 °C)		Subtropical (12–20 °C)		Temperate (7–11 °C)		Arctic (<7 °C)	
Depth (m)	Initial	Final	Initial	Final	Initial	Final	Initial	Final
0–30 m	0.15	0.10	0.17	0.11	0.20	0.13	0.25	0.17
30–100 m	0.12	0.08	0.14	0.09	0.17	0.11	0.20	0.13
100–300 m	0.14	0.09	0.16	0.11	0.19	0.14	0.22	0.17
>300 m	0.18	0.13	0.20	0.15	0.22	0.17	0.22	0.17

Table 6.14 For seawater exposed to steel surfaces recommended mean design current densities (A/m^2), as a function of depth and surface water temperature.

Depth (m)	Tropical (>20 °C)	Subtropical (12–20 °C)	Temperate (7–11 °C)	Arctic (<7 °C)
0–30	0.070	0.080	0.100	0.120
30–100	0.060	0.070	0.080	0.100
100–300	0.070	0.080	0.090	0.110
>300	0.090	0.100	0.110	0.110

Based on DNV, for aluminum components, or those coated with either aluminum or zinc, a design current density of 0.010 A/m^2 is recommended for initial/final as well as mean values. For internally heated components, the design current density will be increased by 0.0002 A/m^2 for each °C that the metal/seawater is assumed to exceed 25 °C.

The current density available to structural steel surfaces will be sufficient, at any time during the design life of the protected structure, to achieve the required potential range and maintain a calcareous deposit. Locally determined maintenance current densities, if applied, will be based upon well-documented monitoring of cathode protection current densities. Locally determined values will be provided by the principal and documented in detail.

6.6.6.3 Coating breakdown factors for CP design

The use of nonmetallic coatings is very important, as it will reduce the CP current demand of the protection object and, accordingly, the required anode weight will be less.

With a structure that is sensitive to weight with a long design life, the combination of a coating and CP provides the most cost-effective corrosion control. In some structures that require very long design lives, CP combined with coatings will be mandatory.

The use of coatings should be considered for applications where the demand for CP of bare metal surfaces is known or expected to be high. The clear example of that is for deep water, as the formation of calcareous deposits may be slow. It should further be considered for surfaces that are partly shielded from CP by geometrical effects.

For large and complex structures like, for example, multiwell subsea production units, extensive use of coating is required to limit the overall current demand and to ensure adequate current distribution. To overcome this, the design coating breakdown factors to be used for CP design will be conservative to guarantee that enough total final current output capacity is installed. Therefore, any calculations of the electrolytic voltage dropping away from the anodes, for example by means of finite or boundary element analyses, and using these coating breakdown factors will generate excessively high cathodic protection in terms of the estimated protection potential. This primarily applies to structures with long design lives in case of the calculated coating breakdown; the current demands and electrolytic voltage drop increase exponentially.

The application of coatings is not practical in some submerged structures that require regular inspection for fatigue cracks, for example the critical welded nodes of jacket structures.

Metallic coatings on zinc or aluminum bases are compatible with galvanic anode CP. However, compared to organic coatings, they have not provided any advantage in decreasing the current demand for CP. Zinc-rich primers are unsuitable for application with CP due to susceptibility to low electrical resistivity, which leads to high CP current demand.

The coating breakdown factor, f_c, presents the reduction in cathodic current density due to the application of an electrically insulating coating. When $f_c = 0$, the coating is 100% electrically insulating, thus decreasing the cathodic current density to zero; $f_c = 1$ so the coating has no current reducing properties.

The coating breakdown factor is different than coating degradation, as the coating degradation is defined by visual examination. Coating showing much damage may still retain good electrically insulating properties. On the other hand, a perfect coated surface may allow a significant current passage.

The coating breakdown factor is a function of coating properties, operational parameters, and time. As a simple engineering approach, f_c can be represented by the following formula:

$$f_c = a + bt \tag{6.2}$$

where t is the coating lifetime in years, and a and b are constants that are dependent on coating properties and the environment.

The marine growth effect is highest in the upper 30 m, where wave forces may further contribute to coating degradation. Another factor is periodic cleaning of marine growth in this zone. The owner should preferably specify constants a and b for calculation of coating breakdown factors based on practical experience of specific coating systems in a particular environment. When the owner does not specify any such data, the default values in Table 6.13 will be used.

Once a and b are defined, mean and final coating breakdown factors, f_{cm} and f_{cf}, respectively, to be used for CP design purposes are to be calculated by introducing the CP design life, t_f (years):

$$f_{cm} = a + 0.5bt_f \tag{6.3}$$

$$f_{cf} = a + bt_f \tag{6.4}$$

For certain protection objects, with large uncoated surfaces, the initial coating breakdown factor, $f_{ci} = a$, may be applied to calculate the initial current demand to include coated surfaces.

If the calculated value exceeds 1, $f_c = 1$ will be applied in the design. When the design life of the CP system exceeds the actual calculated life of the coating system, f_{cm} may be calculated as:

$$f_{cm} = 1 - \frac{(1-a)^2}{2bt_f} \tag{6.5}$$

To account for the effect of a coating system on coating breakdown factors, three coating categories have been defined for inclusion in Table 6.15.

The criteria for the three categories are as follows:

Category I: One layer of epoxy paint coating is not less than 20 μm nominal of dry film thickness.

Category II: One or more layers of marine paint coating by epoxy, polyurethane, or vinyl based on total nominal of the dry film thickness b not less than 250 μm.

Category III: Two or more layers of marine paint coating by epoxy, polyurethane, or vinyl based on total nominal of the dry film thickness being not less than 350 μm.

Note that category I includes shop primer coatings. For categories II and III, the coating materials supplier should provide his qualification document to verify his performance, or it should be done by test. For all three categories, all coating work should be done based on the manufacturer's

Table 6.15 Recommended constants *a* and *b* for calculation of paint coating breakdown factors.

	Recommended *a* and *b* values for coating categories I, II, III		
Depth (m)	**I ($a = 0.10$)**	**II ($a = 0.05$)**	**III ($a = 0.02$)**
0–30	$b = 0.10$	$b = 0.025$	$b = 0.012$
>30	$b = 0.05$	$b = 0.015$	$b = 0.008$

recommendations and surface preparation should include clean by blast to SA 2.5 as a minimum, according to ISO 8501. Also based on the manufacturer's recommendation, the surface roughness will be controlled.

If the coatings were applied without blast cleaning surfaces, coating break-down factors of f_{cm} and f_{cf} will be taken equal to 1, while the initial current demand follows category I.

Published data on the performance of coatings on cathodic protected structures are scarce, for long service lives. The data in Table 6.15 is presented as conservative engineering judgments. For any coating system not covered by the preceding three coating categories, and with major potential effect on the overall current demand, the designer should specify or accept the applicable constants *a* and *b*.

The values of the factors *a* and *b* for a depth of 30–100 m in Table 6.15 are applicable to calculations of current demands of flooded compartments and of closed compartments with free access to air.

The constants in Table 6.15 do not account for significant damage to paint coatings during fabrication and installation. Therefore, if such damage is anticipated, the affected surface area is to be estimated and included in the design calculations as a bare metal surface.

From the technical practice of Shell, Design and Engineering Practice (1995), category I paint coating is not applied to structure nodes, weld surface areas, and up to 0.25 m adjacent to the welds, which will be assumed to be bare and included for the calculation of current requirements.

For each zone, the coated surface average and final coating coverage and breakdown factors will be defined.

6.6.6.4 Galvanic anode material design parameters

The anodes materials should follow specifications. Table 6.14 presents the compositional limits for alloy and impurity elements for Al- and Zn-based anodes.

The CP design parameters related to anode material performance are:

- design electrochemical capacity, ε (Ah/kg)
- design closed-circuit anode potential, E_o a (V)

Both factors, the design electrochemical capacity ε (Ah/kg) and design closed-circuit anode potential $E_o a$ (V), are used to calculate the design anode current output and the required net anode mass by using Ohm's and Faraday's Laws, respectively.

The design values for electrochemical capacity ε (Ah/kg) in Table 6.16 will be used for the design unless otherwise specified or accepted by the

Table 6.16 Recommendation for Al-based and Zn-based anode material composition limits.

Alloying/Impurity element	Zn-based	Al-based
Zn	rem.	2.5–5.75
Al	0.10–0.50	rem.
In	na	0.015–0.040
Cd	≤0.07	≤0.002
Si	na	≤0.12
Fe	≤0.005	≤0.09
Cu	≤0.005	≤0.003
Pb	≤0.006	na

supplier and the designer. The data are applicable for ambient temperature of seawater, which means temperature as a yearly mean value up to 30 °C.

The anode electrochemical efficiency from laboratory examinations has a value close to the theoretical limit (e.g., ≥2500 Ah/kg for Al-Zn-In material). This is due to the relatively high anodic current densities that are utilized for testing. Such data will not replace the recommended design values for electrochemical capacity at sea ambient temperatures.

Long-term testing should be done if we use electrochemical capacity greater than the default values based on Det Norske Veritas DNV (2005) in Table 6.17.

Note that this testing will tend to result in slightly nonconservative values, as the testing time is still relatively short and the anodic current density relatively high compared to the working conditions of real anodes. The design values for closed-circuit anode potential, $E_o\ a$ (V), in Table 6.15 will be used for the design. The data are applicable for all ambient seawater temperatures, which is a maximum of 30 °C yearly average.

Table 6.17 Recommended design electrochemical capacity and design potential voltage for anode materials.

Anode material type	Environment	Electrochemical capacity (Ah/kg)	Closed-circuit potential (V)
Al-based	Seawater	2000	−1.05
	Sediments	1500	−0.95
Zn-based	Seawater	780	−1.00
	Sediments	700	−0.95

6.6.6.5 Anode resistance formulas

Unless otherwise agreed, the anode resistance, R (ohm), will be calculated using the following formula that is applicable to the actual anode shape. Calculations will be performed for the initial anode dimensions and for the estimated dimensions when the anode has been consumed to its utilization factor.

1. For long slender standoff $L \geq 4r$

$$R = \frac{\rho}{2\pi L}\left(\text{In}\frac{4L}{r} - 1\right) \tag{6.6}$$

   where L is the anode length, ρ is the seawater resistivity (ohm m), and r is the effective anode radius.

2. Short slender standoff $L < 4r$

$$R = \frac{\rho}{2\pi L}\left[\text{In}\left\{\frac{2L}{r}\left(1 + \sqrt{1 + \left(\frac{r}{2L}\right)^2}\right)\right\} + \frac{r}{2L} - \sqrt{1 + \left(\frac{r}{2L}\right)^2}\right] \tag{6.7}$$

   The equation is valid for anodes with minimum distance 0.30 m from the protection object. For an anode to object distance less than 0.30 m but a minimum of 0.15 m, the same equation may be applied with a correction factor of 1.3.

3. Long flush-mounted as L more than or equal to 4 times width and also thickness.

$$R = \frac{\rho}{2.S} \tag{6.8}$$

   For noncylindrical anodes; $r = c/(2\pi)$ where c (m) is the anode cross-sectional periphery. S is the arithmetic mean of anode length and width.

4. Short flush mounted, bracelet and other types.

$$R = \frac{0.315\rho}{\sqrt{A}} \tag{6.9}$$

   where A is the anode surface area.

6.6.6.6 Seawater and sediment resistivity

The seawater resistivity, ρ (ohm m), is a function of the seawater salinity and temperature. In the open sea, the salinity does not vary significantly and temperature is the main factor.

For seawater, the marine sediments resistivity is higher by a factor ranging from 2 for very soft clays to approximately 5 for sand. If there is no available sediment data, the highest factor will be considered for calculation of the buried anodes resistance.

In regions with annual average surface water temperature from 7 to 12 °C, resistivity of 0.30 and 1.3 Ωm are recommended, which is reasonably conservative for the calculation of anode resistance in seawater and marine sediments, respectively, independent of depth.

If it presents a lower value, it should be verified by actual measurements, considering any seasonal variations in temperature.

When the offshore structure is in seawater with a temperature higher than 5 °C for a depth interval, the depth interval will be divided into separate zones that cover no more than 5 °C per each interval. The depth with averaged temperature will be used for each created zone. The owner may use seawater resistivity (ρ) from the field basis of the design or it can be obtained by doing resistivity measurements based on the average annual seawater temperature conditions.

6.6.6.7 Anode utilization factor

Anode shape and core insert configuration have a major influence on the utilization factor of any particular anode design. This factor represents the maximum volume of cast anode alloy that can be consumed before the anode can no longer deliver the current required, and it has to take into account the reduced size of the anode alloy and disbandment of anode alloy from the core at the end of life.

The anode utilization factor, u, is the fraction of anode material of an anode with a specific design that may be utilized for calculation of the net anode mass required to sustain protection throughout the design life of a CP system.

When an anode is consumed to its utilization factor, the polarizing capacity (as determined by the anode current output) becomes unpredictable due to loss of support of anode material, or rapid increase of anode resistance due to other factors.

The utilization factor is dependent on the anode design, particularly its dimensions and the location of anode cores. Unless otherwise agreed on, the anode utilization factors presented in Table 6.18, based on DVV RPB401, will be used for design calculations. Based on BS, an anode utilization factor of 0.90 for the long slender stand-off type will be used.

Table 6.18 Recommended anode utilization factors for CP design calculations.

Anode type	Anode utilization factor
Long slender stand-off $L \geq 4r$	0.90
Short slender stand-off $L < 4r$	0.85
Long flush-mounted $L \geq$width $L \geq$thickness	0.85
Short flush-mounted, bracelet, and other types	0.80

6.6.6.8 Current drain design parameters

As a design basis, the current drain calculation will consider all components that are expected to become electrically connected to a CP system.

For offshore structures, there are temporary or permanent components that are not considered to require CP, but will drain current from the CP system. Examples are the mooring systems for floating installations or secondary structural members such as piles and skirts, as they do not need corrosion protection for structural purposes due to their large wall thickness relative to expected corrosion rates.

Calculations of current drain will use the design current densities and coating breakdown factors for items requiring CP. Calculations of surface areas and current demands will be carried out.

The design current densities and coating breakdown factors, respectively, are applicable for calculation of current drains of components that are not considered to need CP, but will be or is expected to be electrically connected to the protection object by the CP system being designed.

With buried surfaces such as mud mats, skirts, and piles, a current drain value of $0.020\,A/m^2$ will be accounted for, based on the outer external surface area exposed to sediment. In some cases, and based on the owner specification for parts of steel skirts and piles to be buried in sediments, a current density (initial/final and average) of $0.025\,A/m^2$ will be used. The current drain to open pipes will cover an internal surface area equivalent to 10π times their respective diameter.

For open pile ends, the top internal surface will be included for 5 times the diameter and will be regarded as seawater exposed. Internal surfaces of piles filled with sediments do not have to be included.

The current drain of (5 A/well casing) will be included in current drain calculations unless the owner has a different requirement.

Subsea wells that are cemented have reduced current drain compared to platform wells that are not cemented. But subsea wells may become exposed to significant current drain during installation and work-over activities.

As per DNV, the current drain for the anchor chains will be considered for 30 m of chain for systems with mooring points topside only; for mooring points under the water level, the seawater exposed section above this point will be considered.

6.6.7 CP calculation and design procedures

For platform substructures, the detailed design of a CP system is normally started after finalizing the conceptual design activity. In the conceptual design, the type of anodes and fastening devices should be selected, considering forces applied on the anodes during installation and operation. Moreover, any coating systems to be applied to specific areas or components will be specified, so it can be considered in the preliminary current demands calculation for CP and the corresponding total anode net mass.

The following information and any optional requirements will be included in the tender package:

- Conceptual CP design report, including the design life of the CP system.
- The project design basis main information, such as salinity and temperature as a function of depth for anode resistance calculation, seawater level, and mud line for platform substructures, and environmental and installation parameters affecting forces applied to anodes.
- The structural drawings and coating systems data as required for calculation of surface areas to be protected, including temporary or permanent current drain.
- Definition of any interfaces to electrically connected components or systems with self-sufficient CP systems, such as pipelines.
- Definition of specific requirements for CP design parameters to be applied, for example coating breakdown factors and current drain to wells.
- Definition of any specific requirements for anode material and design.

The issued engineering package to the contractor should be in the final revision of drawings and specifications affecting the calculation of current demand for CP and location of the anodes are available.

It will be ensured that all necessary information is provided for the contractor to calculate the overall current demand, such as production platform conductors and production control system for subsea valve trees.

CP design involving large or complex objects will be divided its protection to unit areas.

6.6.7.1 Current demand calculations

To calculate the current demand, I_c (A), to provide adequate polarizing capacity and to maintain cathodic protection during the design life, the individual surface areas, A_c (m^2), of each CP unit will be calculated and multiplied by the relevant design current

$$I_c = A_c \, I_c f_c \tag{6.10}$$

Design current density, I_c (A/m^2), and the coating breakdown factor, f_c, if applicable: I_c and f_c are then to be selected, respectively.

It is practical to apply some simplification when calculating surface areas for objects with complex geometries. However, it will be ensured that the overall result of any such simplification is conservative. For major surface areas, an accuracy of from −5% to +10% is adequate. For smaller components, the required accuracy may be lower depending on whether or not a coating will be applied to such items and to the major surfaces.

The designer will verify the design and check that all components affecting CP current demand are included in the surface area calculations. A significant current demand is required for subsea production systems and production control equipment fabricated from uncoated stainless steel.

Moreover, some components like valve blocks and hydraulic cylinders may have coating applied directly to machined surfaces, increasing the coating breakdown factor to be used for the design.

For items with major uncoated metal surfaces, the CP current requirements for both initial polarization and at the end of the design life, I_{ci} (A) and I_{cf} (A), respectively, will be calculated, together with the mean current demand required to maintain cathodic protection during the design period, I_{cm} (A).

The initial current demand can be ignored in the design calculations for steel members with current demand primarily associated with coated surfaces.

6.6.7.2 Selection of anode type

The anode types of stand-off, flush-mounted, or bracelet may be specified by the owner, taking into account the effects of, e.g., sea current drag and interference with subsea interventions. If the anode type has not been specified by the owner, then the contractor will select the anode type, taking into account, e.g., net anode mass to be installed and available space for location of anodes. The anode type further affects the anode utilization factor and the anode current output in relation to weight.

Long stand-off type anodes are usually preferred for relatively large anodes (about 100 kg and more) to be installed on platform substructures and subsea templates. A flush-mounted anode with the same net anode mass will have a lower anode current output and lower utilization factor.

6.6.7.3 Anode mass calculations

The total net anode mass, Ma (kg), required to maintain cathodic protection throughout the design life, t_f (years), is to be calculated from I_{cm} (A) for each unit of the mean current demand including any current drain:

$$M_a = \frac{8760 I_{cm} t_f}{u\varepsilon} \tag{6.11}$$

6.6.7.4 Calculation of anodes number

After selecting the anode type, then the number of anodes, (N), anode dimensions, and anode net mass, ma (kg), are selected to comply with the requirements for:

(1) initial and final current output, I_{ci} and I_{cf} (A) and
(2) anode current capacity, C_a (Ah)
which relate to the CP current demand, I_c (A), of the protection object.

The available anodes on the market control the preliminary sizing of the anodes.

The individual anode current output, I_a (A), required to meet the current demand, I_c (A), is calculated from Ohm's law:

$$I_c = NI_a \frac{N(E_{co} - E_{ao})}{R} = \frac{N\Delta E_o}{R} \tag{6.12}$$

where I_a (A) is the l anode current output, E_{ao} (V) is the design anode material closed-circuit potential, and R (ohm) is the anode resistance.

The initial and final current output, I_{ai} and I_{af}, are to be calculated using the initial and final anode resistance, R_{ai} and R_{af}, respectively.

For calculation of anode resistance, E_{co} (V) is the design protective potential, which is −0.80 V. ΔE_o (V) is termed the design driving voltage.

As the design driving voltage in Eq. (6.13) is defined using the design protective potential for steel, the initial and final design current densities that define the anode current output capacity, and hence the driving voltage, refer to the required anode current output at this potential. Hence, the initial and final design current densities are based on a protection potential of −0.80 V. This means that they will always be used for calculations according to Eq. (6.13) in combination with this potential, and also if a more negative protection potential (e.g., ≤−0.90 V) is aimed at the individual anode current capacity, C_a (A h), is given by:

$$C_a = m_a \varepsilon u \tag{6.13}$$

where m_a (kg) is the net mass per anode. The total current capacity for a CP unit with N anodes thus becomes N · Ca (A h).

The calculations will ensure that the following requirements are met:

$$C_{at} = NC_a \geq 8760 I_{cm} t_f \tag{6.14}$$

$$I_{ati} = NI_{ai} \geq I_{ci} \tag{6.15}$$

$$I_{atf} = NI_{af} \geq I_{cf} \tag{6.16}$$

C_{at} in Eq. (6.14) is the total anode current capacity. $I_{cm}/I_{ci}/I_{cf}$ in Eqs. (6.14), (6.15), and (6.16) are the current demands of a CP unit, including any current drain; 8760 is the number of hours per year. I_{ai} and I_{af} in (6.23) and (6.24) are the initial and final current output for the individual anodes.

If anodes with different size and, hence, anode current capacity, C_a (Ah), and current output, I_a (A), are utilized for a CP unit, N Ca and N I_{ai} /N · I_{af} will have to be calculated for each individual size and then added for calculation of the total anode current capacity (C_{at}) and total anode current output (I_{ati}/I_{atf}).

If these criteria cannot be fulfilled for the anode dimensions and net mass initially selected, another anode size will be selected and the calculations repeated until the criteria are fulfilled.

Optimizing the requirements in Eqs. (6.22), (6.23), and (6.24) is an iterative process using a simple spreadsheet. In general, if Eq. (6.22) is fulfilled, but not Eq. (6.23) and/or (6.24), a higher number of smaller anodes, or the same number of more elongated anodes, can be used. On the other hand, if N·I_a in Eqs. (6.23) and (6.24) is much larger than I_{ci} and I_{cf}, fewer and/or more compact anodes may be applied.

High initial current capacity is required mainly for protection of uncoated steel surfaces. In general, the installed anodes have the same current output (I_a) from the anode to net anode mass (m_a) ratio.

Small anodes that have a high ratio of anode current output to net mass will be consumed more quickly rather than large anodes with a higher ratio, which causes a shortage in total anode current capacity at the design life end.

For anodes that have the same resistance and current output, but a major difference in net anode mass, which depends mainly on different geometry of the anodes, the lowest net anode mass will be consumed first. Similarly, for anodes with the same net anode mass but with major differences in anode resistance and, hence, anode current output, the anode with the lowest anode resistance will be consumed first.

From a practical point of view the following factors will be used:

$$\text{Design protective potential}\,(E_{oc}) = -0.800\,\text{V Ag/AgCl/seawater.}$$

$$\text{Design closed} - \text{circuit potential}\,(E_{oa}) = -1.050\,\text{V Ag/AgCl/seawater.}$$

The anode current output will be calculated for the initial and final life of the CP system.

In the latter case, anodes of 10% remaining mass will be used.

The number of anodes will be computed for the structure's initial and final currents, and the structure's current capacity, by zone, as follows:

$$\text{Anodes for initial current needs}\,(n) = I_c\,(\text{initial})/I_a\,(\text{initial})$$

$$\text{Anodes for final current needs}\,(n) = I_c\,(\text{final})/I_a\,(\text{final})$$

$$\begin{aligned}\text{Anodes for current capacity}\,(n) &= M\,(\text{mm})\\ &= [\text{Mass of anode material per anode}]\end{aligned}$$

To define the anode number, the number of anodes provided within each zone will be greater than the required numbers as calculated, by zone, and greater than the minimum number of anodes required to assure adequate current capacity and proper current distribution.

The total number of anodes required will be the sum of those needed for the three zones.

6.6.7.5 Calculation of anode resistance

The anode resistance, R (ohm), will be based on the applicable Eqs. (6.6) and (6.7), using the actual anode dimensions and specific resistivity of the surrounding environment.

To calculate the initial anode resistance, R_i (ohm), the initial anode dimensions are inserted in the relevant anode resistance equations (Eqs. 6.6 and 6.7). The final anode resistance, R_f (ohm), is calculated based on the expected dimensions when the anode has been consumed to its utilization factor, u, as explained earlier.

When the anode has been consumed to its utilization factor, u, at the end of the design life, t_f, in years, the remaining net anode mass, m_f (kg), is obtained by:

$$m_f = m_i\,(1-u) \tag{6.17}$$

The final anode volume to be used for calculation of R can be obtained from the remaining net anode mass, m_f (kg), anode material density. When anode details are not available, their volume can be ignored or estimated on the conservative side.

For long and short slender stand-off anodes consumed to their utilization factor, assume a 10% reduction in length. If the final anode shape is cylindrical, the final radius will be calculated based on this length reduction, and the final anode mass or volume.

For long and flush-mounted anodes, the final shape will be assumed to be a semicylinder and the final length and radius will be calculated as shown previously.

For short flush-mounted anodes, bracelet anodes, and other shapes mounted flush with the protection object, the final exposed area will be assumed to be equivalent to the initial area facing the surface to be protected.

Anode quantities have been calculated to ensure the correct total net mass of anode material is provided based on mean current density requirements and to be capable of providing sufficient current to maintain polarization throughout the 25-year design life of the platforms.

The total mass of sacrificial alloy required to protect each platform and pipeline is calculated using the following formula:

$$W = \frac{t_f E I_{cm}}{u} \tag{6.18}$$

where W is the total net anode weight alloy in kilograms, t_f is the design life in years, E is the consumption rate of anode alloy (kg/year), I_{cm} is the mean current to protect the structure, and u is the utilization factor.

Once the total weight of the anode material has been calculated for each area to be protected, the quantity of anodes is obtained by dividing the specified anode unit net weight into the total weight.

Number of anodes = Total weight/anode net weight.

The number of anodes for each area should also be capable of providing sufficient current throughout the remaining design life of the structure to maintain polarization. This is verified by calculating the initial, mean, and final current output of each anode type used and comparing this with the platform or pipeline current demands based on surface areas and specified current densities.

This equation is valid for anodes with a minimum stand-off distance of 300 mm away from the structure. A correction factor of 1.3 is applied for anodes between 150 and 300 mm from the structure.

a) For the initial anode resistance calculation, the anode resistance equation is function of the initial anode dimensions. The final anode resistance is calculated based on the expected dimensions when the anode has been consumed to its utilization factor.

b) For long slender stand-off anodes consumed to their utilization factor, a length reduction of 10% will be assumed.

Furthermore, assuming that the final anode shape is cylindrical, the final radius will be calculated based on this length reduction and the final anode mass/volume. For long flush anodes, the final shape will be assumed to be semicylindrical.

6.6.7.6 Anode design precaution

The engineering firm will provide tentative dimensions and/or net mass for anodes to be used in the CP design report.

For use of the anode resistance equation in Table 6.17 for stand-off type anodes, the minimum distance from anode to protection object will be a minimum of 300 mm. However, for distances down to 150 mm, the formula can still be used by multiplying the anode resistance with a factor of 1.3.

The detailed anode design will ensure that the utilization factor assumed during calculations of required anode net mass is met. Hence, it will be ensured that the anode inserts are still likely to support the remaining anode material when the anode has been consumed to its design utilization factor. Unless otherwise agreed, anode cores of stand-off type anodes will protrude through the end faces.

Except for stand-off type anodes, a marine grade paint coating (minimum 100 μm DFT) will be specified for anode surfaces facing the protection object.

6.6.7.7 Distribution of anodes

After calculating the number of anodes, N, for a CP unit, they will be distributed to provide a uniform current distribution for the whole structure, considering the current demand of individual members due to different surface areas and if there are any coatings used.

For platform substructures, there are special areas to be considered when distributing anodes, such as nodes, pile guides, and conductor bundles. The designer should deliver a drawing presenting the location of all individual anodes.

If it is practically acceptable, anodes dedicated to CP surfaces buried in sediments should be located freely exposed to the sea.

The anodes' distribution should allow enough spacing between them to prevent interaction effects that reduce the useful current output. The anodes should be located so that their surfaces intended for current output are not near structural members, which would reduce the current output.

Except for very large anodes, the effects of shielding and interference become insignificant at about 500 mm or more.

Ensure that there are no anodes installed for welding on areas with high fatigue stresses. For main structural components, the minimum distance is 150 mm from anode fastening welds to structural welds. On jacket structures, no anodes will be located closer than 600 mm to nodes.

The anode locations will consider restrictions imposed by fabrication, installation, and operation. The anode placement priority should be based upon the following:

(1) Anode distribution should normally begin with placement upon larger members such as legs near nodes and continue to minor members.
(2) Placement will consider future added components, which usually will be new conductors and risers as operations require.
(3) Locations where attenuation effects will reduce good current distribution, such as conductor clusters and complex pile guides, shall require a higher anode concentration.
(4) Anodes will be placed within 15 m of adjacent anodes.
(5) Anodes will be placed within 10 m of surfaces to be protected.

From the preceding guidelines, define the anode preliminary location and then do a series of iterations until reaching the optimum configuration, to determine the best anode placement.

The placement process may result in conservative adjustments to anode dimensions and additions to the calculated total number, to assure adequate current distribution.

Anodes designed to protect buried surface area will be installed on the bottom elevation of the structure. Fig. 6.9 presents the distribution of anodes (Fig. 6.10).

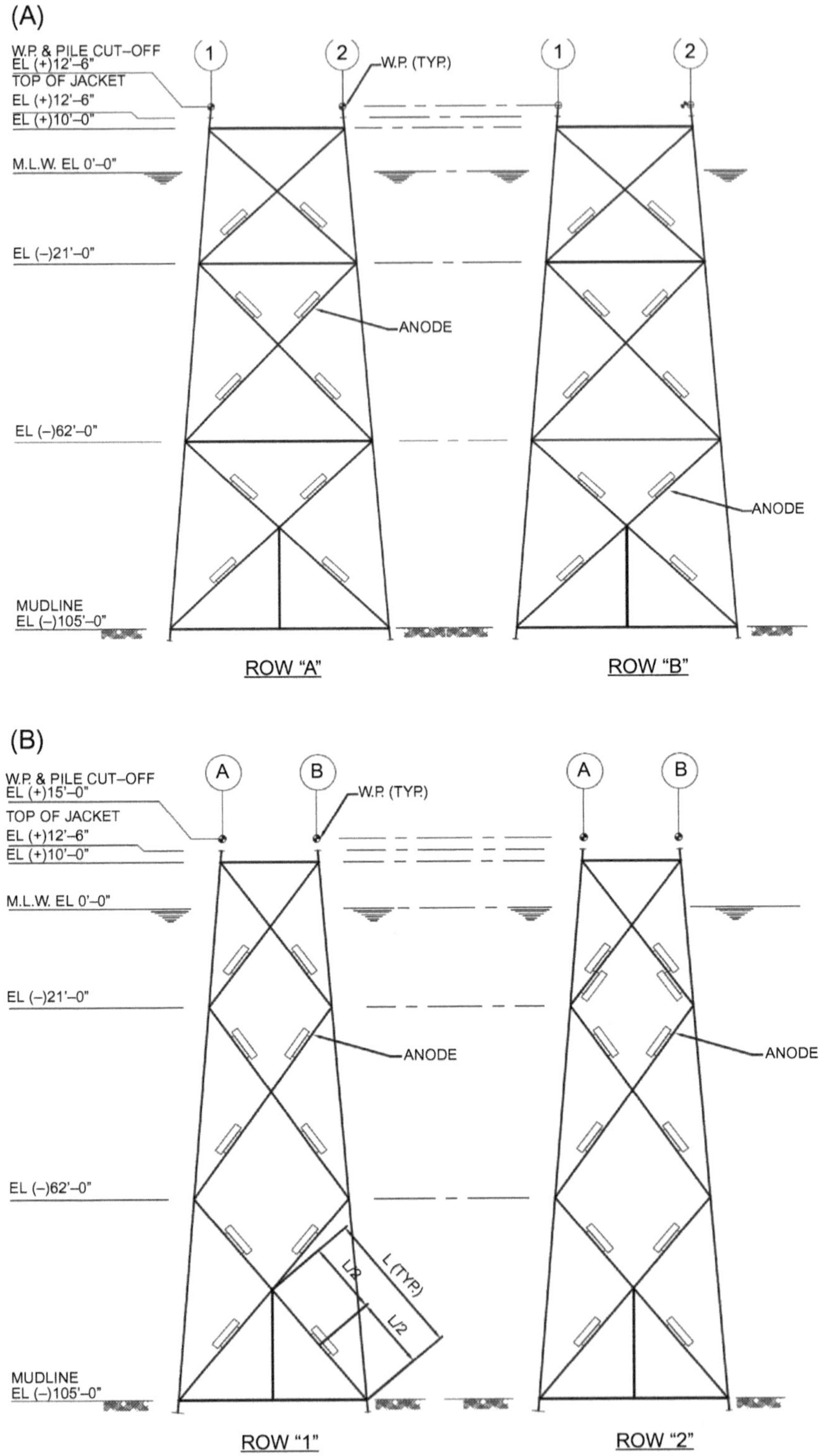

Fig. 6.9 (A) The location of the anodes on 4 leg platform. (B) The location of the anodes on 4 leg platform for rows 1 and 2.

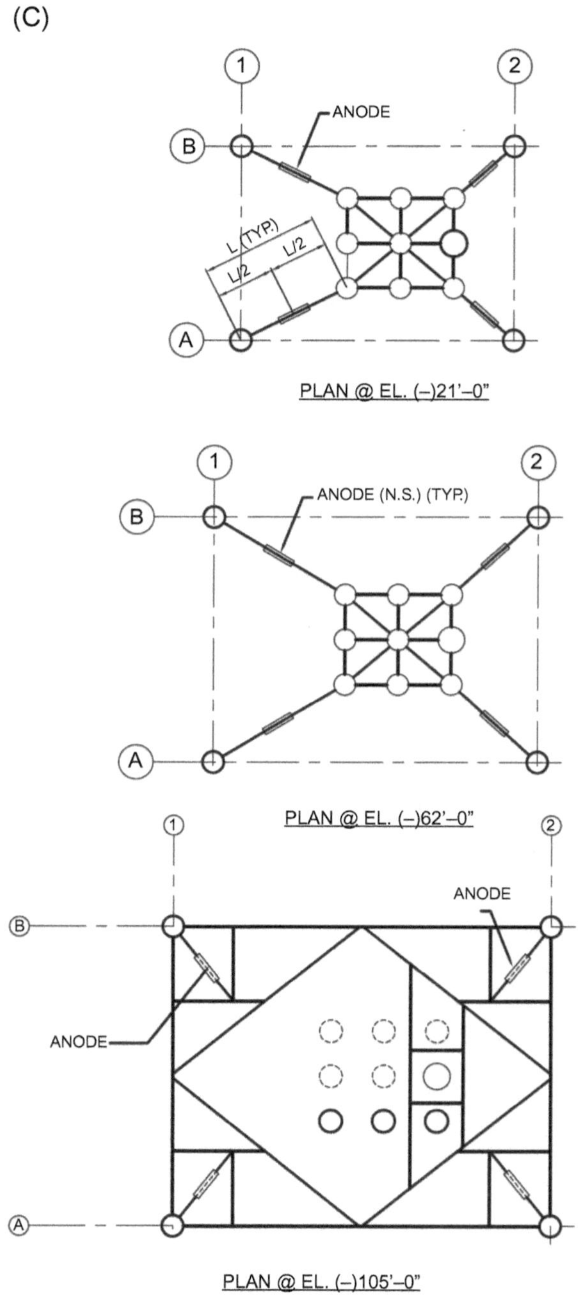

Fig. 6.9, cont'd (C) The location of the anode to 4 leg plat in different horizontal plan.

Fig. 6.10 Location of the anode in the jacket during construction.

6.7 Design example

An example of CP for an offshore structure platform is presented in Table 6.19. This table can be used as a logical step in a design spreadsheet for a CP system.

6.8 General design considerations

Low-alloy steels with minimum yield stress ranging from 275 to 450 MPa are specified for substructures for fixed offshore platforms. Special attention should be given to the use of higher strength steels, particularly those with minimum yield stress above 400 MPa.

The sacrificial anodes will be manufactured from an indium-activated aluminum alloy. The use of other aluminum activators shall be agreed upon by the principal. Tin-activated aluminum alloys shall not be used.

Mercury- or cadmium-activated aluminum alloys may only be considered under the circumstances that the indium-activated aluminum ones are proven to be ineffective in providing protection. However, statutory regulations limiting the use of these alloys may apply.

The core material shall be carbon steel and comply with the specification requirements for tubular structure secondary members and the core diameter and wall thickness will be agreed upon by the principal. Anode core

Table 6.19 Steps of CP design.

Item	Calculation and equations	Input/output data
Coating breakdown		0.350
Jacket HAT[a] to −3000 and risers all depths uncoated at installation		0.35
Jacket −3000 to mud line uncoated at construction		1.00
Jacket coating damaged prior to placement		0.00
Jacket f_c (initial) total factor at placement		1.00
Conductor coating damaged during construction		1.00
Coating breakdown factor per RP B401 paragraph 6.4.4		
Average factor	f_c (average) $= a + bt_f/2$	
	f_c (final) $= a + bt_f$	
Average breakdown factors		
Jacket: HAT to −3.0	f_c (average) $= a + bt_f/2$	0.475
Jacket: −3.0 to −30.0	Bare	1.0
Jacket: −30.0 plus		1.0
Conductors: HAT to −3.0		1.0
Conductors: −3.0 to −30.0		1.0
Conductors: −30.0 plus		1.0
Risers: HAT to −3.0		0.472
Risers: −3.0 to 30.0		0.472
Risers: −30.0 plus		0.472
Final breakdown factors		
Jacket: HAT to −3.0		0.6
Jacket: −3.0 to −30.0		1.0
Jacket: −30.0 plus		1.0
Conductors: HAT to −3.0		1.0

Continued

Table 6.19 Steps of CP design—cont'd

Item	Calculation and equations			Input/output data
Conductors: −3.0 to −30.0				1.0
Conductors: −30.0 plus				0.6
Risers: HAT to −3.0				0.6
Risers: −3.0 to 30.0				0.6
Risers: −30.0 plus				
Current demand jacket				
I_c *(initial)*				
Zone	A_c (m^2)	f_c (initial)	I_c (initial) (A/m^2)	I_c (A)
Jacket: HAT to −3.0	67.3563	0.350	0.035	0.825
Jacket: −3.0 to −30.0	692.6240	1.0	0.035	24.242
Jacket: −30.0 plus	0.00	1.0	0.035	0.00
I_c *(Average)*				
Jacket: HAT to −3.0	67.3563	0.475	0.035	1.120
Jacket: −3.0to −30.0	692.6240	1.0	0.035	24.242
Jacket: −30.0 plus	0.00	1.0	0.035	0.000
I_c *(final)*				
Jacket: HAT to −3.0	67.3563	0.60	0.035	1.414
Jacket: −3.0 to −30.0	692.6240	1.0	0.035	24.242
Jacket: −30.0 plus	0.00	1.0	0.035	0.000

Current demand conductors				
I_c *(initial)*				
Conductors: HAT to −3.0	30.099	1	0.035	1.053
Conductors: −3.0 to −30.0	151.736	1	0.035	5.311
Conductors: −30.0 plus	0.00	1	0.035	0.000
I_c (average)				
Conductors: HAT to −3.0	30.099	1	0.035	1.053
Conductors: −3.0 to −30.0	151.736	1	0.035	5.311
Conductors: −30.0 plus	0.00	1	0.035	0.000
I_c (final)				
Conductors: HAT to −3.0	30.099	1	0.035	1.053
Conductors: −3.0 to −30.0	151.736	1	0.035	5.311
Conductors: −30.0 plus	0.00	1	0.035	0.000
Riser demand conductor				
I_c (initial)				
Risers: HAT to −3.0	11.124	0.35	0.035	0.136
Risers: −3.0 to 30.0	56.076	0.35	0.035	0.687
Risers: −30.0 plus	0.00	0.35	0.035	0.00
I_c (Average)				
Risers: HAT to −3.0	11.124	0.475	0.035	0.185
Risers: −3.0 to 30.0	56.076	0.475	0.035	0.932
Risers: −30.0 plus	0.00	0.475	0.035	0.00
I_c *(final)*				
Risers: HAT to −3.0	11.124	0.60	0.035	0.234

Continued

Table 6.19 Steps of CP design—cont'd

Item	Calculation and equations			Input/output data
Risers: −3.0 to 30.0	56.076	0.60	0.035	1.178
Risers: −30.0 plus	0.00	0.60	0.035	0.00
Pile penetration I_c (initial/final and average)	634.219	1.00	0.01	6.342
Well casing	$4 \times (2.5)$ A/well			10
Total drain current required				16.342
Current demand for jacket buried				
I_c (initial)	49.551	1.00	0.01	0.496
I_c (average)	49.551	1.00	0.01	0.496
I_c (final)	49.551	1.00	0.01	0.496
Total initial current, I_c (initial)				49.092
Total average current, I_c (average)				49.681
Total final current, I_c (final)				50.269
	Mass $(M) = I_c$ (average) $\times t_f \times 8760/(u \times e)$			
Electrochemical anode efficiency				2.0
Utilization factor				0.90
Mass required				
I_c (Ave Ttl)	49.681			
I_c (Ave Ttl) $\times t_f \times 8760/(u \times e)$	Total mass			6044 kg
Anode resistance calculations				
Anode initial physical dimensions				
Anode length				2.420 m
Anode width				0.187 m
Anode depth				0.170
Anode radius				0.114 m

Alloy density		2700 kg/m^3
Net anode volume		0.06 m^3
Anode mass		189.76 kg
Anode final physical parameters	Anode final mass m(final) = m(initial) × (1 − u)	18.976 kg
	Anode final length L(final) = L(initial) − 0.10 × u × L(initial)	2.202 m
Anode resistance, R (initial)		0.0453 Ω
Anode resistance, R (final)		0.0622 Ω
	Anode output $(I_a) = (E_c - E_a)/R$	
Anode (Al) closed circuit potential		−0.95 V
Protected cathode potential		−0.80 V
Anode driving potential		0.15 V
Anode output I_a (initial)		3.310 A
Anode output I_a (final)		2.412 A
Anode number calculation		
Total initial current/I_a (initial)		14.829
Total initial current/I_a (initial)		15.0
Total final current/I_a (final)		20.845
Total final current/I_a (final)	Rounded up to integer	21.00
Anode number		32
Initial current available from anode number above		105.935 A
Total initial current required per calculation		49.092 A
Difference of initial current available and initial current req'	Initial surplus	57 A
Final current available from anode number above		77.172 A
Total final current required per calculation		50.269 A
Difference of final current available and final current required	Final surplus	26.90 A

Continued

Table 6.19 Steps of CP design—cont'd

Item	Calculation and equations	Input/output data
Mass check		
Mass of anodes per distribution		6072 kg
Mass required for life		6044 kg
Difference of mass available and mass required	Mass surplus	28 kg
Average current from mass for design life		55 A
Anode gross weight calculation		
Anode unit gross weight estimate		216
Anode gross weight estimated on numbers above		6912 kg
Total gross weight		6912 kg

[a]HAT is the highest astronomical tide.

attachment by welding will be as specified by drawings and in conformance with the welding specification for the associated structure.

Anodes will be of the "slender stand-off" type, with a minimum stand-off (i.e., the closest distance between the aluminum alloy and the attached steel member) of 150 mm. Anodes will have a cylindrical or trapezoidal shape. The anode core will exit at the anode ends.

The sacrificial anode will be an indium-activated aluminum alloy. Anode material will have a capacity more than 3.6 kg/A years (efficiency = 2420 Ah/kg).

Anode material will have a potential between −1.05 and −1.15 V vs Ag/AgCl/seawater. The number of anodes, their capacity and potential will be enough to reach the polarization and maintenance current requirements for the steel over the design temperature range.

The closed circuit anode potential (E_oa) used in design calculations for determination of indium-activated aluminum anode current output will be −1.05 V Ag/AgCl/seawater.

The objective of the design calculations will be to compute by iteration the number of anodes and anode dimensions to verify the following:

- fulfill the initial current, final current, current capacity, and current distribution requirements of the structure and
- provide the lowest installed cost for the CP system, including cost of anode materials, molds, installation, and coating systems (if used).

To assure adequate and efficient use of the anode material, the submerged structure should be divided into "zones" to be protected. Three zones are normally classified as follows:

- *Zone1*: The zone begins from average sea level to nearly above the first horizontal member below it. This zone depth is the larger of the trough wave depth of 20 m or a 100-year extreme wave. Due to storm waves, the calcareous coatings can be damaged.
- *Zone2*: This zone is under Zone1 and above the sea bed.
- *Zone3*: This zone is below the sea bed.

In this zone the surface area calculation will be compensated for by assuming an allowance of 2.5–3 A per well. This allowance will be provided in addition to the surface area contributions from other components, such as conductors and piles.

The external surface area of the buried part of piles and conductors will be included in the calculations only up to a depth of 30 m into the sea bed.

The current drain to open skirt piles, sumps, and caissons will cover an internal surface area equivalent to 10 times π times their respective diameter.

The internal surfaces of all flooded structural members with unsealed flooding holes will be included in the surface area calculations.

6.9 Anode manufacture

The anodes will be manufactured in accordance with the requirements of the required specification and the supplier will submit detailed anode and steelwork manufacturing drawings, complete with tolerances, as well as manufacturing and testing procedures for the client's approval.

The stand-off supports will be formed from single lengths of carbon steel seamless tube or bar, bent using purpose-built bending equipment of proven performance to provide the minimum clearance specified between the center line of the anode and the nearest jacket member. The tubular supports will have their ends square to the anode axis and to each other within a tolerance of $\pm 2\,\text{mm}$.

The bending method to be used for tubular inserts will be proven to the purchaser's satisfaction to achieve consistent tolerances with a maximum flattening of 10% of the tube outer diameter and all production bends will be checked for general dimensional tolerances and a minimum of 20% of bends will be checked for flattening tolerance.

Inserts formed from flat steel bar will be bent using reduced formers to produce bends of controlled radius. The supplier's proposed procedure will be selected to ensure that the tolerances and acceptance criteria are fully complied with.

6.10 Installation of anodes

The galvanic anodes installation for offshore structures mainly employs welding and sometimes also anode supports clamped to structural steel members. In the latter case, electrical continuity is provided by a copper cable, contacting the anode support and the protection object by brazing, or with special mechanical connections designed to guarantee reliable electrical continuity. Electrical continuity cables can be installed to provide electrical continuity to CP unit components without reliable electrical connection to anodes by welds, metallic seals, or threaded couplings.

Focus should be placed on the anode fastening devices during installation and operation of the applicable structures, to avoid their damage and to meet any special CP design requirements.

For large anodes, the design drawings may include the use of doubler or gusset plates. The following information and any optional requirements will be included in the technical tender package to the supplier:

- anode drawings from detailed CP design, or from manufacturer if completed
- drawings from detailed CP design showing location of individual anodes
- any requirement for preparation
- any special requirements for documentation
- any further amendments to and deviations from this order.

The contractor may specify that the anode installation procedure will be a detailed installation procedure specification document. The minimum information in this document should be as follows:

- Materials specification and its data sheets
- Equipment that will be used, including certificates
- Procedures for receiving, handling, and storage of anodes
- Procedure and specifications for welding and brazing, with welder personnel qualifications
- Performance of anode fastening inspection and testing
- Maintenance of the documentation of materials testing and inspection records.

All welding associated with anode installation will be qualified according to a recognized standard. Installation of anodes will be carried out according to drawings approved for construction, showing locations of individual anodes and any other relevant specifications.

To facilitate installation tolerance in the case of stand-off anodes mounted on structural members, lateral displacement not more than one anode length and circumferentially a maximum of 30 degrees will be accepted.

For welding of anodes to a structure member under high external loads, welded connections will be located at least 150 mm away from other welds, and a minimum of 600 mm from jacket structure nodes.

For coated objects, exposed anode cores will be coated to the same standard. Visual inspection of welding and any brazed connections shall be performed.

For brazed and mechanical connections for electrical continuity, the measurements shall be constructed based on the designated procedure and with an instrument capable of verifying an electrical resistance of 0.1 Ω maximum (Fig. 6.11).

Fig. 6.11 Cow horn anode with 189kg.

6.11 Anode dimension tolerance

In most cases, the anode tubular core will be 80mm NB, Schedule 80 pipe, extending for the full length of the anode casting, having a minimum stand-off height of 300mm, and will be fabricated from ASTM 106 Grade B or similar material. Welding will be in accordance with AWS D1.1, Structural Welding Code. Steel anode cores manufactured from channel sections are not acceptable.

The anodes will be located on the supporting structural members to minimize the impact of wave loads on the structure.

The weight of each anode will be within −0% to +5% surface defect tolerance.

Dimensional tolerances for the long slender stand-off anodes and flush-mounted anodes will conform to the following:

- The smaller of 2.5% on the anode's length or ±20mm, for anodes with tubular stand-off supports.
- The smaller 2.5% on the anode's length or ±10mm, for flush-mounted anodes.
- The smaller of 2.5% on the thickness, width, or diameter or ±5mm, for all anodes.

The straightness of each anode will not deviate by more than 2% from the longitudinal axis over the nominal anode length.

Bends on anode tubular supports will be checked to ensure that flattening (ovality) of the pipe due to bending is within specified limits. Wrinkling of the inside of the bend will not be acceptable.

The core will be centrally located within the casting length. A tolerance of ±5% on the nominal position of the core in the anode is permitted provided the stand-off distance meets the specified minimum anode/member clearance.

The stand-off support fixing centers (to the structure) will be within 20 mm of the nominal dimensions.

Each structural anode will be weighed and will conform to the following weight tolerances:

(a) Net weight of individual anodes will be greater than 97% of the design net weight.

(b) Total anode net weight supplied will not be below the contract net weight.

6.11.1 Internal and external inspection

Verification of anode weight and dimensions will be carried out with the frequency and acceptance criteria specified in NACE RP0387. The anode will comply with tolerances in the manufacturer's drawing and will be checked for a minimum of 10% of the anodes of a specific design and based on the tolerance dimensions as discussed previously.

Note that in any case, the anode surface should be generally free from defects affecting anode efficiency. Inspection for cracks and other surface irregularities will be carried out on all anodes with the acceptance criteria as specified in NACE RP0387. Anodes exhibiting the following defects will be rejected:

- For zinc-based anodes, no cracks visible to the naked eye are acceptable.
- Cracks that are seen to penetrate to anode inserts are not accepted.
- Shrinkage pits or cavities exceeding 10% of depth of anode thickness.
- Cracks exceeding 2 mm in width or greater than 150 mm in length intersecting each other.
- Cracks exceeding 4 mm in width or 300 mm in length.
- Cracks penetrating to the core.
- Apparent slag or dirt inclusions.
- All anodes will be free from slag, dross, and any other inclusions having a particle size of 2 mm or greater.
- Surface irregularity on the top anode surface due to "topping-up" during solidification of the anode will not exceed ±5 mm of the nominal level. Such "topping-up" will not exceed 0.5% of the total anode alloy volume and will be mechanically well-bonded to the main body of the anode.
- Cold shuts or surface laps will not be permitted.

Unless otherwise agreed upon, a minimum of two anodes of each size will be subject to destructive testing to verify absence of internal defects and adequate location of anode inserts. The cutting procedure and acceptance criteria in NACE RP0387 will apply.

At a minimum, electrochemical testing will be performed in the first day production test for purchase orders exceeding 15,000 kg of net anode material and for each further 15,000 kg of production.

The anodes will be transversely sectioned by single cuts at 25% and 50% of the nominal length.

The cut faces, when examined visually without magnification, will not contain more than:

- 0.25% of surface areas as gas holes or porosity with individual pore size limited to 10 mm^2.
- 20% of the insert perimeter with voids porosity adjacent to the insert or any one face and no greater than an aggregate of 10% on all faces.

The cut faces will be dimensionally checked to ensure compliance with the requirements of the core location. All anodes represented by a failed test will be rejected.

6.11.1.1 FPSO protection

FPSO protection follows the same approach as the offshore structure platform, as coating will be used in addition to sacrificial anodes. The sacrificial anodes are mandatory to install inside the tanks of the vessel based on the design.

6.11.1.2 Pipelines and tanks

The production facilities in general can be protected using sacrificial anodes or impressed current. For piping and vessels inside the plant, the impressed current is reasonable and can be monitored appropriately with competent staff.

Pipeline internal coating

The primary reason for applying internal coatings is to reduce the friction of the pipeline and this will improve the pipeline roughness by reducing it, which affects the pressure drop.

In addition, internal coatings will enhance the corrosion protection, precommissioning operations, and pigging operations. In actual pipeline operations, the improved flow efficiency will be observed as a reduction in pressure drop across the pipeline.

Free water in oil pipelines is one of the causes of the corrosion of the inner pipeline; therefore the coating is considered the first barrier against corrosion attack.

During pipeline operations there are frequently pigging activities that over time will reduce the pipeline coating. The coating will reduce the roughness of the pipeline and it also has another benefit in that it will reduce the wear of the pig discs, the wear of which is affected by the pipeline roughness.

The coating specification is governed by the service requirement and the environmental conditions.

In general, the most common internal coatings used for pipelines are epoxies, urethanes, and phenolics. Specific epoxy coating is the most common type used for internal coatings as it has many properties such as sufficient hardness, water resistance, flexibility, chemical resistance, and excellent adhesion.

For protection of the pipeline internally, chemical injection programs add chemicals like oxygen scavengers and cathodic inhibitors, and others, based on the fluid characteristics, will be used periodically to control corrosion inside the pipeline.

6.11.1.3 External coatings

Oil and gas pipelines are protected from corrosion by using coatings in addition to cathodic protection systems. As discussed in the design of the cathodic protection system for platforms, the main benefit of the external coating is that it will reduce the required electric current to protect a steel pipeline from corrosion.

In general, thick coatings are often applied to offshore pipelines to minimize the holidays and defects and to resist damage from handling during transport and installation. High electrical resistivity retained over long periods is a special requirement, because cathodic protection is universally used in conjunction with coatings for corrosion control.

For offshore pipelines, the coatings should have a good bond to the pipe surface to resist degradation due to biological organisms in the seawater. As mentioned before, visual inspection and potential reading will be done periodically to define the efficiency of the cathodic protection system.

The types of wrapping are as follow:

- Tape wrap.
- Asphalt coal tar enamel.
- Fusion bonded epoxy (FBE).
- Wrap polyethylene (PE).
- Extruded thermoplastic PE and polypropylene (PP).

Bay and Bay (2005) mention that fusion-bonded epoxy (FBE) coatings are the most commonly used for offshore pipelines. This coating is a thin film with thickness of 0.5–0.6 mm. They consist of thermosetting powders applied to a white metal blast-cleaned surface by electrostatic spray.

The powder will melt on the preheated pipe with around 230 °C flow and subsequently cure to form thicknesses of between 250 and 650 μm.

Storage tank protection

The bottom of the steel plates rests on the oily sand and the sand will be clean, but API 651 states that practical experience in the industry is varied as to the effectiveness of oiled sand for corrosion control. The presence of oiled sand beneath tank bottoms has not been proven as an effective corrosion control measure and thus does not eliminate the need for cathodic protection. It may even cause cathodic protection to be less effective because of the higher resistivity of the oiled sand.

In some cases, a thick lining is used based on recommendations of API 653. Thick-film, laminated, and corrosion- and chemical-resistant linings are installed in tanks primarily to prevent internal corrosion and are frequently used as an alternative to replacing the steel bottom. The use of a thick-film laminate or other type of internal lining should not be considered as sufficient justification to eliminate the need for external cathodic protection of tank bottoms.

The two types of cathodic protection are sacrificial anode or impressed current cathodic protection systems, as shown in Figs. 6.12 and 6.13, respectively.

When inserting a new bottom on the top of an old bottom, the sacrificial anodes will be inserted in between the new and old bottoms, as illustrated in Fig. 6.14.

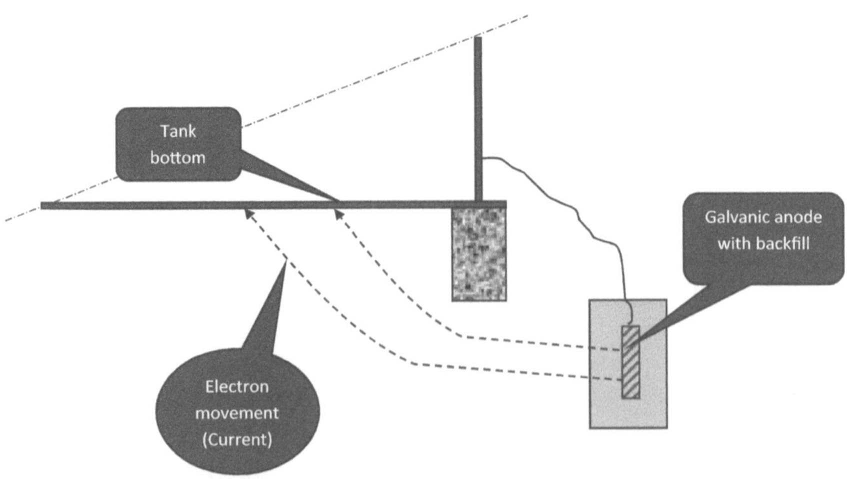

Fig. 6.12 Cathodic protection with sacrificial anode.

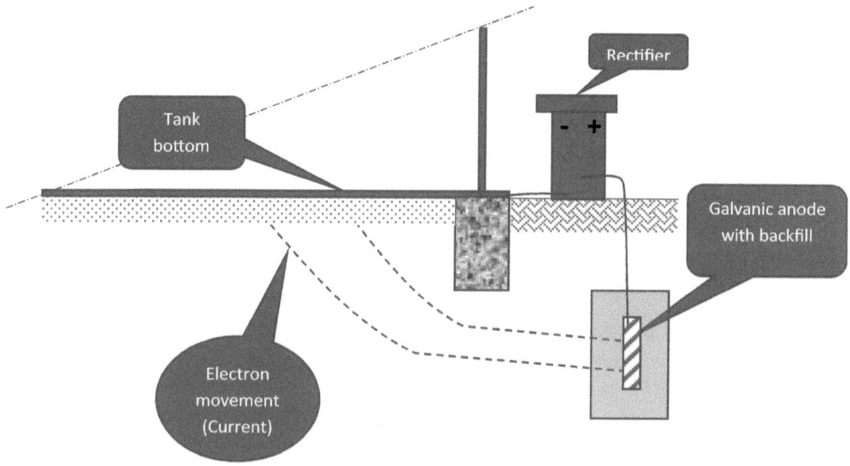

Fig. 6.13 Cathodic protection with impressed current.

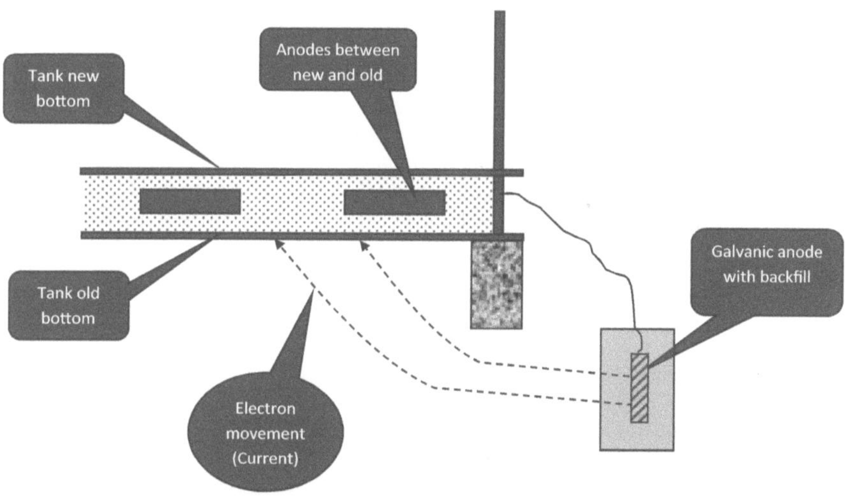

Fig. 6.14 Cathodic protection for new bottom over the old bottom anode.

References

Andrade, C., Holst, J.D., Nurnberger, U., Whiteley, J.J., Woodman, N., 1994. Protection System for Reinforcement, Prepared by Task Group VII/8 of Permanent Commission VII CEB.

Bay, Y., Bay, Q., 2005. Subsea Pipelines and Risers. Elsevier.

Broom Field, J.P., 1995. Corrosion of Steel in Concrete Understanding, Investigation and Repair. E&FN SPON.

Cairns, J., 1992. Design of concrete structures using fusion-bonded epoxy-coated reinforcement. Proc. Inst. Civ. Eng. Struct. Build. 4 (2), 93–102.

Det Norske Veritas (DNV), 1993. Recommended Practice, Cathodic Protection Design, DNV RP B401-1993.
Det Norske Veritas (DNV), 2005. Recommended Practice, Cathodic Protection Design, DNV RP B401-2005. DNV.
El-Reedy, M.A., Sirag, M.A., El-Hakim, F., 1995. Predicting Bond Strength of Coated and Uncoated Steel Bars Using Analytical Model (MSc thesis). Faculty of Engineering, Cairo University.
Miller, J.B., 1994. Structural aspects of high powered electrochemical treatment of reinforced concrete. In: Sawamy, R.N. (Ed.), Corrosion Protection of Steel in Concrete. Sheffield Academic Press, pp. 1400–1514.
Pike, R.G., et al., 1972. Nonmetallic coatings for concrete reinforcing bars. Public Roads 37 (5), 185–197.
Satake, J., Kamakura, M., Shirakawa, K., Mikami, N., Swamy, R.N., 1983. Long term resistance of epoxy-coated reinforcing bars. In: Crane, A.P. (Ed.), Corrosion of Reinforcement in Concrete Construction. The Society of Chemical Industry/Ellis Horwood Ltd, UK, pp. 357–377.
Shell, Design and Engineering Practice, 1995. Design of Cathodic Protection Systems for New Fixed Offshore Steel Structures.
Yeomans, S.R., 1994. Performance of black, galvanized and epoxy-coated reinforcing steels in chloride-contaminated concrete. Corrosion 50 (1), 72–81.

Further reading

ACI Committee 318, 1988. Revisions to building code requirements for reinforced concrete. Am. Concrete Inst. Struct. J. 85 (6), 645–674.
Aluminium and Aluminium Alloys, 2020. Castings, Chemical Composition and Mechanical Properties, BS1706.
Andrade, M.C., Macias, A., 1988. Galvanized reinforcement in concrete. In: Wilson, A.D., Nicholson, J.W., Prosser, H.I. (Eds.), Surface Coating-2. Elsevier Applied Science, pp. 137–182.
Andrade, C., Holst (convenor), J.D., Nurenberger, U., Whiteley, J.D., Woodman, N., 1992. Protection Systems for Reinforcement, CEB Information Bulletin No. 211, Switzerland.
Anon., 1994. The Use of Galvanized Reinforcement in Concrete, Current Practice Note 17. Concrete Institute of Australia.
API 651, 1997. Cathodic Protection System of Aboveground Petroleum Storage Tanks.
Arnon, B., Diamond, S., Berke, N.S., 1997. Steel Corrosion in Concrete: Fundamental and Civil Engineering Practice. E&FN SPON.
Ashworth, V., 1994. In: Shreir, L.L., Jarman, R.A., Burstein, G.T. (Eds.), Corrosion, vol. 2. Corrosion Control, third ed. Butterworth-Heinemann, Oxford, p. 10.15.
Butler, G., Mercer, A.D., 1975. Corrosion of steel in potable waters. Nature 256, 719–720.
Clear, K., Virmani, Y., 1983, June. Corrosion of non specification epoxy-coated rebars in salty concrete. Public Roads Mag. 47.
Clifton, J.R., Beeghley, H.F., Mathey, R.G., 1975 August. Nonmetallic Coatings for Concrete Reinforcing Bars, Building Science Series 65. US Department of Commerce, National Bureau of Standards.
Gustafson, D.P., Neff, T.L., 1994. Epoxy-coated rebar, handled with care. Concrete Construct. 39 (4), 356–369.
Humble, H.A., 1949. Corrosion 5, 292–302.
Kendell, K., Daily, F.S., 1999, June. Cathodic protection for new concrete. Concrete Int. Mag. 21 (6).

Kessler, R.J., Powers, R.G., Lasa, I.R., 1995. Update on Sacrificial Anode Cathodic Protection of Steel Reinforced Concrete Structures in Sea Water' Corrosion 95, Paper 516. NACE International, Houston, TX.

Marine Technology Directorate Limited, 1990. Design and Operating Guidance on Cathodic Protection of Offshore Structure Subsea Installation and Pipeline. The Marine Technology Director Limited.

National Association of Corrosion Engineers (NACE) RP 0176-94, 2021. Corrosion Control of Steel Fixed Offshore Platform Associated with Petroleum Production.

National Association of Corrosion Engineers (NACE) RP 0387-99, 2021. Metallurgical and Inspection Requirements for Cast Sacrificial Anodes for Offshore Applications.

Pourbaix, M., 1974. Atlas of Electrochemical Equilibria in Aqueous Solutions. NACE International, Houston, TX.

Rowlands, J.C., 1994. "Sea Water" Corrosion: Metal/Environment, third ed. Butterworth-Heinemann, Oxford, pp. 2:60–2:71.

Anon., 1969. Zinc Coated Reinforcement for Concrete, BRE Digest 109. Building Research Establishment, United Kingdom.

CHAPTER 7

Integrity management system

7.1 Introduction

The previous chapters cover the preparatory steps that should be carried out before implementing the integrity management system.

Fig. 7.1 presents the concept of the integrity required to establish a strong proactive system with many barriers, including guaranteeing good design, performing inspection and maintenance, hiring competent persons, and other factors. After an event such as a structure collapse, oil spill, or fire occurs, other barriers will arise related to the firefighting or other actions, which will be a different scenario. As a rule of thumb, if such an event happens, there is a lack in the integrity management system.

The main key of the proactive barrier is implementation of the integrity management system. So, it is very important to perform the previous tasks delineated, from data collection, inspection, assessment, and evaluation, and then to establish a database of the facilities with the risk ranking, as will be discussed.

Most oil and gas plants and even refineries and petrochemical facilities have a fleet of assets with different structures and facilities. So, first of all a database covering all the facilities in the plant is needed: for example, the platforms, tanks, pressure vessels, etc.

This database should include all the technical data and all the history, and then the risk ranking should be added on when completed.

This chapter objective is to illustrate the risk assessment methodology based on current industrial practice. In addition, methods of subsea inspection and methods for implementing inspection programs are illustrated at every inspection level.

The complete cycle of the integrity management system is shown in Fig. 7.2. and it is found that the first step is to collect the data for the facilities included.

After collecting the data and performing the technical assessment and evaluation, it is important to run the risk assessment methodology. There are two basic approaches to risk assessment: quantitative risk assessment and qualitative risk assessment.

Asset Integrity Management for Offshore and Onshore Structures
https://doi.org/10.1016/B978-0-12-824540-8.00007-7

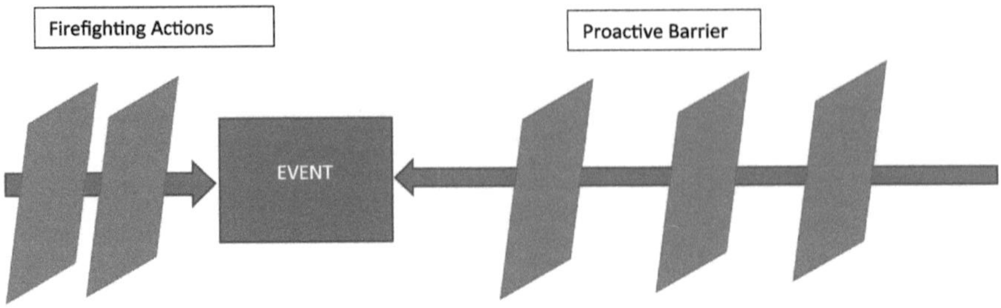

Fig. 7.1 Integrity management concept.

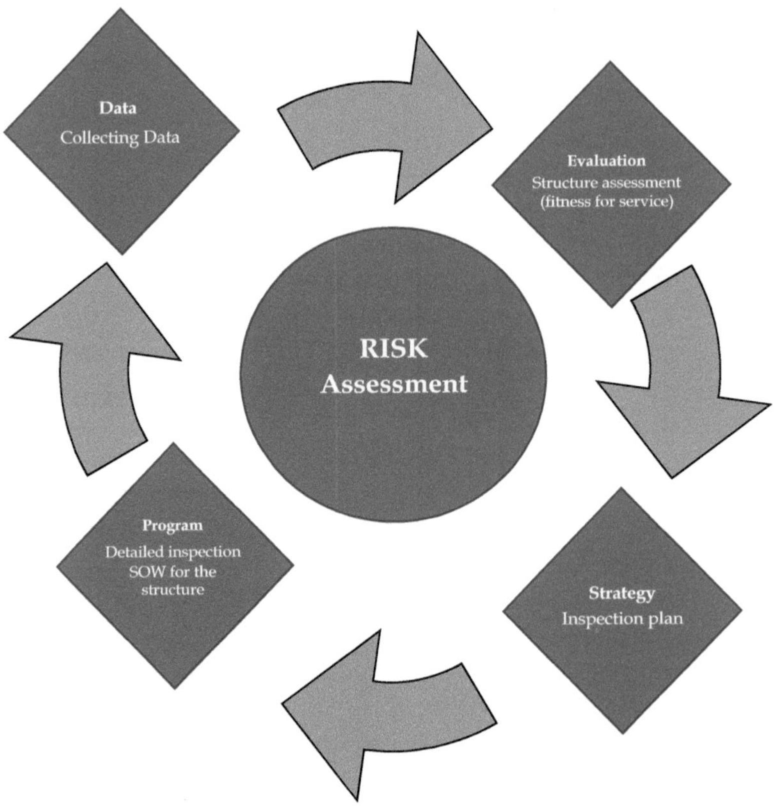

Fig. 7.2 Structure integrity management main steps.

For quantitative risk assessment, it is required to define the value of the structure probability of failure or to predict its lifetime, as discussed in previous chapters. This approach is mainly applied to certain critical structures like offshore structure platforms, to perform pushover analysis, and define the probability of failure, as discussed in Chapter 6. It is run for a specific structure, as it cannot be applied to the whole of a fleet.

For concrete structures, the structure lifetime and its probability of failure can also be determined. From a practical point of view, there are many offshore structures and a lot of foundations under different equipment, in addition to storage tanks, so the quantitive risk assessment is not a practical solution. In this case, the direction is to use the qualitative risk assessment, as it depends on the quality of the data used in the risk assessment process.

7.2 Offshore structure integrity management (SIM) system

The methodology and approach to be applied to the offshore structure, as described in this section, can also be applied for pressure vessels, concrete structures, and FPSO. The following steps should be carried out to implement the integrity management system.

The first and main step is to collect all the available data and perform an evaluation of these data on the structure asset for which the integrity management system is to be implemented. For offshore structures, it is important to separate the structure systems: for example, bridges or flare supports should be considered as separate structure assets. For over 20 years of the structure lifetimes there was no electronic system for archiving and no AutoCAD or even management of change policies. Therefore it is important to collect all available data and to review it to locate the drawings, calculations, specifications, or other items that present the as-is condition.

For offshore structures the collected data will include the following as a minimum:

- Design date
- Construction date
- Water depth
- In-place calculation report
- Construction specifications and drawings
- As-is condition or as-built drawings
- Records for any management of change in case of adding risers, conductors, equipment, or deck extensions, or adding more load or any change in platform configuration
- Existing number of risers and conductors
- Pile design and target depth
- Pile driving records
- The current metocean data
- Data from the last underwater and topside inspection
- Geotechnical study report

- Reports on any structure assessment or modifications or others
- Data from any accidents or fires or other events affecting the structure

For offshore pipelines, the following data are mandatory:

- Design date
- Construction date
- Water depth
- Pipeline subsea calculation report
- Construction drawings
- As-built drawings
- Number of clamps
- The last survey
- Its original service and current service
- Record for any accident
- Mode of operation and flow parameters

Data collection is not an easy step, because many visual inspection checks are required. Sometimes it is required to use a new technique like laser scan to scan the platforms to define their geometric dimensions. The thickness of the steel members can be determined by applying ultrasonic thickness measurements, if unsure about the information on the drawings. The data collection is sometimes considered a separate project due to the huge amount of activity for many structures. In addition, it is required to define the as-is condition of the equipment on the platforms deck. Usually, machines or static equipment can be found on the deck that are not recorded on the drawings and do not exist in the original drawings. This is very important for implementing the integrity system for the static equipment. At the same time, this should be updated to define the weight change on the platform and to assist in issuing the up-to-date weight report that directly affects the structure integrity.

Next, it is required to perform the evaluation of the structures and to calculate their fitness of service. With control of the data achieved, it is thus simple to implement the structure's risk assessment ranking. Based on the risk rankings for the piping, the risk for the pipeline, static equipment, and other components can be implemented. For the offshore structures in the fleet, the overall topside and subsea inspection philosophy, strategy, and plan for all the structures can be obtained.

Fig. 7.2 shows the closed loop of the integrity management system in general. After the plan is defined, the scope of the inspection is then determined and the inspection program will be executed. Then the evaluation of the asset is performed, and a suitable action is determined, recorded, and started. So, this will be a closed loop for infinity, to guarantee that the asset is reliable during its lifetime.

7.2.1 Structure integrity management (SIM)

The asset integrity loop is the same for any asset, but in these sections the focus is on structural integrity. The offshore structure is more complicated than the static equipment, pipeline, or piping. Based on the previous discussion, the offshore platform is a chain of parallel and series elements, so its reliability is more complicated. In the case of static equipment or a pipeline, one failure, such as a hole in the pipeline, causes a leak, so it is a series chain.

The concept of asset integrity is the same for all types, so in these sections offshore structures are considered as an example of the asset integrity approach. The objective for a structure integrity management system is to keep the structure reliable during its lifetime.

For asset integrity in general, and SIMS specifically, the steps are presented in Fig. 7.2. The first step is to collect the data on all the existing offshore fixed platforms and store the platform data in software, as a database for the fleet.

The next step is to evaluate the data and perform a structure assessment and define the structure reliability or structure probability of failure.

The inspection and repair plan is then based on the outcomes from evaluating the available data. Next, a detailed close visual inspection of the topside and a subsea inspection program should be carried out by using a special skilled inspection team for the topside and divers or operating vehicles (ROVs) for the subsea inspection.

The outcomes of the inspection will be reviewed and evaluated and then recorded into the already established database. At this stage, you have all the available data that represents the as-is condition of the structures. Based on these data, a structure assessment can be done, and the condition of each structure is now known. At this time, there is full control over all of the fleet and qualitative or quantitative risk assessment can now be carried out, to make a decision on the next action. The next action is to define the priorities of inspection and repair. This can assist in future decision making, as management can define the required budget for the following years to maintain the SIM cycle.

As mentioned earlier, there are two main ways to do a risk assessment: quantitative or qualitative. In the case of quantitative risk assessment, it is required to calculate the probability of failure for each structure. This is illustrated in Chapter 6 by using pushover analysis and reliability methods. However, this method is very costly and cannot be applied for hundreds of platforms or other assets in general. The most traditional method is to use the qualitative risk assessment method.

The qualitative risk assessment depends on the quality of the data that is obtained and analyzed, so the team needs to be determined to define the required data. This method can be applied for any asset. It will be applied here on the offshore structure's integrity.

For this method, it is required to define the factors that affect the structure likelihood of failure and calculate the impact in case of failure.

In general, the likelihood of failure can be the result of a combination of the strength and the affected loads. The following factors represent the strength and load for offshore structures:

- The construction dates
- The structure system (the jacket is X-bracing, K-bracing, or others)
- The number of legs on the platform
- The thickness of marine growth
- Date of the last inspection

The impact due to the structure depends on whether:

- The platform is manned or unmanned
- Distance from the platform to the shoreline
- Production losses in case of failure

The main target for any integrity management system is to implement an effective inspection, repair, and maintenance plan according to risk-based techniques, by applying this system to be proactive in maintaining the facilities and the structure reliable during its lifetime, with minimum cost.

The risk is calculated by using qualitative risk assessment, which is most practical in the case of a huge number of facilities and a fleet of offshore structures.

It is required to establish a team with members taken from the integrity team, engineering, and operations. The target should be clarified to the team; the integrity team leader is responsible for managing the discussion through workshops. There will be more than one workshop, according to the volume of facilities and the number of offshore structures. This committee is responsible for the first step, which is to define the factors that indicate strength deterioration and extreme loads that may be applied to the structure or the facility, in the case of the offshore structure platform, and the factors that impact the business in general.

7.2.2 Qualitative risk assessment for platforms

In general, the method to rank the offshore structure jackets is based on a risk-based approach. For qualitative assessment, the risk is calculated as a combination of the likelihood of failure occurrence and the impacts on the investment, fatalities, and other factors in the case of platform failure. The risk is equal to the likelihood of failure multiplied by the impact due to this failure.

The likelihood of failure is the definition of the probability of failure, but it depends on a practical calculation, not a statistical calculation of the probability. The consequence of failure is the combination of all the losses that can occur as a result of the offshore structure failure.

Classification factors that do not require any analytical assessment are presented that enable the likelihoods and consequences of failure to be quantitatively evaluated or qualitatively categorized.

For an offshore structure that consists of all the structures in addition to the platforms, such as bridges, for example, the likelihood of structural failure is due to two main factors, which are the structure strength or capacity, and the maximum loads that affect the offshore structure. The likelihood identification system defines the platform factors that decrease or increase the platform strength and loads, respectively.

For instance, all the factors that decrease the platform strength and its capacity to carry the applied load will increase the likelihood of platform failure. Factors that contribute to increasing the applied load on the offshore structure will also increase the probability of failure.

The overall failure consequence is sometimes called the failure impact and it is calculated by a summation of three main components: environmental losses, economic losses, and injuries or any safety-related losses. The impact of each of the three consequences will be reflected by its monetary value, and all these effects are summed to obtain the value of the overall consequences.

One of the quantitative risk analysis methods is to categorize each platform structure in terms of likelihood of failure and the associated consequences of this failure. Then, the categories are inserted into the risk matrix to establish the overall relative risk ranking. Every company has its own risk matrix; there are various types, such as 4 × 4 matrix or 5 × 5 matrix. Color coding is typically used in the risk matrix, with high, moderate, low, and insignificant risk associated with colors red, yellow, green, and white, respectively.

In parallel with the risk assessment, the structures are also grouped into "families" that represent platforms with a similar structure system, such as number of legs, bracing systems, and water depth, and that are in the same area. Hence, after identifying and ranking the high-risk platforms from the fleet of offshore structures, then the high-risk platforms can also be identified within each family. This can help to identify the priority for the development of the maintenance plan. Note that maintenance is a general term, as it includes inspection, repair, or major repairs. So, by using the risk matrix, the target of having a realistic maintenance plan that matches with the structure risks can be achieved.

7.2.3 Likelihood of failure factors

This section illustrates the factors that influence the likelihood of jacket structural failure, noting that for qualitative risk assessment, the likelihood is equal to the probability of failure in quantitative risk assessment. The factors consist of two groups, those that relate to the strength of the platform, listed in Table 7.1, and those that relate to the loadings applied to the structure platform, listed in Table 7.2.

Table 7.1 Strength factors influence platform probability of failure.

Item	Factor	Influence of the factor	Weight	Score
1	Design codes	The modification in the new revision of codes and specification for improvement in design and fabrication practice.	5	0–50
2	Legs number and bracing type	The effect of the number of legs with the bracing system on the structure likelihood of failure.	15	10–150
3	Pile foundation	The pile depth and the pile record data.	10	0–120
4	Risers and conductors	The number of original conductors and risers and compare with the existing numbers.	9	0–90
5	Boat landings	To account for non-designed hydrodynamic loads from boat.	3	0–30
6	Grouted piles	If the annulus is grouted or not as affects the probability of failure.	4	0–40
7	Ineffective member	The number of damaged and cutting members and missing members and their influence on the platform probability of failure.	25	0–250
8	Splash zone	Condition of the corroded member in this zone.	8	0–80
9	Flooded members	The influence of the flooded members on the platform probability of failure.	8	0–80

Table 7.1 Strength factors influence platform probability of failure—cont'd

Item	Factor	Influence of the factor	Weight	Score
10	CP surveys and anode depletion	The efficiency of the CP system. If there is a lack of CP, the probability of failure will increase.	6	0–60
11	Inspection history	The inspection history of the platform. No inspection means high risk.	6	0–60
12	Remaining thickness	The percentage between the remaining to the nominal wall thickness and its relation to the number of corroded members and their location.	8 × BL	0–80
Strength only—subtotal				10–1000

Table 7.2 Load factors influence on the structure likelihood of failure.

Item	Factor	Effect of factor	Weight	Score
13	Design loading	The influence that design loading has on the probability of failure. This differs from the design factor, such as the design details of the structure, like joints and framing schemes.	8	0–80
14	Marine growth	The influence of increasing marine growth thickness.	5	0–50
15	Scour	Scour effect on the platform likelihood of failure.	5	0–50
16	Topside weight change	Additional loads on the deck by adding machines, equipment, or deck extension and their influence on the platform likelihood of failure.	11	0–110
17	Additional risers, caissons, and conductors	Adding risers, caissons, or conductors to the platform.	10	0–100
18	Wave-in deck	The distance between the maximum storm wave crest and lower deck and its influence on the platform probability of failure.	25	0–250
19	Seismic	Whether seismic is considered in design or not, and the seismic zone.	8	0–80
Load only—subtotal				0–700
Total score				10–1700

The factors should be defined separately for different locations, based on the operational experience of the company and the regional data related to the factors that can affect the structures. In addition, as more detailed inspections are undertaken, the number of factors can be increased. Thus it is very important to highlight that the values stated in Table 7.2 and the consequence tables are presented only as guidelines for this approach and they should be examined for your structure fleet based on the conditions of operation, environment, and the back history, as well as the data collected.

For example, for a specific structure, if no member thicknesses were measured to check for corrosion, member thickness is not included. However, once thickness measurements become available in the future, this factor should be considered.

A. The examples of strength factors are as follows:
- Design code
- Year of design
- Number of platform legs (3 or 4 or 6, etc.)
- Structure bracing system (K or X, etc.)
- Pile foundation depth
- Risers and conductors
- Condition of boat landings
- Grouted pile
- Damaged, missing, and cutting members
- Splash zone
- Flooded members
- Cathodic protection survey and anode depletion
- Inspection history
- Remaining wall thickness

B. Load factors are as follows:
- Design loading
- Thickness of marine growth
- Scour
- Change in topside weight
- Additional risers, conductors, and caissons
- Wave-in deck
- Seismic load

Some of the factors are dependent, but others are independent. There are very strong relations between the bracing type and the number of legs on the platform, and this will be discussed regarding the structure system effect. It is important to highlight these factors and their numbers can be used as a guideline only. For example, some companies have platforms with three legs, while others do not have any platforms with three legs, so the factors and criteria will be varied. Therefore the risk-based inspection in the integrity management loop should be different from one company to another and it should be tailored to the company asset characteristics.

The interaction between the factors was discussed by De Franco et al. (1999), who mentioned that the interactions between different platform characteristics might make it impossible to develop a risk-based inspection system that does not need to go through a detailed structure analysis. The guidelines for the interaction of each pair of factors and the resulting relations between the likelihood factors and their interactions are shown in the matrix given in Table 7.3, where H, M, and L mean high, moderate, and low interactions, respectively.

As mentioned, the matrix is considered as a guideline only and should be customized to your fleet, as it represents the relation between the different parameters. As a clear example of this, the design year of the platform implies that certain factors of safety and certain detailing practices were followed in the platform's design, so in this case the design year of the platform shall be a conjunction with the design practice in those years, as it has a strong relation to the applied design load.

As another example, platform redundancy is affected directly by the number of legs and the bracing system together, which thus predicts the possibility of platform damage. On the other hand, the effect of marine growth on the loads is considered small.

Table 7.3 Members effect interfaces.

	Bracing type	Defects location	Legs Number	Grouted piles	Sensitivity to corrosion	Splash zone	Damaged & Flooded members	Deck elevation	Pile Scour	Inspection History	Remaining wall	Change in topside weight
Design year	L	M	M	H	M	L	H	H	M	L	L	M
Bracing type		H	H	M	H	L	H	H	H	M	H	L
Defects location			H	L	H	L	H	L	H	H	M	L
Legs number				M	H	L	H	L	M	L	L	H
Grouted piles					L	L	M	L	L	L	L	L
Sensitivity to corrosion						H	H	L	L	H	H	M
Splash zone							M	L	L	H	H	M
Damaged & Flooded members								L	L	H	M	L
Deck elevation									L	L	L	L
Pile Scour										H	L	L
History of Inspection											H	L
Remaining wall thickness												L

The relation and interaction between different factors should be discussed in detail among the integrity team members, and the interaction in this chapter can be considered as a guideline for the offshore structure platform. The same approach can be considered in other offshore structures like FPSO or tension leg platforms (TLPs).

For other cases, for example in the case of corrosion, the effect of its position on the structure, which is presented in the structure analysis calculation, is considered to be more important than its interaction with the type of structural system, so this relation is omitted.

7.2.3.1 Likelihood of failure calculation

Strength factors

The parameters that affect the platform strength mainly depend on the characteristics of the number of offshore platforms to be maintained, as some factors exist in some companies but don't exist in other companies.

The presentation of these factors here is just an example of the effect of each factor, which should be tailored to each company and its location in the world. Therefore these factors should be defined by holding a workshop, inviting the integrity team, engineering, and the operations and maintenance team.

This is carried out by collecting the data for the whole structure and then a comprehensive study is required to evaluate the role of each factor. For example, one owner may have a platform constructed in 2005, while another owner may have a platform from 1970, so the weights and the scores will be different. On the other hand, some regions in some countries are more strongly affected by seismic load than other countries. In some regions, the ice load is mandatory, such as Russian regions, and other countries have no snow load at all, such as the Middle Eastern countries.

Based on this, the factors and their weight and score in Tables 7.1 and 7.2 are just guidelines and should be tailored based on the nature of the structure fleet and it should be discussed case by case.

The normal procedure for qualitative risk assessment is that a team, consisting of the owner, the operator, and the engineering group, with expertise, holds a meeting and uses brainstorming or other techniques to define the interaction between the factors and their weight.

Design codes There are usually modifications and developments in the offshore structures design and fabrication practice in every revision of the API RP2A. Thus it is required to define the codes that were used during the design of the platform. Considering that the year of construction is different from the year of the design, more investigation is always needed in the stage of collecting the required data.

For example, there is usually a change in tubular joint design due to its complexity and there are many studies and research projects carried out to predict the actual strength of this joint with different parameters. In addition, the joint was established after the 1970s so the increased strength of the joint can be seen in recent structures and it may not exist in older structures. Many constraints and recommendations have been changed over time.

In 1969, the first version of API RP2A was issued, which was considered the first good guideline for platform design. In 1975, a major update to RP2A was done for the joint design practices and in 1979, API RP2A was updated to the point of the API Section 17 Committee on assessment of existing platforms.

The design codes score is presented in Table 7.4; the score is chosen from 0 to 10 for the platform year of design, a factor that estimates the level of confidence that can be placed on it due to the state of design practices prevalent at that time. There was a major enhancement in the code after 1970 so if the platform was designed after 1979 its score is 0, because the platform is considered safer. For platforms designed before 1970, the score is 10, because these platforms have the highest probabilities of failure. Considering this, the overall weight of the total score is 5.

From technical practice, if the design year isn't known, especially for old platforms, you can find the construction or installation year in any document; in most cases the design year can be considered 2 years prior to the installation year.

The design year is very important information, as it is related to the revision of API that it was designed by and the common design methods in this time period. There is a relation between the platform reliability and its design code year or its design year. In addition, this factor is an interface

Table 7.4 Design code factor.

Design year		
≥	≤	Score
	1970	10
1971	1974	7
1975	1979	4
1979		0

and affects the other factors like grouted piles, load, and resistance in general, as the Cm and Cd are different from one code version to another, as discussed before.

Number of legs and bracing system The platform structure redundancy depends on the number of legs and the bracing system. From the pushover analysis and experimental studies that were done, it was mentioned that the X-bracing has a higher redundancy than K-bracing. By rule of thumb, when the number of legs is higher it will provide a higher redundancy as well.

Therefore, if the number of damaged members is the same for two platforms, one of them with 3 legs and a K-bracing system, and the other with 8 legs and X-bracing, the first platform has less reliability or a higher probability of failure than the second platform with 8 legs.

In high seismic zones where required redundancy and ductility are important, it is preferred that the system have X-bracing with 8 legs.

The relation between the bracing system and the legs number on the structure probability of failure is presented as a guideline in Table 7.5. This table should be taken only as a guideline, as possibly in your fleet there is no K-bracing, for example, so it should be eliminated from the table and the score redistributed. Thus this table will be tailored based on the characteristics of the platform fleet in the company assets.

In general, from Table 7.5, the higher score is for K-bracing systems and three-legged platforms and the lowest score is for the eight-legged platform with X-bracing. The scoring system is in the range from 1 to 10 to represent the highest redundancy to the lowest redundancy, respectively. In 1994, Puskar et al. presented that Hurricane Andrew in 1992 caused failure in some eight-legged K-braced platforms, but numerous four-legged X-braced platforms survived without any failures. From this study, the score for eight-legged K-braced structures is slightly higher than the score for 4-legged X-braced structures.

Table 7.5 Score for bracing system and number of legs.

	Number of legs			
Bracing System	**3**	**4**	**6**	**8**
K	10	8	6	5
Diagonal	7	6	4	3
X	5	4	2	1
Unknown	10	7	5	4

This criterion is required to account for the effect that the number of legs and conjunction with the bracing system have on the likelihood of failure. Therefore the legs number and the bracing structure system need to be known.

In some cases, the platform has two systems in different directions, so it is considered that the higher likelihood of failure is on the conservative side, since this is a qualitative method. For example, for a platform with K-bracing in the transverse direction and diagonal bracing in the longitudinal direction, in defining the score it is taken as a K-bracing platform.

As a rule of thumb, increasing the number of legs increases the redundancy of the platform. It is found that the X-bracing type is the most redundant system, while K-bracing is the least efficient. For this factor, the overall weight of the total score is 10.

In case of cutting, damaged, and flooded members, there are significant interaction factors between the bracing system and number of legs.

Piles system This factor accounts for the pile system effect on the platform probability of failure. Uncertainty will exist due to unavailable records relating to the design and installation of the piles and to potential damage to the piles from soil disturbances due to the resting of jack-up rigs near the piles. Obtaining the score for this factor relies on answering these three questions:

Question 1: Is a pile calculation report available (Yes/No)?

Question 2: Do the pile penetration records exist (Yes/No)?

If the answer is yes for these questions, then:

Question 3: Does actual penetration meet target requirement (Yes/No)?

From the answers, select the appropriate score in Table 7.6. The overall weight of the total score is 10.

Note that the accuracy factor of the foundation design itself has not been included in this table.

Risers and conductors Previous factors directly affect the strength and redundancy of the platform and how the particular issue affects the platform's ability to withstand extreme events, such as waves and seismic events. For risers and conductors, it affects the platform's likelihood of failure and needs to be included in the qualitative risk calculation, since inspection of these items is also typically included in an underwater inspection.

The caissons used in fire-control water or other functions are considered to be risers. Other appurtenances such as boat landings, walkways, and barge bumpers can be excluded from the risk study, as they have little effect on the platform risk compared to the risers and conductors.

Table 7.6 Pile strength factor.

Criteria	Answers to questions Q1	Q2	Q3	Score for platform design before 1976	Score for platform design after 1976
Existing pile penetration reach to target	Yes,	Yes,	Yes	2	0
Existing pile penetration known but not target	No,	Yes,	(N/A)	3	2
Target pile penetration known but not existing	Yes,	No,	(N/A)	7	3
Existing pile penetration does not meet target	Yes,	Yes,	No	8	7
Target and existing pile penetration unknown	No,	No,	N/A	10	10

Risers and conductors that contain hydrocarbon gas or oil, and specifically gas, can cause an explosion or fire that can severely damage the platform. The score increases directly with increasing the number of risers or conductors, as there is an increased likelihood of damage as a result of increasing their number.

In terms of weight, the risers have a higher value than conductors in this guideline; assume that one riser is equal to five conductors. This difference between the risers and conductors, because the conductors are less likely to fail than the risers are, is due to the following reasons:

- Conductors are often located inside the platform area and therefore are less vulnerable to boat impact or other external damage.
- Conductors have many pipes inside them as well as the drill string, which decrease the possibility of a complete failure.
- In most cases, the typical conductors have subsurface safety valves that shut off quickly in the event of failure and this should be reviewed during

the integrity session and the gathering data phase, to determine whether these valves are working properly or not.

- The conductor and its inside pipes penetrate inside the seabed and work like a pile, so it will provide more strength that compensates for the additional hydrodynamic load effect. This is different from risers, as risers have no strength increase but just add load to the platform.
- Both increase the wave load on the platform and that will increase the probability of failure for the whole structure.

Also, damage to risers may cause loss of containment, which could result in a major incident, significantly increasing the likelihood of failure.

Therefore, during the gathering data phase the number of existing risers, number of conductors, and their number in the original design should be determined. In addition to that, define the type of hydrocarbon that it is carrying, whether oil or gas, and if it is carrying gas then whether it is high or low pressure.

The main caissons, such as firewater caissons, should be counted and considered as oil risers.

Table 7.7 presents an example of a score in the range from 0 to 10 for both risers and conductors depending upon their contents. These scores are then multiplied by a weighted factor, which is considered to be 1 for risers and 0.2 for conductors, and then the sum of the two scores gives the total score. The total score can vary from 0 to a maximum value of 13.6, considering the overall weight of the total score for this factor is 5.

Risers and conductors do not have any interdependence with others, as the appurtenances do not directly contribute to the redundancy or strength

Table 7.7 Effect of risers and conductors.

	Type		
	Risers and caissons		
Number	**Oil and water**	**Gas**	**Conductors**
0	0	0	0
<3	0	1	0
4–7	1	3	1
8–14	2	6	2
14–20	3	8	2
>20	4	10	3

Score = Riser score × 1.0 + Conductor score × 0.2.

Table 7.8 Additional boat landings.

Number of boat landings	Score
0	0
1	5
2	10

of a platform. On the other hand, riser failure increases the overall risk of a platform.

Boat landings In the case of increases in the marine growth thickness and additional boat landings, which are not considered in the original design of the platform, the global loading on the structure will increase, and consequently likelihood of failure will increase. On the other hand, corrosion and uncontrolled modifications may result in boat-landing failure, creating a dropped-object hazard to the jacket (Table 7.8).

For this factor, the weight of the total score is 5. The number of boat landings attached to the platform shall be defined, because the score for this factor depends only on this number, which ranges from 0 to 10

Grouted piles With induced vibration on the platform or if the platform strength is low, the main solution is to insert a grouting to fill the annulus between the leg and pile to increase the stiffness of the leg. In this case the leg and pile with the grouting work act as a combined section and this will increase the capacity of the joints of the braces that frame into the leg.

The 1975 API RP2A design parameters precautions at that time revealed the design output to have a better joint design, hence a lower score, since the joint may be already well designed and the grouting will not provide much improvement. North Sea platforms have better joint design and therefore grouting will have less effect on them.

Grouted piles will increase the structure strength and will reduce the likelihood of failure.

The score of the non-grouted pile is dependent on the year of design of the platform. From a practical point of view, this score can be reduced by around 60% if the platform location is in the North Sea, as joint design tends to have been much better there than in other areas of the world, hence grouting would have a much lower effect. With a design before 1975, the weight factor is considered to be 10, but after 1976 it is considered to be 4.

As evident from the table for this factor, grouted piles will depend on the design year of the platform, and with North Sea platforms, as the platform joints have historically been better designed, the weight factor can be reduced.

Damaged, missing, and cut members After establishing the first plan for risk-based underwater inspection (RBUI), it should be checked, as the failure probability was estimated primarily based on previous studies from joint industry projects and from research by Puskar et al. (1994) relating the reserve strength ratio (RSR) to the failure probability for a number of locations of members in platforms worldwide. One of the main factors is the number of damaged members, as they can have a significant impact on the strength of the platform.

In 1988 an integrity management project evaluated a platform with numerous common damaged conditions. This study considered several scenarios of damaged members ranging from one to three or more. The damaged members were studied in three different locations in the jacket: near the waterline, at the mid-depth, and near the seabed. This study revealed that there was little reduction on the platform strength with one damaged member, but numerous damaged members, five members or more, will cause a dramatic reduction in platform strength. The scoring range of the damaged members takes into account the results of this study.

Puskar et al. (1994) and Gebara et al. (1998) also stated that K-braced platforms are more sensitive to damage than X-braced platforms. Based on this, it was decided that the K-braced platforms would be 10 times more sensitive to damaged members than X-braced platforms.

The number of damaged, missing, or cut members directly affect the platform likelihood of failure. To be more specific, damaged members are those that remain in their location but have dents, holes, cracks, bends, or other defects that reduce their strength. During an underwater inspection, it will be easy to define the number of missing, cut, or damaged members. If there is no available underwater survey information, the score will depend on platform type, as well-head protectors are particularly vulnerable to having members cut out.

Note that if the inspected platforms have been only partially surveyed, the numbers of damaged, missing, or cut members should be extrapolated from the available survey data.

Tables 7.9 and 7.10 present the score system for platform members damaged or missing. The platform's damage tolerance is accounted for by

Table 7.9 Damaged members.

Number of damaged members	Score
1	1
2–4	2
5–9	3
More than or equal to 10/no data	5

Table 7.10 Missing or cutting member.

Number of missing members	Score
1	3
2–4	7
5–9	9
More than or equal to 10/No data	10

Total score (limited to 10) = (tabulated score for damaged members + tabulated score for missing members) × leg-bracing score/10.

multiplying the tabulated scores by the "legs-bracing" score and dividing by 10. The resulting score is limited to a maximum of 10. The overall weight of the total score is 20.

The damaged member factor is considered one of the strongest factors representing the current condition of the platform. The input data of the inspection result provide a more sensitive factor that affects the overall weight in the scoring. The goal of the qualitative risk technique is to examine the effect of different inspection policies on risk.

The number of damaged members also relates to the bracing type in any given platform. An X-braced jacket with eight legs has a higher redundancy and will not be as affected by damaged or missing members as a K-braced jacket with three legs, as the latter's strength will be reduced dramatically with any missing or damaged members due to its lower redundancy.

The total weight given to damaged members is 10.5. This factor should be checked by revisiting the existing database, which showed that this level of weight was required to increase the risk of platforms with significant damage. As an example, if the platform has four damaged members, the platform rank should be changed to a rank that enables it to do a more frequent underwater inspection.

It is very important to highlight that this score number is only a guideline, and this should be done by trial and error for the fleet of platforms in the same location, to define reasonable weight and score numbers.

Table 7.11 Relation between corrosion condition and topside weights.

Case condition	Original topside weight	
	Measured or calculated	Estimated
Measured or calculated/expected weight increase less than or equal to 10%	5	7
Topside load increase more than 10% or no data	8	10

Splash zone corrosion and damage The splash zone corrosion is considered a major factor affecting the structure integrity. Thus the above-water and underwater survey data of the splash zone members should be defined through inspection.

But if the inspection information is not available, the score depends on topside weight conditions. As presented in Table 7.11 the overall weight of the total score is equal to 10. With light or no corrosion, the score is 0; for moderate corrosion of around 50%, the score is 4; for higher corrosion or no data, the score is 10.

In some practices, the relation between corrosion conditions and the topside weight changes are considered, as shown in Table 7.11.

The relationship with the topside weight factor is essential when the pile-leg annulus is not grouted and the jacket therefore hangs from the crown weld, so that the highest utilized leg members are in the splash zone.

Flooded members An underwater inspection can determine if there are flooded members and their number and locations in the platform. Any damage or crack in the member or the joint can cause water to enter the member and to convert it to a flooded member. This flooded member provides an indication of the current condition of the member and the possible existence of damage or defect, since most platform members are designed to be buoyant.

In terms of the scores assigned for the number of flooded members, the same approach that was used for damaged members applies to the flooded members. In general, the flooded member have a less impact on the structure integrity rather than the cutting or missing members. If there are many flooded members, it will have an impact on the structure integrity due to reduction in strength. The same concepts of the damaged members, the number of legs, and the type of framing structure system will affect the score.

Table 7.12 Flooded member effect.

Flooded members number		
≥	≤	Score
1	3	4
4	9	9
10/no inspection data		10

Total score = tabulated score × leg-bracing score/10.

If the member is flooded due to damage, it will be considered as a damaged member, but if the member is flooded and the reason is not clear, it will be considered a flooded member.

Table 7.12 is similar to that applying to the damaged member. It also provides a score between 0 and 10, where a score of 0 represents no flooded members and hence no increase in the likelihood of failure, and a score of 10 represents 10 or more flooded members, which significantly increases likelihood of failure. The platform's damage tolerance is accounted for by multiplying the tabulated scores by the "legs-bracing" score and dividing by 10. The resulting scores range from 0 to 10. The overall weight of the total score is 6.

Sometimes, to reduce the cost, only a percentage of members are inspected for flooding as opposed to all the platform members. For example, if a survey of 30% of the members detects one flooded member, then a total of three flooded members should be expected for this platform. In general, the existing condition of the platform will be determined based on the number of flooded members compared to the last underwater survey.

As discussed previously, the number of flooded members is also influenced by the bracing system of the jacket. For example, a platform with three legs having a K-bracing type will have few redundant load paths and is thus low redundancy, so it will be affected by any missing or flooded members. On the other hand, a platform with eight legs and X-bracing will not be as affected if there are missing or flooded members.

Cathodic protection and anode depletion If the cathodic protection (CP) system does not work, it will result in corrosion propagation, which will increase the likelihood of failure.

The platform grade in general will be obtained by taking the average from the anodes survey results of 100% of the structure's anode grades.

The overall weight of the total score is 6. For grade 1 less than 10% has been consumed, so the score is neglected; for grade 2 with less than 50% consumed, its score is 3; and for grade 3 with more than 50% consumed, the score is 7; grade 4 indicates 100% depletion and the score is 10. Interpolation of the score between grade 3 and grade 4 can be used. If there is no data, the score is considered 10.

Inspection history The inspection history consists of two major factors. The first is the time period between the inspections, or the date of the last inspection. When there is a time gap between the last inspection and today, it means that the platform likelihood of failure will increase, as there is a high probability that unpredicted local failures have occurred during the time period from the last inspection until now. The second main factor is the level of underwater inspection, because the more detailed the inspection, the less likely it is that an undetected local failure or deficiency in strength will occur. The well-inspected platform is a platform that has been checked for flooded members and most of the joints have been cleaned and inspected and should be able to go longer without an inspection update.

Another factor is the quality of the inspection. This also affects the ranking and it is known from the experience of the integrity team. Assume two platforms with the same risk, but from the history of the last inspection one of them is Level II and the other is Level III. Both will have the same risk immediately after the inspection, but over time there are deviations. It is assumed that both levels of inspection will be able to determine the current condition of the platform with the same level of confidence. With close visual inspection (Level III) it is assumed that hidden deterioration can be detected with higher accuracy, greater than the Level II. Therefore, after 4 years from last inspection, the score of the platform that was inspected with a Level II approach will increase to 8; while the score for the Level III platform increases to 5, because the integrity system considers that the predictive capabilities of the Level III inspection are significantly better than those of the Level II inspection.

As per API RP2A the Level I inspections are defined as above-water walk-around inspections and they will be excluded, since this should be done every year for all platforms.

As presented in Table 7.13, the input data will be the year of last level II, III, and IV inspections. The overall weight for the inspection history factor in the total score is 5.

Table 7.13 Score for inspection data.

	Level of last inspection		
Years from last inspection	**Level II**	**Level III**	**Level IV**
1	0	0	0
2	2	0	0
3	4	2	1
4	8	5	3
5	9	7	5
6	9	8	6
7	10	9	8
8	10	10	9
9	10	10	10

For each inspection level, calculate tabulated scores: *SII* for Level II, *SIII* for Level III, and *SIV* for Level IV. Score = min (*SII*, *SIII*, *SIV*).

For each inspection level II through III, apply the following formula:

$$\text{The years since inspection} = \min\left(9, \text{this year} - \text{inspection year}\right)$$

The inspection history is a stand-alone factor that does not interact with any other factors. The inspection history mainly depends on the operations philosophy and the maintenance plan, which largely depend on the budget constraints.

Remaining wall thickness The remaining wall thickness is represented as a percentage of the nominal wall thickness. Determining the weight of the remaining wall thickness takes a similar approach to that of damaged and flooded members, as the number of corroded members has a major effect.

It is found that the reduction in the thickness does not have much effect until it reaches 20% or more, when it will dramatically affect the strength of the member.

Corrosion is typically localized on the member and this has been accounted for in the approximation, noting that the score would be higher if the entire member were corroded. If the corrosion reaches above 50% of the section, it is considered an extreme case with the highest score.

The score is between 0 and 10, where a score of 0 corresponds to no reduction in wall thickness and a score of 10 is a reduction in wall thickness higher than or equal to 50%. This score is then multiplied by another factor determined by the number of members thinned. The resulting scores range from 0, representing no likelihood of failure, to 10, which represents an

Table 7.14 Remaining wall thickness.

Remaining wall (%)		
>	**≤**	**Score**
	50	10
50	80	8
80	99	3
100		0
Number of members		
Less than	**More than or equal to**	**Weight**
	4	0.3
5	9	0.7
10		1

Total score = tabulated score × weight.

increase in the likelihood of failure; the overall weight of this factor in the total score is 7.5.

The remaining wall and the extent are both inspection results. Consequently, the corrosion factor is taken as a strong indicator of the current condition of the platform (Table 7.14).

The reduction in wall thickness does not depend on the bracing system or number of legs, but when the corrosion increases until a crack or hole appears, it is considered a damaged member and, in this case, the bracing system and number of legs will be taken into consideration.

Loads factors

Design loading A study stated that in the Gulf of Mexico the first structure platforms constructed were designed for a hydrodynamic loading (wave and wind) that was more than today's loads by about 55%. In the North Sea, the earlier platforms of the late 1960s and early 1970s were designed for loading that was about 70% more than today's design load. It was stated also that, in the years between 1980 and 1986, the metocean data produced design loads that were about 45% higher than the present loads. As a conclusion, the metocean data subsequently changed over time due to weather and environmental changes.

The design loading factor differs from the design practice factor, which accounts for the level of confidence in the design of structural components such as joints and members. From a practical point of view, the year of design can be considered 2 years before the year of installation, if there is no definite date of design year.

The scores for this factor are assumed to vary from 1 to 10 to cover different ranges of years designed. Since this is different from company to company depending on the age of the fleet, it should be tailored by the integrity committee.

Low scores represent the higher values of design loads and thus low likelihood of failure, but high scores indicate smaller values of design loads and thus increase the structure probability of failure.

The data in Table 7.15 can be reasonable for any location worldwide. The scores vary from 1 to 10 for different ranges of year designed. The overall weight in the total score is 5.

In some cases one design is repeated as the standard for the same region with the same water depth. So, in this situation, the design year will be taken as the year the first platforms were designed.

It is evident from Table 7.15 that the year-designed parameter is related to the location of the structures, as the North Sea is different due to its harsher environment than other locations worldwide. The scores are based on the design hydrodynamic loading adopted from the metocean data based on practice in the year of design.

Marine growth Another factor that can increase the load is the marine growth. The wave load calculation mainly depends on the diameter of the structure member based on the Morison equation for inertia and drag force calculations. The existence of marine growth will increase the member diameter and this will increase the wave load correspondingly. This factor depends on the actual measured marine growth (mm) with respect to the marine growth thickness that is considered in the design.

The marine growth score factor is shown in Table 7.1. This score is in the range of 0 to 10, with a higher score reflecting a higher likelihood of

Table 7.15 Designed year.

	Year designed		Score
	≥	≤	
Any location worldwide		1970	10
	1971	1974	7
	1975	1978	4
	1979		1

structure failure. An increase in the measured marine growth thickness higher than the designed thickness of the marine growth means an extra load affecting the structure. The table suggests a score value of 10 represents measured marine growth exceeding the design value by 50% or more, and if there is no data at all the score is 10 also. If the marine growth thickness increases from 110% to 135% the score is 3, and if the increase is in the range between 135% and 150%, the score is 8, and the score is considered 10 if there is no available inspection data or the marine growth thickness have a thickness over 150% than the design marine growth thickness.

This factor should be changed based on the inspection and it is easy to be defined in any inspection level. As per APIRP2A, the marine growth thickness in design is considered 50 mm but for recent revisions of API it depends on the engineering office judgment. Thus it is important to return to the design year and the basis of design for the new platform.

The marine growth factor has no interaction with any other factors. The measured marine growth should be included in the inspection scope. The thickness of marine growth will be determined from underwater inspection and should be compared with the original design if there is an excessive increase in the thickness over the design value, as it means there is an extra load on the platform.

Note that the increase of the marine growth has a direct effect on increasing the load on the platform and specifically the global base shear. But there are other major factors that affect the increase of the horizontal load, so the sensitivity is not high and it appears after an increase in its thickness 50% over the design thickness. In some locations and specifically in tropical areas with hot weather, the marine growth is very thick and increases rapidly, so this scope may be changed.

Scour The scour problem, if it occurs, will reduce the design depth of the pile, which reduces its capacity. As per API RP2A, the scour should be considered in the design, assuming the scour is 1.5D where D is the pile diameter.

The scour should be included in the scope of the inspection to have a reliable risk assessment.

Although scour is an inspection-related input, the parameter to cause failure is the resultant stress. With deep scour when the depth exceeds the design depth, this will cause an increase in the stress level on the piles and the lower portion of the jacket, which may increase the likelihood of failure.

The scour factor is the percentage between the measured scour depth and the designed value. The scour score is between 0 and 10, with the highest value meaning a higher increase in the probability of failure.

If the existing scour depth exceeds the design value, it will increase the stresses on the pile, therefore increasing the likelihood of failure for the platform. The maximum score value is 10, which is used if the scour depth is over the design value by 150% or there is no data. If the scour measurement has a value more than the design values by 110% to 120%, the score is considered 3; if it is between 120% to 150%, it is 8; if the value of the measured over the design is 150%, or there is no data from the inspection, the value is considered 10.

If there has been no inspection and there is no record of an inspection of a platform near this platform, the score is considered 10. Sometimes, the inspection is not available for the platform but there are available scour inspections for platforms near by, so this can give an indication of the score value to use.

Scour affects the pile capacity but its sensitivity is not high, as there are many factors affecting the pile design. In this guideline table, the highest score will be after a 50% increase over the design scour. On the other hand, if the existing scour is less than design scour depth by 10%, the score will be 0. However, it will be an opportunity as the structure is conservative in this case.

We recommend that this value can be negative to be considered safer, to compensate for any other high score.

The scour has no interdependence with other factors or parameters in the likelihood category.

Topside load change Topside load changes usually occur in normal practice, when adding new equipment or new machines or performing deck extensions for any reason.

The main problem with old platforms is that the topside weight data in most cases do not exist, as in the past there has been no management of change policy or electronic drawing by AutoCAD or pdf as is done currently, or a computer archive to facilitate storing and recording any change.

This can pose a challenge, so Table 7.16 presents many outcomes from this condition. If you do not have any record of data, whether the load is increased or not, this case will represent a higher probability of failure, so

Table 7.16 Topside load change.

Criteria	Score
Not known if Load Increase or a reduction	2
Estimated believed load increase ≤(5%)10%	5
Estimated load increase >(5%) 10%	9
If no data available	10

the score will be taken as 10. If the is increased by over 10%, it will mean a higher risk, as it is important to do a global analysis as a rule of thumb. But if the load increases no more than 10% and it is certain there is no local deformation for any member, it is simply a reduction from the platform conservativity, so the score will be smaller. Some studies designate 5% instead of 10% regarding the uncertainty in load increase, so the team should decide whether to use 5% or 10%.

Note that the load factor is the percentage of the existing loads on the deck as a percentage of the design load on the deck.

Risers, caissons, and conductors factor From a practical point of view, it is normal practice to increase the number of conductors or risers to match with increasing production.

Therefore, the additional risers, conductors, or caissons over the design numbers should be considered, as they will affect the platform probability of failure.

The score system mainly depends on both the number of risers, caissons, and conductors in the design and the additional number that exists currently, as presented in Table 7.17. This factor has an overall weight of 5.

Table 7.17 Additional risers or conductors.

Additional number of risers, caissons, and conductors	Original number of risers, caissons, and conductors in design		
	1–10	11–20	>20
2	2	1	0
4	4	3	1
6	6	4	2
8	8	5	2
>8	10	7	3

Table 7.18 Additional risers or conductors.

Sum of diameters for existing riser/sum of diameter of original design	Score
1–20%	2
20%–50%	6
50%–80%	8
80%–100% or more	10

The data should be available from the petroleum engineers following or working on the platform. If the integrity team during the investigation finds that there is uncertainty in the data of the original design, a conservative estimate of the number of additional risers, caissons, and conductors will be used.

Based on this, another method of calculation is shown in Table 7.18, which is calculating the percentage of the summation of the existing risers and conductors to the summation of the diameter of the design. The score is presented against every ratio range for the increase in the summation of the diameters.

Wave in platform deck The wave-in-deck factor directly affects the likelihood of the platform failure due to storm wave crests that strike the platform's lowest deck. This factor may be affected if the offshore platform is not designed well or if a change occurs in metocean data; based on this, the lowest deck elevation fails to satisfy the minimum air gap requirement as per API. Therefore, in the case of an extreme event, the maximum wave may strike the deck levels and cause catastrophic damage to the platform. Thus the consequence of this factor is very high, as it directly affects persons and the structure integrity. Therefore it is essential that this situation be highlighted and reported to the operating and integrity management team. This situation is not a rare occurrence, so if it is recorded, it should be highlighted in order that a decision can be made. Thus this factor will occur or not; if it has not occurred, then this factor will be neglected, but if it has occurred, it will receive the higher score, which is 10. The overall weight in the total score is 20.

Seismic load Some locations worldwide, such as Trinidad, lie in a high seismic zone. Recent platforms have been designed based on their seismic zone location. However, most of the old platforms did not consider seismic loads in their original design, and there have been many updates in seismic design over time. A weakness in seismic calculation will increase the platform likelihood of failure when compared to platforms that have been designed for seismic zones, or are located in non-seismic zones.

ISO codes present zone 1 equal to 0.1 g, Zone 2 equal to 0.2 g, and so on.

It is known that offshore platforms that are designed for waves only can normally withstand ground accelerations of 0.15 to 0.2, provided that they were initially designed for a reasonable wave height and have good design and detailed joints. Ground accelerations higher than this can cause problems, however. ISO states that Zone 4 or higher requires specific seismic design features for ductility, such as special member slenderness coefficient values, extra-heavy cans, redundant framing schemes, and others.

As a first step, it is important to define the seismic zone of the platform location. The score of the seismic load effect is presented, as for seismic zones 3, 4, and 7, the scoring is 4, 7, and 10, respectively.

This factor has no interdependence with other factors or parameters in the likelihood category. It is formulated with the seismic zone and seismic design (yes or no) parameters from the design category of the data sheets.

7.2.3.2 Likelihood categories

In general, in this guideline most factors have a score between 0 and 10 for simplicity. Factors like bracing legs and year-load location have values ranging from 1 to 10 and scores for appurtenances range from 0 to 12.

A score value of 0 means that this factor has no or a negligible effect on increasing the platform likelihood of failure. A score value of 10 means that the factor has a maximum effect on the platform likelihood of failure. The factors' weights indicate how strongly the effect assessed by each factor affects the overall likelihood of failure of the platform.

The total score is calculated by the formula $\sum_i w_i s_i$ where w_i is the weight of the ith factor and s_i is the score. From a theoretical point of view, mathematically, the score range can be from 15 to 2255, but from a practical view, the score ranges from about 40 to 750.

The total weighted scores are converted to a likelihood category according to the ranges in Table 7.19. These ranges have been developed from the ranges presented in the DNV report.

Then after calculating the scores for all platforms under study, the cumulative percentage can be calculated and, using statistical data, the cumulative distribution function can be obtained. Because the category limits identify the best and worst 5% of the platforms, it is assumed that 5% of the platforms have a high likelihood of failure and 5% of the platforms have an approximately minimal likelihood of failure. This distribution of the category limits assigns two

Table 7.19 Likelihood category.

Category	Range
1	≤S1
2	S1–S2
3	S2–S3
4	S3–S4
5	>S4

category ranges for platforms that have better than average likelihood scores and three ranges for platforms that have worse than average scores.

After the platform ranking is done, then the platforms should be revisited to determine if the rankings are logical or not. If something is not logical, try to find the reason. An optimum verification will include two or three platforms having their probability of failure calculated to evaluate the ranking, and the weight and score system can be reevaluated or redistributed based on that. Sometimes other platforms in the same regions or other regions worldwide can be checked.

The previous values of the categories S1, S2, S3, and S4 can be obtained from the cumulative density function.

7.2.4 Impact or consequence factors

The terms impact and consequence factor are the same, but they are used for qualitative and quantitative risk assessment, respectively. In general, the impact of any structure or facility failure is the sum of three main factors, where the components account for:

- Loss in Environmental Condition (I_E)
- Losses in Business (I_B)
- Losses in safety and human life, such as injuries or fatalities (I_S)

Each of the impact factors in this section is converted to a monetary value in any unit currency (UC) whether USD, GB, or others, and the three factors are then summed to determine the overall impact. This method can also be used for the quantitative risk assessment, but it will be considered as a guideline when applying the qualitative risk assessment, as our qualitative approach will be used as a score system in obtaining the likelihood of failure.

While the summation of the monetary values obtained does not represent the total amount of money that could be lost due to a failure, the concept of a monetary value for consequences has been adopted to measure and sum the effect of safety, environmental, and business losses.

Each type of consequence factor can be measured in units of money like UC or others, with different factors determining the value of each type of loss. When possible, the consequence calculations are quantitative and are related to actual economic values during the time of calculation. One of the main factors in calculating cost is the price of oil and gas and the cost of offshore reinstatement.

The total cost will be calculated as follows:

$$I_T = I_E + I_B + I_S \tag{7.1}$$

where I_T is the total impact value, I_E is impact due to environment, I_B is impact value due to business losses, and I_S is an impact value due to safety loss

Each category will be calculated separately in the following sections

7.2.4.1 Environmental impact losses

Environmental losses are calculated based on the amount of hydrocarbon that will be spilled, which can be calculated based on the daily production or the amount of hydrocarbon release. Considering this, there is a fixed cost that includes the mobilization of personnel and equipment to perform the cleanup in addition to the regulatory cost.

The variable cost varies depending on the volume of the oil spill. The fixed and variable costs are different from one country to another, depending on many factors. In some countries, the waves in their region have a strong wave action that will quickly break up, dissipate, and volatize spilled oil, reducing the overall environmental impact of an oil release.

The cost should be adjusted by multiplying the calculated cost by a factor to consider the distance of the platform to the shore area. The distance from shore is important because both the clean-up cost and the damaging effect of an oil spill will increase if oil is washed up on shore. In addition, this oil spill can affect rare coral reefs or can destroy tourism, as in Egypt the oil fields are near the tourist area.

Estimated environmental loss calculations are presented in Eq. (7.2):

$$I_E = f(d) \times \{F_C + V_C \times \min(D_P, R)\} \tag{7.2}$$

where

$$f(d) = 1 \quad d > d_m \tag{7.3}$$

$$f(d) = 1 + \left(\frac{d_s - d}{d_s}\right)^2 \quad d \le d_m \tag{7.4}$$

where

I_E is the environmental loss, in UC;
V_C is the variable cost, in UC/bbl;
F_C is fixed environmental cost, in UC;
R is minimum released oil volume, in bbl;
D_P is daily production, in barrels of liquid, bpl;
d is distance offshore, in km; and
d_s is the maximum significant distance offshore, which is the distance from the shoreline in km.

The calculations are followed only when the environmental loss, I_E, has not been determined for a given platform. The estimate is based on the assumption that the environmental losses are proportional to the daily production. The fixed cost, F_C, and the marginal cost, V_C, depend on the location to account for the variations worldwide in the costs associated with clean-up, regulations, and other factors.

Eq. (7.2) also accounts for the increased environmental effects of near-shore spills where a spill is more likely to threaten beaches and other ecosystems. In these locations, cleanup and regulation costs are likely to be higher than in the open sea, where wave action and weather can dissipate and volatize a large portion of a spill.

The value of the significant distance should be defined in each region separately based on the geographic location.

Note that, for input variables that are unknown, the variables can revert to default values.

The environmental factor alone made the oil spill that happened in the Gulf of Mexico in 2010 a disaster for BP: the cost could reach to 60 billion UC according to new analysis from the rating agency Moody's. In addition, there is also an effect on the reputation of the company. The reputation factor should be considered another consequence of platform failure.

7.2.4.2 Impact on business losses

To assess the business impact due to any structure or facilities failure, first there is the replacement cost of constructing a new platform. Second, there is the impact of deferred production, which reflects the cost of not being able to produce for the time it takes to replace the platform.

With special tower failure in gas plants or in water and gas injection platforms, and with failure of jackets supporting vent lines and flares, this value should reflect the loss of production.

Eq. (7.1) estimates the total business loss component of the failure consequences. There are two main factors that influence the business loss cost, which are as follows:

- Replacement, as in the case of failure there will be a cost of constructing a new structure or building a new facility.
- Deferred production due to the facilities or structure failure completely or partially, as this is the cost due to stopping the production and shutting off the well until replacement, or doing a major repair for the failed structure or facility unit.

In this calculation the oil and gas reserve are modeled as an investment present value (PV). The value of the deferred production is equal to the difference between the present value and the value of the reserve discounted for the time it takes to return it to production after a failure at a given interest rate.

The owner company should define the estimated cost of the structure or the facility if there is no given replacement value for the platform.

The deferred production losses are a function of the production rate. The production rate can be expressed as either the value of a month's production or the volume of oil and gas produced in a month. When the volume is given, the value of the production is estimated using default oil and gas prices that vary with location to reflect the actual value of the oil and gas in any given region of the world.

The value of all the oil in the reserve should be known by the petroleum engineer staff and can be included in the model of the net present value by multiplying the monthly or yearly production by oil price or its expected value.

If the reserve lifetime is unknown, it is assumed to be infinite and the value of the reserve is the "capitalized cost" of the reserve. This is equal to monthly production in dollars divided by the monthly interest rate M/I. This is the value of the reserve if production is never interrupted. But if the facility or the structure fails and production must be deferred, the value of the reserve decreases because monthly income will be suspended from the date of shutdown until the date of the start up after repairing or building a new platform.

The deferred production loss is the difference between the value of the uninterrupted production and the present value of the interrupted production. The day production resumes, the value of the reserve is equal to the value of the reserve based on the uninterrupted assumption. This is converted to a net present value by discounting the value of the reserve for the interruption period. By default, this is calculated at an interest rate of 1% per month.

In determining the business loss for collapse of the structure, if C_{DP} is not given, then determine its default value by calculating the net monthly production from the following equation:

$$P_M = (D_{P(oil)} \times P_{oil} + D_{P(gas)} \times P_{gas}) \times (30\ \text{days}) \quad (7.5)$$

where

D_P = daily production in bpd or MSCFD, in UC;

P_{oil} is oil price/bbl, in UC/bbl;

P_{gas} is gas price/mscf, in UC/MSCF;

P_m is the net monthly production.

The present value will be as shown in the following equation as r_m is the monthly return investment equal approximately to (r/12), which will be obtained by defining r, which is the annual rate of return on investment.

$$PV = P_M / r_m \quad (7.6)$$

$$C_{DP} = P_m \sum_{k-1}^{n} \left(\frac{1}{1+r}\right)^k \quad (7.7)$$

where n is calculated from N/30 where N is the total downtime per day.

Note that this equation assumes the volume of oil is fixed for the period of shutdown, which is not really true, as in most cases there is a decline in production that depends on the reservoir characteristics.

$$I_B = C_{DP} + C_R \quad (7.8)$$

where C_R is cost of replacement; C_{DP} is cost due to deferred production; I_B is cost of business loss. The value of deferred production is based on daily production from oil and gas. If this is not given, PV is estimated from the monthly production in dollars by assuming an infinite reserve life.

The values for deferred production, replacement, and total business loss can be input into the system from a variety of sources. The values here are default values for these losses when no additional information is available. The defaults for net oil price, expected return on investment, etc., can be different from one country to another.

The loss from failure of the structure is the sum of the replacement cost and production deferral. If the replacement cost is unknown, it can be estimated using the approximate weight of the structure.

7.2.4.3 Safety impact loss

To assess the safety impact in case of a platform failure, two terms are used. The calculation is simply equal to the product of the number of persons and a marginal safety cost per person. Because there is an increased hazard potential when the platform handles gas, therefore the cost will increase 20% for a platform producing gas or in the case of gas compression platforms.

For platforms that have a production with high H_2S, the consequence value is recommended to be higher by around 50%. If the platform is bridge-linked to another platform, the consequence value is reduced by around 50%.

The default safety loss is proportional to the average number of people on a platform. The estimated safety loss is calculated by multiplying the safety loss cost per person and the average platform persons. As in other consequences, the default number of persons can vary from one company to another, so the resulting consequences reflect the relative differences in hazard exposure costs worldwide.

The safety loss cost will be reduced if the platform can be evacuated during severe storms. The expected consequence is increased if the platform handles gas production, because gas production implies that fire could result from an underwater structural failure. For the unmanned platforms, a minimum safety loss is assumed.

Regional factors, such as workplace safety regulations, fines, or other legal costs, are accounted for by using regionally specific default mean per capita safety losses.

The safety-related losses, I_S, are proportional to the number of people exposed, the location, and the marginal safety loss per person as a result of failure:

$$I_S = C_{ex} \times N \times G \tag{7.9}$$

where N is the number of persons on a platform; G is equal to 1.2 in the case of gas platforms and equal to 1 for others.

The input data for these equations depend on the information that will be collected about the number of persons on the platform and its nature, its location, and the company operations approach, which is different from one country to another and from one company to another, so it should be collected during the phase of gathering data.

7.2.4.4 Consequence categories

In quantitative risk assessment, the estimated loss in dollars is converted to consequence categories. In order to determine the ranges for the

Table 7.20 Relationship between estimated consequence and consequence category.

Category	Percentage of the platform fleet	Consequence (UC $M)
A	<5%	<C1
B	5%–50%	C1 to C2
C	50%–75%	C2 to C3
D	75%–95%	C3 to C4
E	>95%	>C4

categorization for a particular fleet of platforms, it is necessary to calculate the total loss in all the platforms and then the cumulative distribution of the estimated total losses is plotted on probability by using the appropriate software. The results are then used to estimate the consequence against 5%, 50%, 75%, and 95% probability. The four values C1, C2, C3, and C4 are obtained and will be used as the bounds to define the consequence categories, as shown in Table 7.20.

In order to reflect local differences in the factors that affect the consequences, Table 7.20 shows which values can be changed for the various locations worldwide, as the factors C1 to C4 should be defined on each category, as the values will different from one region and one company to another.

The environmental, business, and safety losses are summed to calculate an overall loss in terms of dollars or any currency depending on the country. This overall loss is converted to a consequence category according to the ranges illustrated in Table 7.20.

As with the likelihood of failure, the consequence category limits are intended to identify the best and worst 5% of the platforms. The intermediate category limits are intended to represent the 50th and 70th percentiles. This distribution of the category limits assigns two category ranges for platforms that have better-than-average likelihood scores and three ranges for platforms that have worse-than-average scores.

7.2.5 Overall risk ranking

After each platform has been categorized in terms of likelihood (1 to 5) and consequence category (A to E), the categories are represented through the risk matrix, as in Fig. 7.3, to establish the overall relative risk ranking. The overall rankings are high (red; dark gray in print version), moderate (orange; light gray in print version), low (green; dark gray in print version), and insignificant (white).

Consequence

Likelihood	A	B	C	D	E
1	Orange	Orange	Orange	Red	Red
2	Green	Green	Orange	Orange	Red
3	White	White	Green	Orange	Red
4	White	White	Green	Green	Orange
5	White	White	Green	Green	Orange

Fig. 7.3 Risk matrix proposed by API for the categorization of qualitative risk.

The highest-risk items are located in Category E1 and the lowest-risk items fall into category A5. The different colors show that the relative risks are grouped into high, moderate, low, and insignificant risk levels.

Where guidance in terms of ranking within each of the four risk categories may be useful, a preliminary assessment can be made by multiplying the total likelihood score by the total loss in abstract dollars. This approach is far from rigorous, so the resulting listing should be treated as indicative only.

Fig. 7.3 shows the qualitative risk matrix proposed by the API Committee for Refinery Equipment and based on Bea et al. (1988) for the assessment of refinery and petrochemical plants. It is the same 5 × 5 matrix that has been adopted for the risk-based underwater inspection system (Fig. 7.4).

Puskar et al. (1994) delivered other risk matrices that have been used for offshore risk assessment. Many of these systems are based on each company's practical experience. The 5 × 5 matrix elements shown in Fig. 7.3 are preferable, as they present enough range in likelihood and consequence that it will be easy to differentiate between the platforms in a fleet.

Note that API RP2-SIM recommends the matrix in Table 7.21 to represent the risk level of the platform.

High consequences are E5, D5, E4, and E3 and medium consequences are A5, B5, C5, C4, D4, D3, E2, and E1, whereas the low-risk rankings are the remaining other cells. Some references, such as API RP2-SIM, have only three categories, as shown in Table 7.21.

7.3 Risk assessment outcomes

After defining the risk ranking for all facilities or fleet structures onshore and offshore, the next step is to define the future maintenance plan, which

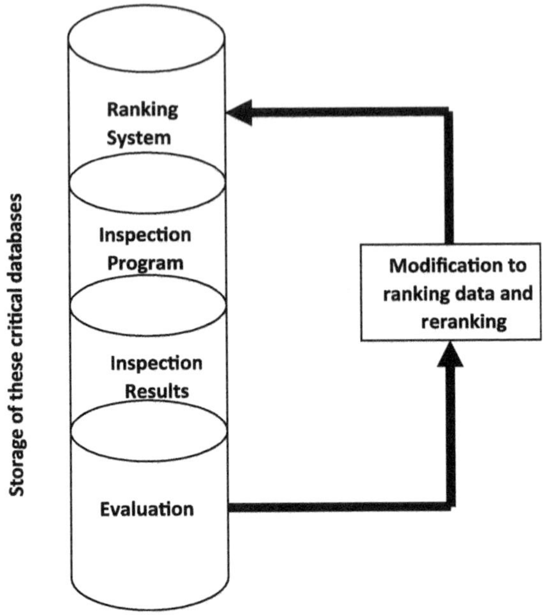

Fig. 7.4 Risk-ranking procedure.

Table 7.21 Risk matrix based on API SIM.

		Likelihood of Failure		
		Low	**Medium**	**High**
Failure impact	High	Medium Risk	High Risk	High Risk
	Medium	Low Risk	Medium Risk	High Risk
	Low	Low Risk	Low Risk	Medium Risk

includes a repair and inspection plan. For the offshore structure platform, this will be the main tool to determine the priority list for the subsea inspection, as it is the most costly inspection since it requires marine vessels and special tools and skills like ROV and divers.

For the offshore platform, the periodic underwater inspection should be carried out at intervals consistent with the SIM strategy adopted by the owner or the operator. Inspection intervals are provided based on platform consequence of failure; they are default intervals in lieu of intervals developed based on risk of failure. Periodic underwater inspection intervals may also be influenced by operational considerations, such as riser importance and condition, seabed movement, CP system condition, etc.

The main objective of the routine underwater inspection is to collect data on the as-is condition of the platform to assist in assessment of its condition and to take any remedial action in a reasonable time.

Some oil and gas companies use time-based maintenance, so they schedule the interval of inspection based on time, regardless of the existing condition of the structure or the platform. With our case and based on the SIM strategy as per API, there are two approaches; the time interval is based on the risk category, as shown in Table 7.22.

The other approach is using the time interval based on the consequences of failure. With high consequences, the inspection is every 3 years, but with medium and low consequences the inspection interval is 5 years, so the maximum inspection interval is 5 years.

However, these two scenarios do not match up with the structure integrity approach, because the structure could deteriorate faster than the time period and thus be under risk, or the structure could be new and have no major changes, so frequent intervals could be a waste of money and time. Therefore the risk-based inspection or maintenance in general is an optimum solution.

The risk-based inspection interval should not exceed 5 years for high-consequence platforms where the consequence category is driven by the presence of pipeline risers at the platform.

After the qualitative risk assessment is performed based on the methodology applied before for offshore platforms, finally the list of the platforms will be as shown in Table 7.24.

If the structure integrity management is not implemented or in progress, the inspection interval should be carried out as per Table 7.22, considering the consequence in the case of a platform failure. These default intervals are based on historic industry practice that has resulted in satisfactory in-service performance of the platform, cathodic protection system, and appurtenances.

Note that the timing of the first underwater periodic inspection is determined from the date the baseline inspection was completed.

Table 7.22 Risk-based inspection interval ranges.

Risk category	Inspection interval ranges
High	3–5 years
Medium	6–10 years
Low	11 years or more

After going through the previous steps to implement the general ranking for the whole fleet of platforms, it will appear as seen in the example in Table 7.23.

After implementing the platform ranking as in Table 7.23, then all the platforms are divided into three categories, with an equal number of

Table 7.23 Data for platforms constructed.

Platform ID	Design year	Total score
PLTF 49	1985	410
PLTF 50	1985	470
PLTF 51	1984	465
PLTF 52	1986	473
PLTF 53	1985	399
PLTF 54	1985	394
PLTF 55	1985	394
PLTF 56	1985	323
PLTF 57	1985	405
PLTF 58	1985	416
PLTF 59	1985	391
PLTF 60	1984	418
PLTF 61	1987	423
PLTF 62	1981	483
PLTF 63	1984	348
PLTF 64	1986	463
PLTF 65	1983	439
PLTF 66	1984	458
PLTF 67	1984	463
PLTF 68	1987	420
PLTF 69	1985	503
PLTF 70	1985	333
PLTF 71	1982	463
PLTF 72	1983	473
PLTF 73	1989	407
PLTF 74	1989	449
PLTF 75	1989	444
PLTF 76	1984	458
PLTF 77	1989	421
PLTF 78	1982	503
PLTF 79	1985	463
PLTF 80	1985	645
PLTF 81	1981	458
PLTF 82	1981	662

platforms from red, yellow, and green, as presented in Table 7.24. For example, with 150 platforms, each category will contain 50 platforms and the priority of the inspections and follow-up on each platform is based on its number. The qualitative risk assessment method is thus simple to use to manage a large number of offshore structure platforms.

Table 7.24 Risk ranking with cost for fleet of offshore structures.

S/N	Platform ID	Design year	Total score	Cost of maintenance (inspection and repair)
1	PLTF 82	1981	662	
2	PLTF 80	1985	645	
3	PLTF 78	1982	503	
4	PLTF 69	1985	503	
5	PLTF 62	1981	483	
6	PLTF 52	1986	473	
7	PLTF 72	1983	473	
8	PLTF 50	1985	470	
9	PLTF 51	1984	465	
10	PLTF 64	1986	463	
11	PLTF 67	1984	463	
12	PLTF 71	1982	463	
13	PLTF 79	1985	463	
14	PLTF 66	1984	458	
15	PLTF 76	1984	458	
16	PLTF 81	1981	458	
17	PLTF 74	1989	449	
18	PLTF 75	1989	444	
19	PLTF 65	1983	439	
20	PLTF 61	1987	423	
21	PLTF 68	1987	420	
22	PLTF 60	1984	418	
23	PLTF 58	1985	416	
24	PLTF 49	1985	410	
25	PLTF 73	1989	407	
26	PLTF 57	1985	405	
27	PLTF 53	1985	399	
28	PLTF 54	1985	394	
29	PLTF 55	1985	394	
30	PLTF 59	1985	391	
31	PLTF 63	1984	348	
32	PLTF 70	1985	333	
33	PLTF 56	1985	323	
34	PLTF 77	1989	421	

Most engineers focus on the technical issues, in this case the structure probability or likelihood of failure; on the other hand, the top management usually focus on the business; therefore the consequences should be presented in the risk equation.

In this regard, the risk categories are based on the consequence or impact and the likelihood of failure or probability of failure. To be clear, in this case the maintenance plan is business-oriented and not technically oriented only. Sometimes the direction may be not to spend any money on a platform that has a higher probability of failure but with very low consequence cost, and that has a very low reserve. In this case, you should predict that this platform will fail. But if the direction is to implement a maintenance and inspection plan to prevent any failure, then you should rank the platforms from the probability-of-failure point of view only.

As shown in Table 7.24, there are 34 offshore platforms divided between the three categories. The advantage of this technique is that you have a risk ranking under each category and you can insert the column with estimated cost of repair and maintenance. From this, the operator or the owner can determine the required budget to do the required maintenance as per the risk category. This approach is applied for all the facilities such as tanks, pipeline, piping, and others.

7.3.1 Cathodic protection

Cathodic protection is very important to the platform and pipeline. The anodes for the platforms are usually depleted over time and replacing these anodes is costly. This will be discussed in the next section on the economy of the platform. Therefore the anodes are evaluated, based on the last inspection, from the condition of the anode and its reading. As shown in Table 7.25, the anodes are placed into four grades and every grade represents the percentage of depletion: grade 4 is completely depleted. Sometimes the anode is actually depleted, but still a small electrical reading is present, as

Table 7.25 Anode condition.

Grade of anode	Anode depletion (%)
Grade 4	100 (red cell)
Grade 3	>50 (yellow cell)
Grade 2	<50 (yellow cell)
Grade 1	<10 (green cell)

Table 7.26 Cathodic protection (CP) readings.

CP readings (mV)	Protection condition
<850	Unprotected (red cell)
850–900	Marginal protection (yellow cell)
900–950	Satisfactory (yellow cell)
950–1000	Good (yellow cell)
>1000	Very good (green cell)

shown in Table 7.26. It is preferred to have a color code also for both the anode condition and its reading. As a case study, at the end the operator or the owner has a sheet, as shown in Table 7.27, from which they can determine any platforms that need rapid remedial action to replace the anodes.

7.3.2 FPSO risk-based inspection

With FPSO the inspected areas are mainly on the topside and the hull, so it is required to do a topside inspection, underwater inspection for the hull, and inside-tanks inspections.

Most FPSO follow the class rules for the inspection, but the new direction nowadays is to go through risk-based inspections. There is a lot of research currently devoted to this topic, since class rules for inspection can be fit for purpose for a tanker, but they are not tailored in some ways to the FPSO. For example, the tanker can go to the dry dock, but the FPSO does not have this chance due to the nature of its operations.

For FPSO the likelihood is expressed as a lifetime or annual probability of failure. The calculation of the probability of failure for FPSO is rarely explicit. This is because the calculation is computationally intensive and the estimate, in any case, is "notional" rather than "actual."

Lee et al. (2006) calculated the likelihood of failure by considering the term "design ratios." The simplest expression of likelihood is by using a matrix in which the likelihood is characterized as high, medium, and low, as in Table 7.21, in terms of design ratio. For example, the likelihood of plate yielding failure is high if the design ratio of plate yielding strength is greater than 0.85. If the design ratio is in the range of 0.7–0.85, the likelihood of plate yielding failure is "medium." If the design ratio plate yielding strength is less than 0.7, the likelihood of plate yielding failure is "low." The precise numerical range is the subject of current investigation and

Table 7.27 Maintenance plan for anode retrofit.

Structure name	WD (ft)	RA Rank	Last inspection	Inspection reading (mV)	Reading condition	Anode condition	Anode installation date	Budget cost	Color code
PLTF 1	123	2	2000	691–708	Un-Protected	Grade 4			RED
PLTF 2	123	5	2000	684–690	Un-Protected	Grade 4			RED
PLTF 3	123	6	2001	686–699	Un-Protected	Grade 3–Grade 4			RED
PLTF 4	113	11	2000	811–907	Un-protected /marginal	Grade 3–Grade 4			RED
PLTF 5	280	16	2002	871–906	Marginal	Grade 4			RED
PLTF6	127	20	2001	721–739	Un-Protected	Grade 4			RED
PLTF7	155	23	2003	1008–1014	Very good	Grade 1			GREEN
PLTF8	130	26	2003	995–1012	Very good	Grade 2			YELLOW
PLTF9	134	27	2001	1004–1020	Very good	Grade 2–Grade 3			GREEN
PLTF10	122	32	2001	850–938	Marginal–satisfactory	Grade 4			YELLOW
PLTF11	130	39	2003	954–980	Good	Grade 4			YELLOW
PLTF12	250	40	2003	941–962	Good–Very good	Grade 1			GREEN
PLTF13	123	44	2001	716–719	Un-Protected	Grade 4			RED
PLTF14	123	45	2001	714–725	Un-Protected	Grade 4			RED
PLTF15	123	46	2001	707–719	Un-Protected	Grade 3–Grade 4			RED
PLTF16	132	47	2003	919–924	Satisfactory	Grade 2			YELLOW
PLTF17	150	104	2002	664–675	Un-Protected	Grade 4			RED

calibration. The impact calculation is the same as that presented previously regarding impact on persons, environment, and business loss.

7.3.3 Strategy of inspection and repair

The inspection and repair strategy should be determined based on the probability of failure and the economics of inspection and minor and major repair. El-Reedy (2003) presented the optimal inspection and repair strategy.

Inspection alone does not improve reliability unless it is accompanied by corrective action if there is a defect present. There are some other tactics used in practice in offshore structures if a crack is found, such as:

- Monitoring the crack until its depth reaches a certain percentage of the original thickness and then perform repair;
- Immediate repair upon finding the crack;
- Repair at a fixed time period, for example, 1 year after detecting the crack;
- Repair by welding.

Some oil and gas companies follow a strategy of doing inspections at constant time intervals, as shown in Fig. 7.5, as some companies use a preventive maintenance approach with periodic maintenance and constant inspection intervals to facilitate planning. Inspection intervals are selected by using risk assessment techniques to minimize the expected cost of inspection, repair, and failure.

"No repair will be made unless the damage is detected" is a very important statement. Most offshore structure inspection techniques are visual or use NDT. The ability to detect damage mainly depends on the inspectors'

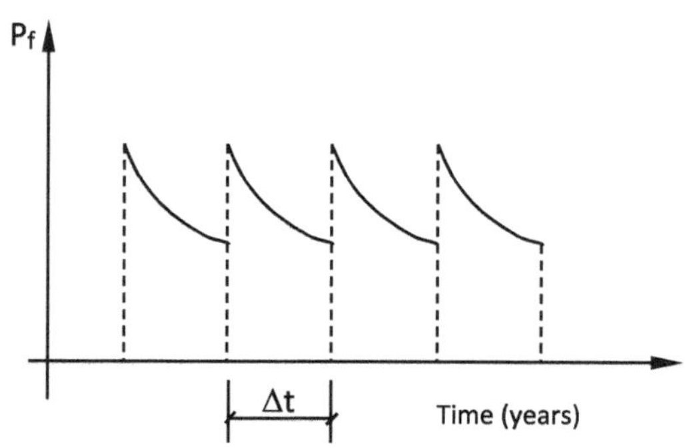

Fig. 7.5 Inspection strategy.

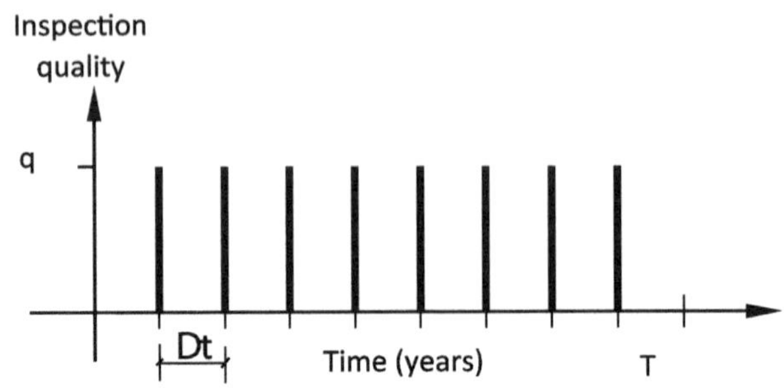

Fig. 7.6 Offshore structure performance.

experience and the accuracy of their tools. A higher-quality inspection method will provide a more dependable assessment of damage.

Higher-quality inspection may lead to higher-quality repair, which could bring the reliability of the structure closer to its original condition (although the reliability of the structure decreases with age). After inspection and repair, the structure's capacity should be returned to its original capacity based on the design, as seen in Fig. 7.6.

7.3.3.1 Expected Total cost

To calculate the expected cost, it is required to determine the service life of the structure. Assume the service life is 50 years and the routine maintenance (inspection and minor repair) is scheduled to be once every 2 years. It starts at $t = 2$ years and continues until $t = 48$ years. Based on this, the preventive maintenance will be performed 25 times during the life of the structure. The cost of maintenance during the lifetime will be:

$$C_{FM} = \text{C}_{\text{m2}} + \text{C}_{\text{m4}} + \text{C}_{\text{m6}} + \ldots + \text{C}_{\text{m48}} \tag{7.10}$$

If the total expected cost during the lifetime (T) is based on the present value, according to Turan (1990) the expected lifetime preventive maintenance cost becomes

$$C_{IR} = C_{IR2}\frac{1}{(1+r)^2} + C_{IR4}\frac{1}{(1+r)^4} + \ldots\ldots + C_{IR74}\frac{1}{(1+r)^{74}} \tag{7.11}$$

where r is the net discount rate of money.

In general, for a strategy involving n number of inspections during the lifetime, the total expected inspection cost is

$$C_{ins} = \sum\nolimits_{i=1}^{n} C_{ins} + C_R\frac{1}{(1+r)^{T1}} \tag{7.12}$$

where C_{ins} is cost of inspection as per the inspection method, and C_R is repair cost.

Finally, the total expected cost, C_{ET}, is the sum of the structure capital cost C_T, the expected cost due to periodic maintenance, which is the cost of the inspection, the minor repair, and maintenance, where C_f is the cost due to consequence of structure failure, P_f is the structure probability of failure, C_{in} and C_{re} are the consequent costs of inspection and repair. Accordingly, C_{ET} can be calculated from the following formula:

$$C_{ET} = C_T + (C_{in} + C_{re})\left(1 - P_f\right) + C_f \bullet P_f \tag{7.13}$$

The main objective is for C_{ET} to be a minimum value and in the meantime the structure should be reliable during its lifetime.

7.3.3.2 Optimization strategy

The target is to obtain the optimum maintenance strategy plan during the platform lifetime. The following problem must be solved:

$$\text{Minimize } C_{ET} \text{ subject to } P_{f\ life} \leq P_{max}$$

where P_{max} = maximum acceptable probability of failure during its lifetime or, considering the minimum accepted reliability index,

$$\beta = \phi^{-1}(1 - P)$$

where ϕ is the function for the standard normal distribution. The optimum inspection strategy during the platform lifetime is obtained by solving the following mathematical problem:

$$\text{Minimize } C_{ET} \text{ subject to } \beta_{life} \geq \beta_{min}$$

By formulating an optimization equation, the optimal inspection strategy plan with respect to cost can be obtained.

Eq. (7.5) represents the objective function (C_{ET}), which is affected by the periodic inspection cost and minor repair costs for the joint or member if it exists, as well as the cost as a consequence of failure occurrence, which includes the cost of major joint or member repair. The inspection periodic time (Δt), which is the optimization variable in this equation, is constrained by the minimum index β I, which is also called a reliability index, that is specified by the code and the maximum periodic time.

The optimization problem may be mathematically written as: find the Δt that minimizes the objective function

$$C_{BT}(\Delta t) = (C_{IR})(1 - P_f(\Delta t))\left(\frac{(1+r)^T - 1}{\left((1+r)^{\Delta T} - 1\right)(1+r)^T}\right) + C_f P_f(\Delta t)\left(\frac{(1+r)^T - 1}{\left((1+r)^{\Delta T} - 1\right)(1+r)^T}\right) \quad (7.14)$$

subject to $\beta(t) \geq \beta_{min}$, $\Delta t \leq T$, where C_{IR} is the periodic inspection and minor repair cost per inspection, C_f is the cost for major repair, r is the discount rate, and β_{min} is the minimum acceptable reliability index. C_{IR} and C_f are assumed to be constant with time, but these values can escalate with time factor, which requires future research. Due to code requirements, it is often necessary to place a constraint on the reliability index.

The preceding method is matched with the quantitative risk method for determining the probability failure and its consequences. However, from a practical point of view, it is easy and not very costly to use a qualitative risk assessment approach for a fleet of offshore structure platforms.

7.3.3.3 Onshore facilities and structure integrity

The same approach and way of thinking and implementing can be applied to onshore facilities and structures. The FPSO will be divided into the topside and hull, and also the tanks for ballast and storage, and the data should be collected based on that.

For pipelines, the same methodology will be followed as for offshore structures and the efficiency of the CP will aid in determining reduction in thickness. However, the redundancy in pipelines is less, as any reduction in wall thickness directly affects the pipeline integrity.

For the tanks there are regular inspections every 5 to 7 years for settlements. The tanks are divided into parts, such as bottom, shell, and top, and also there may be a fleet of tanks, so the same methods should be applied. For onshore structures like underground concrete tanks, pipe racks, or others, there should be a list of structures and their rankings as presented in the offshore structures section.

References

Bea, R.G., Puskar, F.J., Smith, C., Spencer, J.S., 1988. Development of AIM (assessment, inspection, maintenance) programs for fixed and mobile platforms. In: Offshore Technology Conference, OTC Paper 5703 Houston, TX, USA.

De Franco, S., O'Connor, P., Tallin, A., Roy, R., Puskar, F., 1999. Development of a risk based underwater inspection (RBUI) process for prioritizing inspections of large numbers of platforms. In: Offshore Technology Conference, OTC10846, Houston, TX, USA.
El-Reedy, M.A., 2003. Life-cycle cost design of deteriorating offshore structures. In: Offshore Mediteranian Conference OMC2003, Ravenna, Italy.
Gebara, J.M., Westlake, H., DeFranco, S., O'Connor, P., 1998. Influence of framing configuration on the robustness of offshore structures. In: Offshore Technology Conference, OTC Paper 8736, Houston, TX, USA.
Lee, A.K., Serratella, C., Wang, G., Basu, R., Spong, R., Energo Engineering, Inc., 2006. Flexible Approaches to Risk-Based Inspection of FPSOs. OTC 18364,.
Puskar, F.J., Aggarwal, R.K., Cornell, C.A., Moses, F., Petrauskas, C., 1994. A comparison of analytically predicted platform damage to actual damage during Hurricane Andrew. In: Offshore Technology Conference, OTC Paper 7473., Houston, TX, USA.
Turan, G., 1990. Engineering Economy for Engineering Managers. John Wiley and Sons, USA.

Further reading

API, 1996. Base Resource Document—Risk-Based Inspection—Preliminary Draft, API 581. American Petroleum Institute, Washington, DC.
API RP2A-SIM, 2006. Structural Integrity Management of Fixed Offshore Structures. American Petroleum Institute, Washington, DC.
API RP2A-WSD, 2000. Recommended Practice for Planning, Designing, and Constructing Fixed Offshore Platforms, Twentieth Edition, Supplement 1. American Petroleum Institute, Washington, DC.
Det Norske Veritas (USA) Inc. (DNV), 2002. Risk Based Underwater Inspection—Phase II, for Amoco, Houston. Document Number 221-8823, 8 February.
Gonen, T., 1990. Engineering Economy for Engineering Managers. John Wiley and Sons, Hoboken, NJ.
PMB Engineering, Inc, 1988. AIM—Assessment, Inspection and Maintenance, Phase III Final Report. PMB Engineering, Inc., San Francisco, CA.
Stahl, B., 1986. Reliability engineering and risk analysis. In: McClelland, B., Reifel, M.D. (Eds.), Planning and Designing of Fixed Offshore Platforms. Van Nostrand Reinhold, New York.
Stahl, B., De Franco, S. n.d. Internal Presentation Material Developed by Amoco, Amoco WE&C, Houston, TX.
Tallin, A.G., Puskar, F.J., Matos, S., 1998. Risk-Based Underwater Inspection, Report for Amoco WE&C, DNV Materials Engineering Consulting Report 232–8507.

Index

Note: Page numbers followed by *f* indicate figures and *t* indicate tables.

FSC
www.fsc.org
MIX
Papier | Fördert gute Waldnutzung
FSC® C083411

Kolibri 360 GmbH
Ferdinand-Jühlke-Straße 7
99095 Erfurt, Deutschland
produktsicherheit@kolibri360.de